# Differential Equations
## as Models in
## Science and Engineering

# Differential Equations as Models in Science and Engineering

## Gregory Baker
### The Ohio State University, USA

**World Scientific**

NEW JERSEY · LONDON · SINGAPORE · BEIJING · SHANGHAI · HONG KONG · TAIPEI · CHENNAI · TOKYO

*Published by*

World Scientific Publishing Co. Pte. Ltd.

5 Toh Tuck Link, Singapore 596224

*USA office:* 27 Warren Street, Suite 401-402, Hackensack, NJ 07601

*UK office:* 57 Shelton Street, Covent Garden, London WC2H 9HE

**Library of Congress Cataloging-in-Publication Data**
Names: Baker, Gregory R. (Gregory Richard)
Title: Differential equations as models in science and engineering / by Gregory Baker
     (The Ohio State University, USA).
Description: New Jersey : World Scientific, 2016. | Includes index.
Identifiers: LCCN 2016001687| ISBN 9789814656962 (hardcover : alk. paper) |
     ISBN 9789814656979 (pbk. : alk. paper)
Subjects: LCSH: Differential equations--Textbooks. | Differential equations--Numerical solutions--
     Textbooks. | Science--Mathematical models--Textbooks. | Engineering--Mathematical models--Textbooks.
Classification: LCC QA371 .B25 2016 | DDC 515/.35--dc23
LC record available at http://lccn.loc.gov/2016001687

**British Library Cataloguing-in-Publication Data**
A catalogue record for this book is available from the British Library.

Printed in Singapore

Engineering students have strongly influenced this book. By paying attention to their questions and discussing their frustrations, I realized what they wanted: to see and appreciate how mathematics is beneficial in science and engineering. They are attracted to the power of mathematics rather than its beauty. This book, then, is dedicated to them.

# *Preface*

Change occurs around us all the time. Indeed, our very notion of time emanates from our daily experience of life. There is always, before, now and next. Much of human endeavor is directed to gaining better and better understanding of swirling changes around us, presumably because that understanding will aid our survival.

Some change is very gradual, some very rapid and others take about the time of a few heart beats. The universal unit of change is the second, and has been so since the dawn of civilization, perhaps because it is a measure of a heart beat. Connected to the idea of time is that of motion, the movement of things. Some things move very fast (the speed of light), some at a moderate pace (walking people), and others extremely slowly (the rock in the garden). There appears to be no limit to the possible ranges in time. But our sense of time depends very much on the capabilities of our eyes. The changes we observe seem for the most part to be a continuous variation of events. What really happens is that our brains receive a very fast sequence of snapshots that is interpreted as continuous. That is why we can go to the theatre and watch a movie made from a fast sequence of images.

Some changes are predictable, for example, falling objects, the ticking of a clock, sunrise and sunset. Others occur regularly but with some randomness, like the crashing of waves at the beach. In some cases, we can predict for a while but not for too long, like the tracking of hurricanes. In our search to provide valuable predictions, we have come to rely more and more on mathematical models that can describe the patterns underlying change; and for continuous behavior the mathematical models are primarily differential equations, equations that connect change (time derivatives) with the current state of affairs.

Much of the current research and design activity of scientists and engineers is geared around the solution to differential equations. In the not too distant past, the only tools to treat differential equations were those of analysis and approximation. Many clever techniques have been developed, especially for those equations that are of fundamental importance to science and engineering. The solutions found by these techniques provide the insight that allows us to conceive what happens under much more complicated circumstances. Today we can add computer algorithms to our tool box and the power of modern computers is opening up doors in almost all subjects, but certainly in the solution to differential equations. Yet we all start the journey to understanding differential equations with simple, but important examples from science and engineering. It is only with a basic understanding of differential equations and the properties of their solutions that the journey can continue into the more complex nature found in modern science and engineering.

Given that differential equations started their central position in science and engineering in the middle of the 17th century, it is not surprising that a large body of knowledge has been created about the nature of differential equations and their solutions and the techniques for constructing them. More recently (since the 20th century), more abstract interpretations have arisen that extend the type of possible solutions to cases where the functions are not even continuous. All this progress has enriched our understanding of differential equations, but it also means that it takes some considerable time to learn all the complexities of differential equations, the huge variety of solution techniques, and the importance of properly stated additional conditions, in particular, initial and boundary conditions. On the other hand, there are fundamental properties of differential equations that are extremely important in science and engineering, and these fundamental properties are what students need to learn first.

Some organization of the material is necessary for the purpose of guiding a student's conceptual development of differential equations, and there are of course many possible themes. For a science and engineering student, the conceptual view should be at the same time strongly connected to relevant applications. The hope is that students begin to incorporate mathematical activity, including mathematical reasoning, into the natural activities of science and engineering, and at the core of that activity is the ability to translate physical reality into mathematical models that allow analysis and construction of solutions that in turn lead to interpretation and understanding of physical phenomena. The notion is one of mathematical literacy for scientists and engineers: it is more than simply translation, but the capability to identify and understand the role that mathematics plays in the scientific and engineering world. Literacy invokes the mantra "Learn to read so that you can read to learn." Mathematical literacy suggests "Learn mathematics so that you can use mathematics."

The material in this book reflects this perspective. Common problems in science and engineering are introduced as *the motivation* for developing an understanding of differential equations. Students are exposed to the complete process:

- the idealization of a real-life situation, itself a partly mathematical process, for example, we record the location of a particle as a *precise* number when the precision is a convenient idealization but not completely scientifically justified;
- the conversion of this idealization into a mathematical framework, based on the decision of what is important and what is needed (usually reflected on what assumptions are made);

- the mathematical processes and solution strategies that result in meaningful conclusions;
- and the evaluation of the success of the model.

The purpose of the mathematical model should be clear: it should be able to shed light on behavior of the system. Does the solution establish the nature of the long term behavior, whether stationary or steady or unstable? How does the solution make a transition from some initial state to the long term behavior? Subordinate to this process is the identification of important length and time scales. What role do parameters play in the mathematical results? Are there possibilities for control of parameters to optimize design? All these properties are stated in *essentially mathematical* terms and must be incorporated into learning about differential equations. Mastery of these properties demand higher cognitive function and can only be learned through consistent use and practice throughout the material, but once mastered the ability is lost only through lack of use.

There is another very important aspect to the organization of the material in this book, cohesion. There is a central theme that connects all parts of the material. A good starting point is the nature of solutions to linear differential equations because they exploit many of the fundamental ideas behind linear operators, such as linear superposition of homogeneous solutions. They also reveal several other important concepts that run through all differential equations, such as the role of initial and boundary conditions. The student needs to master these concepts before moving on to more advanced topics. The central theme in this book is to introduce these concepts right from the start and then re-enforce them as the material moves from first-order ordinary differential equations, through second-order ordinary differential equations and on to the standard second-order partial differential equations. The fundamental concepts are:

- The solutions to differential equations are functions! While apparently obvious, it takes some adjustment in thinking to shift from solving algebra equations where a single quantity is determined to the idea that differential equations determine a function. Also, there is a shift from viewing differentiation as a process to produce a result (the slope) and integration as a process to obtain a result (the area) to a view where they are operators acting on functions to produce functions.
- The properties of the function that solves the differential equation must reflect the nature of the equation. For example, if there is a jump in one of the terms, especially the forcing term, then an appropriate derivative must have the appropriate jump to ensure the balance of terms in the differential equation.
- Homogeneous linear equations have the property that if a solution is known, then any multiple of the solution will satisfy the differential equation. It is important to find all the independent *homogeneous* solutions because a linear combination of them is needed to solve initial and/or boundary conditions.

- Inhomogeneous equations arise from forcing terms and can be solved by seeking a *particular* solution, one that ensures a balance of the differential equation with the inhomogeneous or forcing term and/or a match with inhomogeneous or forcing effects at boundaries.
- A *general* solution is constructed by adding a particular solution, if needed, with the linear combination of all homogeneous solutions that are independent.
- A *specific solution* is obtained when initial conditions, or boundary conditions, are used to evaluate the arbitrary coefficients in the linear combination of homogeneous solutions. In this setting, science and engineering students learn how initial and boundary conditions are important in the mathematical model to describe completely a physical situation.

This theme is driven home most easily when the differential equations have constant coefficients. This restriction does not weaken the conceptual understanding of differential equations but allows students to construct concrete solutions with important applications in science and engineering. Indeed, the engineering faculty complain that students who have taken differential equation courses are not able to solve even the simplest linear equations with constant coefficients. Emphasizing these equations in this book helps students retain the necessary perspectives and skills needed in subsequent engineering courses without losing important mathematical concepts.

## Course Outline

*Homogeneous solutions* to differential equations with constant coefficients are just exponential functions for the simple reason that any derivative of an exponential is an exponential, thus ensuring all terms in the homogeneous differential equation will balance. At the same time, exponential functions describe the standard ideas behind growth and decay, in other words, the question of stability. For second-order differential equations, and higher-order, the arguments in the exponentials can be complex. The connection between exponentials of complex arguments and trigonometric functions with growing or decaying oscillations extends the concepts of stability and introduces the possibility of oscillatory behavior that occurs frequently in science and engineering.

The next important concept is that *particular solutions* to inhomogeneous differential equations tend to follow the nature of the forcing term. Important examples arise when the forcing term is a combination of polynomials, exponential functions and cosines and/or sines. A particular solution, then, is easily guessed and can be completely determined without "integration formulas." Furthermore, this approach is intuitively appealing. If a system is forced periodically, its response is periodic: if the forcing effect is exponential, the response is exponential. It is closely associated with the method of undetermined coefficients, but has extensions also to partial differential equations. Particular solutions rarely satisfy initial conditions, and so the need to compose a *general solution* with particular solutions and homogeneous solutions that contain unknown coefficients.

During the development of the constructive procedure outlined above, two other aspects of the nature of the solutions to differential equations must be brought out: the mathematical properties of the solution and the need for initial conditions to determine a specific solution completely. Important mathematical properties of the solution follow from the requirement that terms in the differential equation must balance the nature of the forcing term. For example, if the forcing effect is a linear ramp followed by a constant value, then the forcing term has a jump in its derivative, and the term containing the highest derivative in the differential equation must have the same mathematical property. Such considerations help remind students of the nature of differential equations as determining functions and their properties. These properties under the appropriate conditions allow the differential equation to be differentiated and the Taylor series evaluated provided there is sufficient information from the initial conditions to ensure all the initial derivatives can be evaluated recursively. While students dislike the Taylor series, it does bring home the need for appropriate initial conditions to determine a *specific solution* completely.

Besides the mathematical concepts underlying linear differential equations, the applications chosen to illustrate them also introduce the student to subsidiary ideas important in differential equations as models for phenomena in science and engineering. These include the notion of evolutionary processes that are continuous in time and lead naturally to time-dependent differential equations. Certain scientific principles, such as the conservation of quantities, are fundamental to understanding what governs the processes and leads to the derivation of differential equations. The differential equation is not just some mathematical expression but contains important information, and the student needs to be able to recognize that information. The solution to the differential equation will also reveal important information, such as the role of parameters in the model, the eventual state of the process, and the opportunity to design the process in some optimal way. It is the whole picture that counts, and if only solution techniques to differential equations are presented, the student is denied the opportunity to see the whole picture.

Once the student sees how differential equations arise in science and engineering and has grasped the ideas behind constructing particular solutions, homogeneous solutions and general solutions, it is time to move on to forcing effects that have more complicated mathematical nature. The next obvious step is to consider forcing terms that are periodic, for example, as might be found in electric circuits. This is a good time to introduce the Fourier series. Without concepts from linear algebra, it is difficult to motivate the derivation of the Fourier series as a representation in terms of a basis or eigenfunction expansion. Instead, I appeal to the idea that certain averages must balance, for example, if the Fourier series is to represent a periodic function, then the averaged values of the function must match the averaged value of the Fourier series. The important point is that the periodic inhomogeneous or forcing term is replaced by a Fourier series with trigonometric functions that allow the ideas underlying the method of undetermined coefficients to be applied successfully in the construction of the particular solution which also takes the form of a Fourier series.

The journey through the material now takes on the challenge of developing these ideas for two-point boundary value problems. The motivation here is the need to find steady solutions to transport phenomena with spatial variation. Previous ideas extend naturally: particular solutions will take care of inhomogeneous terms in the differential equation and the homogeneous solution will take care of the boundary values. But now the student will see for the first time that solutions do not always exist!

There is a new perspective that can be introduced at this point. Boundary value problems consist of both a differential equation and a set of boundary conditions. The problem is homogeneous if both the differential equation and the boundary conditions are homogeneous. Generally, there is only the trivial solution to the homogeneous boundary value problem, except there might be times when the differential equation has oscillatory solutions that automatically satisfy the boundary conditions so that a unique solution is not determined. If there is a parameter in the differential equation, non-unique solutions may arise for special values of the parameter, akin to the role of eigenvalues.

The stage is now set for the composition of the ideas developed for solutions to initial value problems, boundary value problems and the Fourier series to solve partial differential equations. Here the thread of the previous concepts can be brought together: the multiplicative nature of exponentials for homogeneous problems (both the partial differential equation and the boundary conditions) suggests the method of separation of variables. Separation of variables leads to homogeneous boundary value problems which need non-unique solutions (eigenvectors) for the construction of the solution to succeed. The Fourier series then completes the approach. If the problem is inhomogeneous, then a particular solution must be constructed separately and added with the solution to the homogeneous problem. The solutions to higher-dimensional, second-order partial differential equations rest on the recursive construction of particular and homogeneous solutions, and this process is the most challenging conceptual part of the course.

The last chapter introduces the ideas behind systems of first-order differential equations. It serves a dual purpose of presenting the construction of solutions to linear differential equations and presenting the geometric ideas underlying the interpretation of the global behavior of solutions, in other words, the phase diagram. To draw a phase diagram, equilibrium points must be located and their stability provides the information to start the trajectories in phase space. Stability of an equilibrium point is approached through linearization, so the dual purpose of this chapter is brought together through the importance of linear approximations even in nonlinear models. Here, linear algebra would be a great help. Instead, the systems are mainly restricted to two and three-dimensional where the calculations of the solutions are relatively straightforward.

During these conceptual developments in mathematics of differential equations, connections are continually made to: the origin of differential equations in science and engineering; the importance of initial and boundary conditions to complete the description of the physical models; the role of parameters; and the physical consequences of the solutions. In this way, the natural place of mathematics in the lives of scientists and engineers is placed at the center of this course.

*Gregory Baker*

# *A Note to the Student*

This book, and the course that uses it, will most likely be very different from any mathematics course you have ever taken. The material is dedicated to showing you how mathematics is used in science and engineering. That means ideas in science and engineering must be expressed in mathematical terms and the ability to do so must be learnt along the way of learning how to solve differential equations. As a scientist and engineer, you will have to convert real-life situations into mathematical models and then employ mathematical thinking to find answers. This process is very different from that you embark on when learning mathematics in a traditional course, where problems are stated in mathematical terms and you simply have to follow the procedures given in class or the textbook to find answers.

The usual strategy for a student to "survive" a mathematics class is:

- Attend lecture to gain some idea of what you will need to do to answer assigned problems.
- Attempt homework and, if stuck, scan lecture notes and the textbook to find a similar problem and copy how it is solved.
- Make sure your answer agrees with the one given in the back of the book.
- Do many problems to be sure that you will find one or more of them on exams.
- Be sure to get as much partial credit as you can.

This strategy suffers from several disadvantages: it encourages low-level cognitive behavior such as pattern recognition and mimicking procedures but does not encourage understanding nor the development of a connected body of mathematical knowledge capable of being used whenever needed. It decreases the opportunity to struggle with mathematical reasoning that leads to learning and reduces the opportunity to gain confidence that what you do mathematically is correct.

Instead, a new approach will likely be very successful, not only in this course but in subsequent courses in science and engineering when mathematics is needed. Try to:

- Go over the problems in class and this book in detail and understand how each step is done. The goal is that you can do these problems on your own.

- Ask yourself why did the procedure succeed in answering the problem. What is the "key" idea?
- Use the approach in similar problems to gain confidence in how to find answers.
- Learn to "read" mathematics. Each mathematical statement should make sense to you.
- Check your work carefully. Not only must each mathematical statement make sense but also that subsequent statements follow logically and make sense.
- Pay attention to the units of measurement for all quantities. In particular, the arguments of standard functions, such as $\exp(x)$, $\cos(x)$, must have no units. A mismatch of units indicates the presence of errors.
- Repeat problems every few days to make sure that you remember what is important to solve the problems.
- When you believe that you have the correct answer, write it down clearly. The very process of writing aids in learning because you must pay attention to what is expressed.

The ultimate goal is simple but requires dedication and hard work. If you can solve all the problems on your own and be sure that you are correct, then you will have mastered the material in this book. Just as an athlete must focus and practice, you too must focus and practice, but the payoff is great. When you finally play the game, you will be prepared and confident and relish the thrill of succeeding.

# Contents

# Linear Ordinary Differential Equations

# 1

## 1.1  Growth and decay

A good starting point is to see just how differential equations arise in science and engineering. An important way they arise is from our interest in trying to understand how quantities change in time. Our whole experience of the world rests on the continual changing patterns around us, and the changes can occur on vastly different scales of time. The changes in the universe take light years to be noticed, our heart beats on the scale of a second, and transportation is more like miles per hour. As scientists and engineers we are interested in how specific physical quantities change in time. Our hope is that there are repeatable patterns that suggest a deterministic process controls the situation, and if we can understand these processes we expect to be able to affect desirable changes or design products for our use.

We are faced, then, with the challenge of developing a model for the phenomenon we are interested in and we quickly realize that we must introduce simplifications; otherwise the model will be hopelessly complicated and we will make no progress in understanding it. For example, if we wish to understand the trajectory of a particle, we must first decide that its location is a single precise number, even though the particle clearly has size. We cannot waste effort in deciding where in the particle its location is measured,[1] especially if the size of the particle is not important in how it moves. Next, we assume that the particle moves continuously in time; it does not mysteriously jump from one place to another in an instant. Of course, our sense of continuity depends on the scale of time in which appreciable changes occur. There is always some uncertainty or lack of precision when we take a measurement in time. Nevertheless, we employ the concepts of a function changing continuously in time (in the mathematical sense) as a useful approximation and we seek to understand the mechanism that governs its change.

In some cases, observations might suggest an underlying process that connects rate of change to the current state of affairs. The example used in this chapter is

---

[1] For example, the front? the back? the atom in the middle?

bacterial growth. In other cases, it is the underlying principle of conservation that determines how the rate of change of a quantity in a volume depends on how much enters or leaves the volume. Both examples, although simple, are quite generic in nature. They also illustrate the fundamental nature of growth and decay.

### 1.1.1  *Bacterial growth*

The simplest differential equation arises in models for growth and decay. As an example, consider some data recording the change in the population of bacteria grown under different temperatures. The data is recorded as a table of entries, one column for each temperature. Each row corresponds to the time of the measurement. The clock is set to zero when the bacteria is first placed into a source of food in a container and measurements are made every hour afterwards. The experimentalist has noted the physical dimensions of the food source. It occupies a cylindrical disk of radius 5 cm and depth of 1 cm. The volume is therefore 78.54 cm$^3$. The population is measured in millions per cubic centimeter, and the results are displayed in Table 1.1.

Table 1.1   Population densities in millions per cubic centimeters.

| Time in hours | Temperature | | |
|---|---|---|---|
| | 25 °C | 35 °C | 45 °C |
| 0.0 | 6.68 | 2.67 | 7.88 |
| 1.0 | 7.62 | 3.87 | 7.33 |
| 2.0 | 8.68 | 5.61 | 6.86 |
| 3.0 | 9.89 | 8.09 | 6.38 |
| 4.0 | 11.22 | 11.72 | 5.99 |
| 5.0 | 12.83 | 16.98 | 5.52 |
| 6.0 | 14.56 | 24.59 | 5.20 |

Glancing at the table, it is clear that the populations increase in the first two columns but decay in the last. Detailed comparison is difficult because they do not all start at the same density. That difficulty can be easily remedied by looking at the relative densities. Divide all the entries in each column by the initial density in the column. The results are displayed in Table 1.2 as the change in relative densities, a quantity without dimensions. Since we use the initial density as a yardstick, all the first entries are just 1, and now it is easy to see that the second column shows the fastest change when the temperature is 35 °C.

But looking at data in a table has limitations. It does not show, for example, whether the changes appear regular or random. Is there a smooth gradual change? Instead, we prefer to "see" the data in a graph as displayed in Fig. 1.1. Now it is apparent that the changes appear systematic. The question is "what is the pattern in the data?" The data for temperature at 35 °C looks like it might be a quadratic. One way forward then is to try to fit the data to a quadratic. To proceed, we need

Table 1.2  Relative population densities.

| Time in hours | Temperature | | |
| | 25 °C | 35 °C | 45 °C |
| --- | --- | --- | --- |
| 0.0 | 1.00 | 1.00 | 1.00 |
| 1.0 | 1.14 | 1.45 | 0.93 |
| 2.0 | 1.30 | 2.10 | 0.87 |
| 3.0 | 1.48 | 3.03 | 0.81 |
| 4.0 | 1.68 | 4.39 | 0.76 |
| 5.0 | 1.92 | 6.36 | 0.70 |
| 6.0 | 2.18 | 9.21 | 0.66 |

to introduce some symbols to represent the data before we can test the fit to any choice of function.

An obvious symbol for the measurement of time is $t$, but there are measurements at several different times. We may designate each choice by introducing a subscript to $t$, $t_j$ where $t_0 = 0$, $t_1 = 1$, and so on. Note that the symbol $t$ records a number but with some units in mind, in this case hours.[2] We may, of course, use any symbol to represent time, but it makes good sense to pick one that will remind us what the quantity is. Now we need to pick a symbol for the relative density. My choice is $\rho$ (no units). For each time measurement $t_j$, we may associate a density measurement from one of the columns, for example, $\rho_j$(Temperature = 25 °C). Obviously, we have three choices for the temperature so we can introduce a subscript to a symbol, $T_k$ say, for the temperature with $T_1 = 25$ °C, $T_2 = 35$ °C, and $T_3 = 45$ °C. From the experimentalist's point of view, $t_j$ is the independent variable (measurements

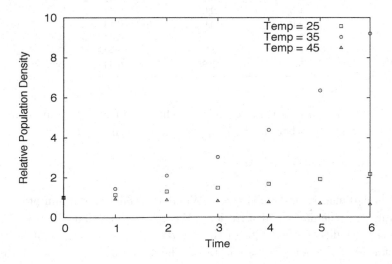

Fig. 1.1  Relative population density of bacteria for three different temperatures.

[2]Whenever a symbol is introduced, an appropriate choice of units will be included in parentheses; thus $t$ (hr).

are made to reveal how the density changes with time), $\rho_j$ is the measurement, and $T_k$ is a control variable held fixed while the bacteria grows. From a mathematical perspective, $t_j$ is the independent variable and $\rho_j$ is the dependent variable, while $T_k$ is a parameter (a fixed constant for each set of results). If we wish to emphasize the role of the temperature on the measurements, we can use $\rho_j(T_k)$, but we generally take it for granted that we know which data we refer to and drop the reference to $T_k$.

Now we are in a position to ask how the increases in the relative density change in time. There are two obvious ways the change in relative density can be measured. One is to record the jump in value (the absolute increase) and the other is to record the relative increase (the ratio of the absolute change to the current value). At each time $t_j$, we calculate the absolute increase by

$$\rho_{j+1} - \rho_j$$

and show its value in Table 1.3 for temperature $T_2$. Also, we can calculate the relative increase,

$$\frac{\rho_{j+1} - \rho_j}{\rho_j}$$

and it is shown as a separate column in Table 1.3.

Table 1.3  Changes in the relative densities during an hour.

| Time in hours | Absolute change | Relative change |
|---|---|---|
| 0.0 | 0.45 | 0.450 |
| 1.0 | 0.65 | 0.448 |
| 2.0 | 0.93 | 0.443 |
| 3.0 | 1.36 | 0.449 |
| 4.0 | 1.97 | 0.449 |
| 5.0 | 2.85 | 0.448 |

From the data, it seems that the relative change during an hour remains very close to a constant. In other words, it appears as though

$$\frac{\rho_{j+1} - \rho_j}{\rho_j} = C \approx 0.45, \quad \text{or} \quad \rho_{j+1} = (1 + C)\rho_j, \tag{1.1.1}$$

where $C$ is a dimensionless constant.[3] When we make this assumption, we are introducing a mathematical idealization. We are stating that there is a definite principle that guides the behavior of the growth of the bacteria, namely, that the relative change of the density is a constant. The advantage is that we can now use the power of mathematics to study the behavior of $\rho_j$ and draw some conclusions.

---

[3]The proper procedure is to include error bars on the measurements taken and record them in Table 1.1. Then we can use statistical methods to ask how likely is $C$ a constant.

Equation (1.1.1) is a simple example of a recursion relation. A solution is generated by applying the recursion sequentially.

$$\rho_0 = 1\,,$$
$$\rho_1 = (1+C)\,\rho_0 = (1+C)\,,$$
$$\rho_2 = (1+C)\,\rho_1 = (1+C)^2\,,$$
$$\rho_3 = (1+c)\,\rho_2 = (1+C)^3\,,$$

and so on. It is very easy to make the guess that

$$\rho_j = (1+C)^j\,. \tag{1.1.2}$$

But is this guess correct? Although it seems intuitive that it must be correct (how can it be wrong?), we should always verify our guess, in this case by direct substitution into Eq. (1.1.1). The right and left hand sides of Eq. (1.1.1) are

$$\rho_{j+1} = (1+C)^{(j+1)}\,, \qquad \text{(left hand side)}$$
$$(1+C)\rho_j = (1+C)\,(1+C)^j = (1+C)^{(j+1)}\,, \qquad \text{(right hand side)}$$

and both sides agree for *any* choice of $j$: thus Eq. (1.1.2) must be the solution for Eq. (1.1.1).

Sometimes it helps to see how a different guess for the solution fails. Suppose we guess that $\rho_j$ falls on a quadratic function,

$$\rho_j = 1 + aj^2\,,$$

where $a$ is to be determined so that the equation is satisfied. The right and left hand sides of Eq. (1.1.1) are

$$\rho_{j+1} = 1 + a\,(j+1)^2 = 1 + a^2 + 2aj + aj^2\,, \qquad \text{(left hand side)}$$
$$(1+C)(1+aj^2) = 1 + C + a(1+C)j^2\,, \qquad \text{(right hand side)}$$

and there is no value of $a$ which will make the left and right hand sides match perfectly, for example, there is a term with $j$ on the left hand side but there is no such term on the right hand side. Clearly, then, $\rho_j = 1 + aj^2$ cannot be a solution to Eq. (1.1.1)

Informed choices for the solution of equations will prove to be a very valuable technique. The idea is that we make a certain guess for the solution, replace the unknown function in the equation by the guess and verify that the equation is always satisfied. If this works then we are sure we have found the solution. For example, we made the guess Eq. (1.1.2) for the solution to Eq. (1.1.1) and verified it satisfies the equation for any choice of $j$.

The behavior of the solution in Eq. (1.1.2) is *exponential*. To see this clearly, use Eq. (A.1.1) to rewrite Eq. (1.1.2) as

$$\rho_j = e^{j\ln(1+C)} \tag{1.1.3}$$

and now it is clear that the density increases exponentially as $j$ increases. A review of the properties of exponential functions is provided in Appendix A.

### 1.1.2  *From discrete to continuous*

The growth of the bacteria at a temperature of 35 °C has proved to be exponential when measurements are made at each hour. Suppose we record the results more frequently, say every $\triangle t$ units of time. If $\triangle t$ is less than an hour, then obviously the relative change must be less. For example, if we now record results every half hour, then we might expect the relative change to be halved. Let us make the assumption that the density recorded at $t_j = j \triangle t$ changes at a fixed relative rate that is proportional to $\triangle t$. In other words, we should replace $C$ in Eq. (1.1.1) by $\lambda \triangle t$. The parameter $\lambda$ must have the units per hour because $\triangle t$ must have the same units as $t$ which is hours.

$$\frac{\rho_{j+1} - \rho_j}{\rho_j} = \lambda \triangle t, \quad \text{or} \quad \rho_{j+1} = (1 + \lambda \triangle t)\rho_j, \qquad (1.1.4)$$

with the solution indicated by Eq. (1.1.3),

$$\rho_j = e^{j \ln(1 + \lambda \triangle t)}.$$

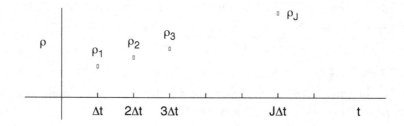

Fig. 1.2   General pattern in the discrete data.

Let us keep halving the measuring interval, and imagine more data points added to Fig. 1.1. Surely, the points will continue to fill in a continuous curve and eventually the relative density becomes $\rho(t)$. Mathematically, the counter $j$ becomes a continuous variable $t$. How this happens is illustrated in Fig. 1.2. If we knew what the function $\rho(t)$ was, then we could easily determine the data points $\rho_j$ by simply evaluating

$$\rho_j = \rho(t_j).$$

The challenge is to determine $\rho(t)$ when all we know is $\rho_j$. Clearly, to connect $\rho_j$ with $\rho(t)$, we must pick a fixed time $t = J \triangle t$ and state $\rho_J = \rho(J \triangle t)$. Note that as we decrease $\triangle t$, we must increase the number $J$ of time increments to remain at the same time $t$. Indeed, $J = t/\triangle t$ and

$$\rho_J = \rho(t) = e^{t\ln(1+\lambda\triangle t)/\triangle t}. \tag{1.1.5}$$

It may seem that we have constructed a solution $\rho(t)$ that is a continuous function of $t$. Not so, because $t$ still jumps in multiples of $\triangle t$ and $\triangle t$ still appears in Eq. (1.1.5). We must take the limit of $\triangle t \to 0$ so that $t$ becomes continuous. Of course we must take this limit with $t = J\triangle t$ kept fixed, which means $J \to \infty$ (it takes infinitely many very small time increments to reach a fixed time). By replacing $J$ with $t/(\triangle t)$ in Eq. (1.1.5) and then keeping $t$ fixed, we avoid the problem of what to do with $J$.

In the limit $\triangle t \to 0$,

$$\frac{\ln(1+\lambda\triangle t)}{\triangle t} = \lambda.$$

One way to obtain this limit is by L'Hospital's rule. Another is to use the Taylor series expansion for $\ln(1+x) = x - x^2/2 + \cdots$ and cancel $\triangle t$ before taking the limit. Either way the result is,

$$\rho(t) = e^{\lambda t}. \tag{1.1.6}$$

The relative density grows continuously as an exponential function!

For the data in Table 1.2 to match with Eq. (1.1.6), we must have at $t = 1$,

$$e^{\lambda} = (1 + C) = 1.45, \quad \lambda = \ln(1 + C) = 0.372,$$

and the solution Eq. (1.1.6), along with the data is shown in Fig. 1.3. We have constructed a continuous approximation to the growth of the bacteria that matches the experimental observations!

We have found a continuous function Eq. (1.1.6) as the limit of the solution Eq. (1.1.5) of the recursion Eq. (1.1.4), but what is the limit for the recursion? Pick $j = J$, and set $t = J\triangle t$. Then the two sides of Eq. (1.1.4) become

$$\rho_{J+1} = \rho\big((J+1)\triangle t\big) = \rho(t + \triangle t),$$
$$(1 + \lambda\triangle t)\rho_J = (1 + \lambda\triangle t)\rho(t),$$

and Eq. (1.1.4) can be written as

$$\frac{\rho(t + \triangle t) - \rho(t)}{\triangle t} = \lambda\rho(t),$$

which has the limit,

$$\frac{d\rho}{dt}(t) = \lambda\rho(t). \tag{1.1.7}$$

How remarkable! The solution Eq. (1.1.6) solves Eq. (1.1.7). To be sure of this, we should check the solution by direct substitution.

$$\frac{d\big(e^{\lambda t}\big)}{dt} = \lambda\big(e^{\lambda t}\big).$$

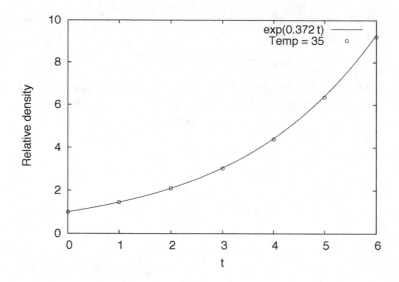

Fig. 1.3    Continuous exponential growth compared to the data.

Note how this statement compares with Eq. (A.1.5)! It is equivalent to the property of differentiation of exponential functions.

Equation (1.1.7) is the simplest example of a differential equation. It is an equation that involves the derivative of the unknown function $\rho(t)$. It is called first-order because only the first derivative is involved: more derivatives, higher order. It is also *linear*. That means the unknown function and its derivatives appear separately and not as the argument of some other function.

---

**Reflection**

The differential equation Eq. (1.1.7) states that the instantaneous rate of change of some quantity is proportional to that quantity. There are many examples of such a relationship and we will consider several of them in what follows.

---

### 1.1.3   *Conservation of quantity*

Suppose we have a drum filled with contaminated water, and we want to flush it out. We plan to pour clean water in until the water reaches a desired level of purity. If we want to study this real-life situation we must be more clear about what is happening, and in bringing clarity we inevitably make assumptions. For example, what is the rate at which the clean water is injected into the drum? We will assume that it is a constant, and that the rate of contaminated water flowing out is the same. The consequence is that the volume of water in the drum remains constant as well. The flow of water in is clean while the flow of water out contains contamination. How much contaminant flows out? If the water in the tank is well stirred, we expect the concentration in the tank to be uniformly constant, and that must then be the concentration in the water flowing out of the drum.

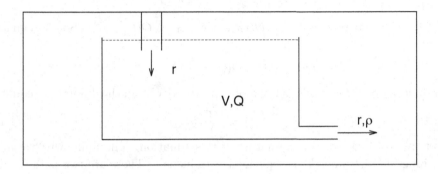

Fig. 1.4   A schematic of the drum cleansing.

We are now ready to develop a mathematical model. What do we need to know? It often helps to draw a "cartoon" of the situation and to consider what quantities are involved. In Fig. 1.4, the drum is represented as a rectangular shape with inlet and outlet pipes. The capacity of the drum is important, and is assumed to be $V = 100$ gal. Also in the drum is the contaminant dissolved in the water. The amount of contaminant is represented by $Q$ (lb), but we anticipate it will change in time, so we write it as a function of time $Q(t)$ where $t$ is measured in minutes. We can set our timer at $t = 0$ min when we start to flush the drum, and we assume we know the amount of contaminant at this moment, $Q(0) = Q_0$. The arrow at the mouth of the inlet pipe in Fig. 1.4 indicates the flow of water into the drum at a rate of $r = 3$ gal/min and we assume that the outflow at the outlet pipe has the same rate, thus keeping the volume of water a constant value $V$. Finally, we assume that the water flowing out carries a concentration of the contaminant that is the same as the concentration in the drum $\rho(t) = Q(t)/V$ (lb/gal). So what do we want to know? We would like to know how much contaminant is left at a later time $t$ (min). This means we must know how the amount of contaminant changes in time. $Q(t)$ is an unknown function, $t$ is the independent variable, and $V$, $r$ and $Q_0$ are parameters (remain fixed for each simulation of drum cleaning).

Now that we have established a mathematical description of the drum cleansing, we need an equation that determines $Q(t)$. Let us consider a small time interval $\triangle t$ (min) and imagine we have made a series of measurements $Q_j = Q(j\triangle t)$. How does the amount $Q_j$ change during the time interval $\triangle t$? The basic principle of conservation says that the amount of contaminant must change according to

$$Q_{j+1} - Q_j = -(\text{amount of } Q \text{ flowing out during } \triangle t). \qquad (1.1.8)$$

The assumption we make is that the concentration flowing out is the same as the current concentration in the drum which is $Q_j/V$ (lb/gal). The volume of water flowing out during $\triangle t$ (min) is $r\,\triangle t$ (gal). The amount of contaminant in this water is the volume times the concentration. Thus Eq. (1.1.8) becomes

$$Q_{j+1} - Q_j = -\frac{Q_j}{V}\, r\,\triangle t\,.$$

Pick $t = j\triangle t$ and replace $Q_j = Q(t)$ and $Q_{J+1} = Q(t + \triangle t)$. So the equation becomes

$$Q(t + \triangle t) - Q(t) = \frac{r}{V}\, Q(t)\,\triangle t\,.$$

After dividing by $\triangle t$ and taking the limit $\triangle \to 0$, we obtain the differential equation

$$\frac{dQ}{dt}(t) = -\frac{r}{V}\, Q(t)\,. \tag{1.1.9}$$

Let us first take note of the nature of this equation. The unknown "variable" is the function $Q(t)$ and the independent variable is $t$. There are two parameters $r$ and $V$ to be considered known for each time we solve the equation. We know how to solve this equation, since it is very similar to Eq. (1.1.7) and the solution must be similar to Eq. (1.1.6). Indeed, we just replace $\lambda$ with $-r/V$.

$$Q(t) = e^{-rt/V}\,.$$

Great, but there is a problem! At $t = 0$, $Q(0) = 1$. That was true for $\rho(t)$ in Eq. (1.1.6), but that is not right for $Q(t)$. The initial amount is given by $Q(0) = Q_0$. The parameter $Q_0$ allows us to study the situation for any initial amount of contamination and we certainly do not want to restrict ourselves to the choice $Q_0 = 1$.

Fortunately, the remedy is simple. We start afresh and assume that the solution has the form

$$Q(t) = C\,e^{\lambda t}\,.$$

The choice made reflects the expectation that the solution is an exponential but we do not know yet what $\lambda$ to pick. We also multiply the solution by $C$ because we will need to satisfy the initial condition. Thus we have two unknown constants in the guess for the solution and we expect them to be determined by requiring the guess to solve the differential equation and the initial condition. First, substitute the guess into Eq. (1.1.9). Since

$$\frac{dQ}{dt}(t) = \frac{d(C\,e^{\lambda t})}{dt} = \lambda C\,e^{\lambda t}, \quad -\frac{r}{V}\, Q(t) = -\frac{r}{V}\, C\,e^{\lambda t}, \tag{1.1.10}$$

the two sides of Eq. (1.1.9) will balance perfectly with the choice $\lambda = -r/V$.

Notice that the key to success is that the derivative of an exponential is proportional to the exponential. The constant $C$ simply cancels on both sides and remains unknown. It is determined by the application of the initial condition, $Q(0) = C = Q_0$ at $t = 0$, and finally the solution is completely determined,

$$Q(t) = Q_0\,e^{-rt/V}\,. \tag{1.1.11}$$

Finally, we can answer how long we must wait until $Q(t) < \alpha$, some tolerance factor. The solution states that the $Q(t)$ decays exponentially, and so there will be a time $T$ when $Q(T) = \alpha$ and that is the first moment when $Q(t)$ begins to fall below the tolerance level.

$$Q_0 \, e^{-rT/V} = \alpha, \quad \text{or} \quad T = -\frac{V}{r} \ln\!\left(\frac{\alpha}{Q_0}\right).$$

The factor $V/r$ is a very important physical quantity. Note first that it has the units of time. For this example, $V = 100$ gal, $r = 3$ gal/min, so $V/r \approx 33$ min. The result gives an estimate for the typical time it takes for $Q$ to change appreciably. To be more specific, note that at time $t = \tau \equiv V/r$,

$$Q(\tau) = Q_0 e^{-1}, \tag{1.1.12}$$

and the quantity $Q(\tau)$ is reduced by the factor $1/e$ of its initial value. For this reason $\tau$ is called the e-folding time. In conclusion, it will take about half an hour to reduce $Q$ by the factor $\alpha/Q_0 = 1/e \approx 1/3$.

---

### Reflection

The instantaneous rate of change of a quantity must be related to how it is added or subtracted. The presence of parameters usually governs the rate of change, and hence the time scales for change to occur. As scientists and engineers, we are very interested in just how long a process will take. Mathematically, we look at the argument of any function in the solution, in the example, the exponential function. The argument of this function can have no units! Imagine the difficulties in evaluating $\exp(3\,\text{min}) = \exp(180\,\text{s})$ since the results should be the same.

---

### 1.1.4 *Simple electric circuits*

So far, two examples have shown ways in which differential equations arise: population growth/decay that led to a phenomenological model, and the conservation of quantity that led to a model for mixing. Another way that differential equations arise is in models for electric circuits. A simple electric circuit is shown in Fig. 1.5. It is composed of a resistor $R$ ($\Omega$) and a capacitor $C$ (F) in series and may have an externally applied voltage $E$ (V). Also shown is a switch that can turn the applied voltage on or off. As a start, suppose there is no applied voltage, but the capacitor is fully charged with charge $Q_0$ (C). The switch is closed and the capacitor simply discharges.

The governing principles for electric circuits are called Kirchoff's laws. For the simple circuit in Fig. 1.5, Kirchoff's laws state:

- The sum of all the voltage changes around a closed circuit must be zero.
- The current is the same at all parts of the circuit.

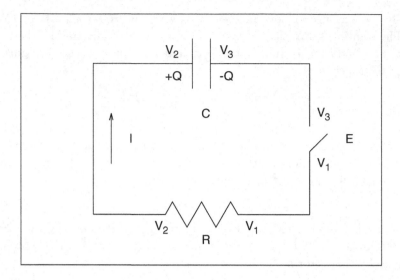

Fig. 1.5   RC circuit.

To apply Kirchoff's laws it is useful to introduce values for the voltage at different parts of the circuit. They are shown as $V_j$ (V) in Fig. 1.5. The usual assumption is adopted, namely, that there is no voltage change along the wire, but only across the resistor and capacitor. Also shown is the current and its direction. The charge on the capacitor is shown in a manner consistent with the direction of the current $I$ (A). Let $V_R = V_1 - V_2$ be the change in voltage across the resistor and $V_c = V_2 - V_3$ be the change in voltage across the capacitor. Then

$$V_R + V_C = (V_1 - V_2) + (V_2 - V_3) = V_1 - V_3 = E = 0. \qquad (1.1.13)$$

Here we acknowledge that no external voltage is being applied. While there is no current through the capacitor, the rate of change of the charge $Q(t)$ must match the current in the circuit.

$$I(t) = \frac{dQ}{dt}(t). \qquad (1.1.14)$$

The voltage drop across the resistor is given by $V_R = RI$ and across the capacitor by $V_C = Q/C$. Thus Eq. (1.1.13) becomes

$$R\frac{dQ}{dt}(t) + \frac{Q(t)}{C} = 0 \quad \text{or} \quad \frac{dQ}{dt}(t) + \frac{1}{RC}Q(t) = 0. \qquad (1.1.15)$$

To solve this equation, assume the solution has the form

$$Q(t) = A\,e^{\lambda t} \qquad (1.1.16)$$

where $A$ and $\lambda$ are unknown constants.[4] Substitute this guess into the differential equation,

$$\lambda A\,e^{\lambda t} + \frac{1}{RC}A\,e^{\lambda t} = A\,e^{\lambda t}\left(\lambda + \frac{1}{RC}\right) = 0. \qquad (1.1.17)$$

---

[4]Since we already have a symbol $C$ representing the capacitor, we pick a different symbol $A$ in the guess for the solution to avoid confusion.

When solving Eq. (1.1.10), we picked $\lambda$ to ensure the balance of the terms on either side of the differential equation because the exponential was common. Here we present a different, but equivalent, point of view. We have factored $A$ and the exponential function from both terms because they are common, and then we should pick

$$\lambda = -\frac{1}{RC}$$

to ensure the differential equation is satisfied.[5] Both approaches Eq. (1.1.10) and Eq. (1.1.17) are equivalent, and will lead to the same result for $\lambda$.

We have found a function that satisfies the equation,

$$Q(t) = A \, e^{-t/(RC)},$$

with an unknown coefficient $A$. The last step is to invoke an initial condition to determine $A$. Initially, the capacitor has a charge $Q_0$. Stated mathematically, $Q(0) = Q_0$. The perspective here is that $Q_0$ is some known number, a parameter that reflects different possible starting charges on the capacitor, but for each discharge, it is a fixed constant. Apply this initial condition,

$$Q(0) = A = Q_0$$

and $A$ is now known. Thus the specific solution is

$$Q(t) = Q_0 \, e^{-t/(RC)}. \qquad (1.1.18)$$

The solution Eq. (1.1.18) is similar to Eq. (1.1.11) except that the time scale is $RC$. In both cases, the concentration Eq. (1.1.11) and the charge on the capacitor Eq. (1.1.18) relax exponentially in time to the state where they become negligibly small.

### 1.1.5 *Abstract viewpoint*

There are three examples of differential equations in this section, Eq. (1.1.7), Eq. (1.1.9) and Eq. (1.1.15). They all have a similar form aside from the choice of symbols. Let $y(t)$ be either $\rho(t)$ or $Q(t)$, then the equations can be written in a standard form,

$$\frac{dy}{dt}(t) = a \, y(t) \qquad (1.1.19)$$

where $a$ is either $\lambda$, $-r/V$ or $-1/(RC)$. The equation is linear.

In addition, we have an initial condition

$$y(0) = y_0, \qquad (1.1.20)$$

which in the first case is $y_0 = 1$ and in the other cases $y_0 = Q_0$.

---

[5] If a product of two factors is zero, then at least one of the factors must be zero, and we certainly do not want $A = 0$ otherwise our solution Eq. (1.1.16) will be just zero.

The solutions we found, Eq. (1.1.6), Eq. (1.1.11) and Eq. (1.1.18), suggest that we try a guess for the solution of the form

$$y(t) = e^{\lambda t}. \tag{1.1.21}$$

There is nothing wrong with making a guess for a solution, but we *must* verify that it is a solution. In the process, we expect to determine the unknown constant $\lambda$.[6] The way this is done is to invoke a simple principle.

## Principle 1

*If we have a guess for a solution, we must replace the unknown in the equation by the guess and verify that the equation is satisfied exactly.*

So we replace $y(t)$ in Eq. (1.1.19) by Eq. (1.1.21) with the result

$$\lambda e^{\lambda t} = a e^{\lambda t}.$$

Our guess will indeed be a solution provided we make the choice $\lambda = a$: the choice ensures that the same expression occurs on both sides of the equation.

An important property of the solutions to Eq. (1.1.19) is that if $y(t)$ is a solution, then so is $C y(t)$, where $C$ is an arbitrary constant. By direct substitution into Eq. (1.1.19),

$$\frac{d}{dt}(C y(t)) = C \frac{dy}{dt}(t) = C a y(t),$$

and $C$ can be cancelled. Solutions with this property are called *homogeneous*; so $\exp(at)$ is a homogeneous solution, and the *general* solution is obtained by multiplying it with an arbitrary constant $C$;

$$y(t) = C e^{at}. \tag{1.1.22}$$

The presence of $C$ is very important because there is also an initial condition (1.1.20) that must be satisfied. Evaluate $y(0)$ in Eq. (1.1.22) and use Eq. (1.1.20) to obtain

$$y(0) = C = y_0.$$

Now we have a completely specified solution,

$$y(t) = y_0 e^{at}. \tag{1.1.23}$$

The solution Eq. (1.1.23) depends on two parameters, $a$ that appears in the equation and $y_0$ which sets the initial condition. When $a > 0$, the solution grows exponentially, and when $a < 0$, the solution decays exponentially. Both behaviors are extremely important in science and engineering because they are associated

---

[6]We often use the same symbols repeatedly. There should be no confusion about using $\lambda$ here which is different from $\lambda$ in Eq. (1.1.6).

with the characterization of instability ($a > 0$, growing solutions that depart from equilibrium) and stability ($a < 0$, return to equilibrium). The time scale for these behaviors is given by $1/|a|$: it is the amount of time for the solution to change appreciably – grows by the factor $e \approx 3$ or decays by the factor $e^{-1} \approx 1/3$.

What happens when $a = 0$? Naturally, the solution to

$$\frac{dy}{dt}(t) = 0 \tag{1.1.24}$$

is just $y(t) = C$, consistent with the choice $\lambda = 0$. It is always reassuring when a procedure produces an expected solution in a special case. Of course, it is questionable whether Eq. (1.1.24) is indeed a differential equation rather than just a specification of a derivative with an obvious anti-derivative.

What else can we say about the nature of the solution to Eq. (1.1.19)? Certainly the solution Eq. (1.1.23) is a very nice function. It is continuous and very "smooth," meaning that it has continuous derivatives at any order. Could we have realized those properties just from the differential equation Eq. (1.1.19)? In an intuitive way we can!

For the equation to have a solution, there must be a function that satisfies it, and since the derivative of the function also appears in the equation, the solution must have a derivative. Thus the function must be continuous. That means the right hand side of Eq. (1.1.19) is continuous, and so the derivative must be continuous since both sides of the equation must satisfy the same properties. So far, we have concluded the solution must be continuous and must have a continuous derivative.

### Principle 2

*The solution to a differential equation is a function that should have the appropriate properties so that any term in the equation makes sense.*

Let us now differentiate Eq. (1.1.19):

$$\frac{d^2y}{dt^2}(t) = a \frac{dy}{dt}(t). \tag{1.1.25}$$

Since the right hand side of Eq. (1.1.25) is continuous, we conclude by Principle 2 that the second derivative is continuous. We simply proceed recursively: differentiate repeatedly,

$$\frac{d^n y}{dt^n}(t) = a \frac{d^{n-1}y}{dt^{n-1}}(t) \tag{1.1.26}$$

and at each step, note that the next derivative must be continuous because the previous derivative is continuous. *All derivatives of the solution must be continuous.* Well, that is certainly true for the solution we found, namely an exponential function.

There is another interesting consequence of Eq. (1.1.26). First note that if we evaluate Eq. (1.1.19) at $t = 0$, we find

$$\frac{dy}{dt}(0) = a\,y_0\,, \tag{1.1.27}$$

since $y(0) = y_0$ is the initial condition. Now use this information in Eq. (1.1.25) to obtain

$$\frac{d^2y}{dt^2}(0) = a\,\frac{dy}{dt}(0) = a^2\,y_0\,.$$

Apply the procedure recursively with Eq. (1.1.26): For the choices $n = 3, 4$, we obtain

$$\frac{d^3y}{dt^3}(0) = a\,\frac{d^2y}{dt^2}(0) = a^3\,y_0\,,$$

$$\frac{d^4y}{dt^4}(o) = a\,\frac{d^3y}{dt^3}(0) = a^4\,y_0\,,$$

and the obvious guess for the pattern is

$$\frac{d^n y}{dt^n}(0) = a^n\,y_0\,. \tag{1.1.28}$$

This guess can be verified by direct substitution into Eq. (1.1.26).

In other words, we know the value of all the derivatives at $t = 0$, and it is precisely this information that we need to evaluate the Taylor series of a function. The solution has the Taylor series,

$$y(t) = \sum_{n=0}^{\infty} \frac{d^n y}{dt^n}(0)\,\frac{t^n}{n!} = y_0 \sum_{n=0}^{\infty} \frac{(a\,t)^n}{n!} \tag{1.1.29}$$

and it should be easy to recognize the result is the Taylor series expansion Eq. (B.3.1) for $\exp(at)$. For a review of the Taylor series, please read Appendix B.

From a practical point of view we are done, but a nagging concern arises. We have constructed a solution, but is it possible for there to be another? There is a clever way to show that another solution is not possible. Suppose there are two solutions $y_1(t)$ and $y_2(t)$ that satisfy both Eq. (1.1.19) and Eq. (1.1.20). Now form the difference $Y(t) = y_1(t) - y_2(t)$ and substitute this as a guess into Eq. (1.1.19) and Eq. (1.1.20). First, use Eq. (1.1.19):

$$\begin{aligned}\frac{dY}{dt}(t) &= \frac{dy_1}{dt}(t) - \frac{dy_2}{dt}(t) \\ &= a\,y_1(t) - a\,y_2(t) \\ &= a\,Y(t)\,. \end{aligned} \tag{1.1.30}$$

So $Y(t)$ must satisfy the same differential equation (1.1.19). In this case, however, the initial condition is

$$Y(0) = y_1(0) - y_2(0) = y_0 - y_0 = 0\,,$$

and if we calculate the Taylor series solution Eq. (1.1.29) for $Y(t)$ we find $Y(t) = 0$.[7] In other words, $Y(t) = 0$ is the only possibility, and that means $y_1(t) = y_2(t)$: there cannot be two different solutions. Very reassuring, since we do not want many different solutions to Eq. (1.1.19) and Eq. (1.1.20) otherwise we would not know which one to use.

Besides building confidence that we have constructed a unique solution to the differential equation and the initial condition, we also observe the importance of an initial condition as revealed by the following principle.

### Principle 3

*Additional information is needed to select a unique solution to a differential equation. If the equation is first order, we need just one more piece of information, typically an initial condition.*

### 1.1.6    *Exercises*

**The following exercises explore the connections between discrete data and exponentials.**

(1) Confirm that the data in Table 1.2 for the temperatures 25 °C and 45 °C can also be approximated by constant relative changes. Then find the exponentials that match the data.

(2) Consider data $f_j$ drawn from an exponential function

$$f(t) = A\,e^{\alpha t}$$

at fixed intervals of time $t_j = j\,\triangle t$. Show that $f_j = f(t_j)$ satisfies

$$f_{j+1} = R\,f_j\,,$$

and find an expression for $R$. What is the relative change in $f_j$?

(3) One way to test whether the data in Table 1.3 matches a quadratic $\rho(t) = 1 + a\,t + b\,t^2$ is to calculate what the absolute difference should be. Let $\rho_j = \rho(j\,\triangle t)$;

    (a) Show that $\rho_{j+1} - \rho_j = a\,\triangle t + b\,(2j+1)\,(\triangle t)^2$.

    (b) Can the data in Table 1.3 match this behavior?

(4) Suppose instead of the solution to (1.1.4),

$$e^{j\,\ln(1+\lambda\triangle t)}\,,$$

---

[7]The result is intuitive but to establish it rigorously we must first recognize that $Y(t)$ is an analytic complex function.

we consider the simpler function,

$$\rho_j = e^{j(1+\lambda \triangle t)}.$$

Does this discrete relationship become a continuous function as $\triangle t \to 0$? To appreciate what might be happening, plot $\rho_j$ for $\triangle t = 0.2, 0.05$ with $\lambda = 1$. What can you conclude from your results?

(5) Suppose that the data in Table 1.3 gave the absolute difference as a constant instead of the relative change. The recursion formula would be

$$\rho_{j+1} - \rho_j = C \triangle t.$$

Determine the solution $\rho_j$ to this recursion and then take the limit $\triangle t \to 0$. What is the resulting continuous function and what differential equation does it satisfy?

**The point of the next exercises is to explore various ways exponentials can arise and how best to "read" them.**

(6) Can you simplify the following expressions?

$$e^{\alpha^2}, \quad \left(e^\alpha\right)^2, \quad \ln\left(e^{\alpha^2}\right), \quad \frac{e^x}{x}, \quad e^{2\ln(x+1)+3}.$$

(7) Is there a value for the constant $\alpha$ so that

$$e^{\alpha x} = 1 + x^2$$

for all values of $x$?

(8) Let us try some other guesses,

(a) $y(t) = \alpha \sin(t) + \beta \cos(t)$,
(b) $y(t) = \alpha + \beta t$,

for the solution to the differential equation,

$$\frac{dy}{dt}(t) + 2 y(t) = 0.$$

Are there values for $\alpha$ and $\beta$ that will give non-zero solutions?

**These are some basic exercises to help develop and improve skill.**

(9) Find the solution to the following differential equations:

(a)

$$\frac{dH}{dz}(z) + \pi H(z) = 0, \quad \text{with } H(0) = 1.$$

**Answer:** $H(1) = e^{-\pi}$.

(b)

$$\frac{dy}{dt}(t) - \sigma y(t) = 0, \quad \text{with } y(1) = \pi.$$

**Answer:** With $\sigma = 1$, $y(2) = 8.54$.

(c)

$$\frac{dx}{dy}(y) = a\,x(y), \quad \text{with } x(0) = 2.$$

**Answer**: With $a = 2$, $x(1) = 14.8$.

**Differential equations can be used to confirm the properties of exponential functions.**

(10) Take two functions $y(t)$ and $z(t)$ that satisfy the differential equations,

$$\frac{dy}{dt}(t) = a\,y(t), \quad \text{and} \quad \frac{dz}{dt}(t) = b\,z(t),$$

respectively. The solutions are:

$$y(t) = y_0\,e^{at}, \quad \text{and} \quad z(t) = z_0\,e^{bt}.$$

Consider the product $w(t) = y(t)\,z(t)$ and determine the differential equation it must satisfy. Determine the solution for $w(t)$: you will also need the initial condition from the product $y(0)\,z(0)$. Then compare your solution for $w(t)$ with the product $y(t)\,z(t)$: you should have verified the identity (A.1.3) without using it in your reasoning.

**Some problems that arise in applications.**

(11) Suppose the contaminant has an initial amount of 2 lb. How long must we wait until the amount given by Eq. (1.1.11) is below 0.1 lb?
    **Answer**: 100 min for $V = 100$ gal and $r = 3$ gal/min.

(12) A worker accidentally spills some contaminant into a drum filled to the top with 10 gal of water. He quickly grabs a hose and begins to pour in water to clear out the contaminant. The water flows through the hose at 2 gal/min and spills out of the tank at the same rate. After ten minutes, his supervisor catches him and stops him. The worker claims he only spilled a small amount. The supervisor takes a reading and finds the concentration of the contaminant is 0.01 lb/gal. How much did the worker spill?
    **Answer**: 0.74 lb.

(13) A worker pours salt into a 100 gal tank at a rate of 2 lb/min. The target is a 0.01 lb/gal concentration of salty water. Derive an equation that determines the rate of change of concentration in time. Determine how long the worker must pour in the salt to reach the desired concentration.
    **Answer**: 30 s.
    Now suppose the worker pours in the salt from a big bag at a rate that slows down exponentially. Specifically the rate is $4\exp(-2t)$ lb/min. How long will it take to reach the desired concentration?
    **Answer**: 0.35 min.

(14) A bacterial culture is known to grow at a rate proportional to the amount present. After one hour, there are 1000 strands of bacteria and after four hours, 3000 strands. Find an expression for the number of strands present in the culture at any time $t$. What is the initial number of strands?

**Answer**: 693 strands.

(15) A boat of mass 500 lb is approaching a dock at speed 10 ft/s. The captain cuts the motor and the boat coasts to the dock. Assume the water exerts a drag on the boat proportional to its current speed; let the constant of proportionality be 100 lb/s. How far away from the dock should the captain cut the motor if he wants the boat to reach the dock with a small speed, 3 in/s say?

**Answer**: 48.75 ft.

(16) According to Lambert's law of absorption, the percentage absorption of light $\triangle I/I$ is proportional to the thickness $\triangle z$ of a thin layer of the medium the light is passing through. Suppose sunlight shining vertical on water loses its intensity by a factor of a half at 10 ft below the water surface. At what depth will the intensity be reduced by a factor of $1/16$ (essentially dark)?

**Answer**: 40 ft.

(17) Carbon dating is based on the theory of radioactive decay. A certain concentration of carbon-14 is always present in the atmosphere and is digested by plants and thus animals. When they die, the carbon in their remains begins to decay radioactively, the rate of decay being proportional to the concentration present. The constant of proportionality, called the decay constant, is about $1.21 \times 10^{-4}$ yr$^{-1}$. The concentration $\rho(t)$ of carbon-14 present after decaying for $t$ (yr) is measured in the laboratory and compared to the concentration $\rho_0$ currently present in the atmosphere, a quantity assumed constant in time.[8] Suppose the measurement of carbon-14 in a wooden sample gives the ratio $\rho(t)/\rho_0 \approx 0.01$. Determine how long ago the tree died.

**Answer**: 38,060 yr ago.

(18) One of the difficulties in carbon dating is the sensitivity of the results. To explore this issue, consider an exponential decay where there is a relative uncertainty $\epsilon$ in the measurement of $\rho(t)/\rho_0$. As a result there will be uncertainty in the calculation of the age. Determine this uncertainty as a function of $\epsilon$. Suppose there is a ten percent uncertainty in the measurement made in the previous exercise for $\rho(t)/\rho_0$. What will be the range in time that can be estimated.

**Answer**: Between 37,272 and 38,930 yr ago.

---

[8]This is not strictly true and calibration curves are determined through comparison with tree-ring data.

### 1.1.7 *Additional deliberations*

Hopefully, the exercises have served to re-enforce the use of exponentials as a technique to solve differential equations. But this technique relies on two properties of Eq. (1.1.21) that should be emphasized and both can be best presented by rewriting the equation as

$$\frac{dy}{dx}(x) - a\,y(x) = 0\,. \tag{1.1.31}$$

In this form, all the terms containing the unknown function $y(x)$ are placed on the left and there are no terms on the right hand side.[9]

The first property, an extremely important property, is that the equation for $y$ is *linear*. In practical terms, that means $y$ appears by itself in each term; it is not the argument of a function such as $y^2$ or $\sin(y)$. More technically, it means that if $y_1(x)$ and $y_2(x)$ are solutions to the equation, then their sum, $y_1(x) + y_2(x)$, is also a solution. Note how this property was used in Eq. (1.1.30).

If the differential equation is nonlinear, for example,

$$\frac{dy}{dx}(x) + y^2(x) = 0\,,$$

then the solution does not take the form of an exponential. If we guess $y(x) = \exp(\lambda x)$ and substitute into the equation, we find

$$\lambda\,e^{\lambda x} + e^{2\lambda x} = 0$$

and there is no way to balance $\exp(\lambda x)$ with $\exp(2\lambda x)$. It is only when the equation is linear that all terms with $y(x)$ will be replaced with the same exponential $\exp(\lambda x)$ and a balance is always possible.

The other important property reflects the presence of a zero on the right hand side of Eq. (1.1.31) and the equation is called *homogeneous*. The consequence is that if $y(x)$ is a solution to Eq. (1.1.31), then so is $C\,y(x)$, where $C$ is an arbitrary constant. By direct substitution into Eq. (1.1.31),

$$\frac{d}{dx}\big(C\,y(x)\big) - a\,C\,y(x) = C\left(\frac{dy}{dx}(x) - a\,y(x)\right) = 0\,.$$

The fact that $C$ can be factored out ensures the $C\,y(x)$ is a solution. This property is used in the solution Eq. (1.1.22), and the presence of the constant is used to satisfy the initial condition as in Eq. (1.1.23). In the next section, we will consider what to do when there is a non-zero term on the right hand side of Eq. (1.1.31).

## 1.2 Forcing effects

Forcing effects occur as inputs to a system. One simple example is the effects of blowing air over the surface of a hot cup of coffee; we see waves generated on the

---

[9]The independent variable $t$ has been replaced by $x$: it is quite common for mathematicians to use $x$ in place of $t$.

surface in response to the forcing effects of the air flow. We pick up electro-magnetic signals, forcing effects, when we listen to the radio. The thrust from jet engines on an aircraft is a forcing effect that drives the motion. Adding chemicals to a reacting unit induces reactions to produce products. Often, forcing effects are under the control of the scientist or engineer, and by studying the response of the system to the forcing effects, we can understand certain properties of the system. Hopefully, this will lead to improved design or changes in operating conditions.

From a mathematical perspective, forcing effects appear as additional terms in a mathematical model, but these terms contain specified functions. The response to the forcing terms is expressed by the solution to the mathematical model. Different forcing terms will produce different responses, and if the system is described by a differential equation, then this means we should expect different solutions to the differential equation. These solutions will be called *particular* or *inhomogeneous* solutions. There is still the evolution of the system from some initial state, which would occur even in the absence of forcing. This *free* response of the system is reflected in the presence of *homogeneous* solutions and we saw several examples in the previous section; see Eq. (1.1.6), Eq. (1.1.11) and Eq. (1.1.18). For linear differential equations, we must combine the particular with the homogeneous solution multiplied by some arbitrary constant to create the *general* solution. The inclusion of a particular solution in the general solution is one of the key features of this section. As before, the application of initial conditions to the general solution will produce a *specific* solution. Several examples will help make these thoughts concrete.

### 1.2.1  *Constant inflow as an input*

The first example supposes that there is a dam fed by a river carrying a soluble contaminant. The volume of water in the dam is held constant by releasing water from the dam at the same rate as the river supplies water. What we want to know is how long does it take the contamination to affect the dam if the dam is initially clean? Obviously, the amount of contamination in the dam changes according to the amount entering through the river and the amount leaving from the dam release. We view the natural flow of clean water through the dam as the "system" and the inclusion of a contaminant in the river as a "forcing effect."

There are some details necessary to state a mathematical equation of balance, and the first step is to draw a "cartoon" of what is happening, for example, as shown in Fig. 1.6. At the same time, it is useful to introduce symbols to represent important quantities. For example, we represent the volume of the dam by $V\,(\mathrm{km}^3)$. Let the rate of flow of the river water into the dam be $r\,(\mathrm{m}^3/\mathrm{s})$; it is also the rate of flow released from the dam. Let the concentration of contaminant in the river water be $c\,(\mathrm{kg/m}^3)$. Let the concentration of contaminant in the dam, assumed uniform through mixing, be $\rho(t)\,(\mathrm{kg/m}^3)$.

The next step is to identify what processes are controlling the situation. Water

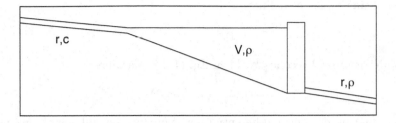

Fig. 1.6   A schematic of the river flow through the dam.

flows through the dam in a way that the amount of water is conserved. Because the amount of water flowing into the dam is the same as the amount of water leaving the dam, the amount of water in the dam remains unchanged; in other words, the volume of water in the dam is constant in time. On the other hand, the amount of the contaminant that enters the dam need not be the same amount leaving, with the consequence that the amount of pollutant in the dam can change in time.

So how do we measure the change in the amount of contaminant in the dam? During a time interval of $\triangle t$ (s), the amount is given by the change in density $\rho(t + \triangle t) - \rho(t)$ multiplied with the volume. Here we must be the careful because the density is measured in kilograms per cubic meter but the volume of the dam is measured in cubic kilograms. Thus we need the conversion factor $\text{km}^3 = 10^9 \, \text{m}^3$ to measure the change in amount during $\triangle t$ as

$$10^9 \, V \left[ \rho(t + \triangle t) - \rho(t) \right].$$

This change can only come about from the difference in the amount entering the dam and the amount leaving it.

$$10^9 \, V \left[ \rho(t + \triangle t) - \rho(t) \right] = \text{flow in} - \text{flow out} . \qquad (1.2.1)$$

Note that the units on the left are kilograms. We need to determine the amount of pollution flowing in with the river in units of kilograms. Since the volume of water flowing in is $r \, \triangle t \, (\text{m}^3)$ and carries with it a pollutant concentration $c \, (\text{kg/m}^3)$,

$$\text{the amount flowing in} = c \, r \, \triangle t .$$

Similarly, the rate of release from the dam is $r$ and carries with it a concentration $\rho(t)$; the concentration in the water leaving is the same as the concentration in the dam,

$$\text{the amount flowing out} = \rho \, r \, \triangle t .$$

As a result, Eq. (1.2.1) becomes

$$\frac{\rho(t + \triangle t) - \rho(t)}{\triangle t} = \frac{10^{-9} r}{V} \left( c - \rho(t) \right)$$

with the obvious limit,

$$\frac{\mathrm{d}\rho}{\mathrm{d}t}(t) = \frac{10^{-9} r}{V} \left( c - \rho(t) \right). \qquad (1.2.2)$$

For convenience, let us define

$$\alpha = \frac{10^{-9}r}{V}.$$                                    (1.2.3)

Its units are reciprocal seconds.[10] Then Eq. (1.2.2) becomes

$$\frac{d\rho}{dt}(t) = \alpha\left(c - \rho(t)\right),$$          (1.2.4)

which is a first-order, linear differential equation for the unknown concentration $\rho(t)$. It contains two parameters, $\alpha$ and $c$, to be regarded as fixed for each attempt to solve the equation. Equation (1.2.4) is similar to the differential equation of the last section (1.1.9) except that there is a source term $\alpha c$, representing the input of contaminant into the dam, and this additional term will affect the solution.

Progress in solving (1.2.4) starts with rewriting the equation so that only terms with the unknown variable $\rho$ appear on the left hand side,

$$\frac{d\rho}{dt}(t) + \alpha\,\rho(t) = \alpha\,c.$$            (1.2.5)

Compare this equation with Eq. (1.1.31) and the presence of the forcing term on the right hand side is clear. The differential equation Eq. (1.2.5) is called an *inhomogeneous* equation because of the presence of the forcing term. While not apparently so, the forcing term is a function, a function that is constant in time and determined by the two parameters $\alpha$ and $c$.

The technique we use in Sec. 1.1 is to assume a solution of the form

$$\rho(t) = e^{\lambda t},$$                                      (1.2.6)

and substitute it into Eq. (1.2.5), but this time the result

$$\lambda\,e^{\lambda t} + \alpha\,e^{\lambda t} = \alpha\,c$$

displays a problem. There is no exponential on the right hand side to balance the exponential behavior in $t$ on the left hand side. Our guess Eq. (1.2.6) fails! So we have to be clever and guess a different solution that will balance the right hand side.

We seek a guess so that the terms on the left hand side of (1.2.5) will be a constant to balance the constant on the right hand side. So let us start with assuming the derivative is a constant. That means $\rho(t)$ will be a linear function in $t$ and the linear function will not balance the right hand side. The next attempt is to assume that the term $\alpha\,\rho(t)$ is a constant. The derivative will be zero and a balance with the right hand side is possible. So a guess that works is $\rho(t) = A$. As required by Principle 1 in Sec. 1.1, we must verify this guess works by direct substitution into Eq. (1.2.5):

$$0 + \alpha\,A = \alpha\,c,$$

---

[10]It is often convenient to define combinations of parameters, here $r$ and $V$, as new parameters to simplify the control of algebra and to reveal the importance of the combination. In this case, $\alpha$ will prove to be a decay rate.

and with the choice $A = c$, the balance is perfect. Curiously, the response revealed by the solution, a constant in time, parallels the forcing term, a constant in time. In other words, constant forcing solicits a constant response! We will find that this is a common pattern; the nature of the response follows the nature of the forcing.

We have found a solution, $\rho(t) = c$, called the *particular* solution $\rho_\mathrm{p}(t)$, which is a direct response to the forcing term. But this solution is too limited. The concentration in the dam cannot be a constant in time; we expect it to be zero initially and then to increase to some value in time. The next important idea is that we compose a solution of two parts. To the particular solution we add another solution, denoted as $\rho_\mathrm{new}$. Our guess for the solution now takes the form,

$$\rho(t) = \rho_\mathrm{p}(t) + \rho_\mathrm{new}(t) = c + \rho_\mathrm{new}(t) \,, \qquad (1.2.7)$$

and we must substitute our guess into Eq. (1.2.5) and insist that it satisfies the equation. The result is

$$\frac{\mathrm{d}\rho_\mathrm{new}}{\mathrm{d}t}(t) + \alpha\,c + \alpha\,\rho_\mathrm{new}(t) = \alpha\,c \,,$$

or

$$\frac{\mathrm{d}\rho_\mathrm{new}}{\mathrm{d}t}(t) + \alpha\,\rho_\mathrm{new}(t) = 0 \,, \qquad (1.2.8)$$

because the forcing term has been cancelled by the particular solution. This is brilliant because we can now solve Eq. (1.2.8) by recognizing that it is a homogeneous differential equation and can be solved by assuming a solution of the form Eq. (1.2.6) just as we did in Sec. 1.1. After substituting Eq. (1.2.6) into Eq. (1.2.8),

$$\frac{\mathrm{d}}{\mathrm{d}t}\left(e^{\lambda t}\right) + \alpha\,e^{\lambda t} = 0 \,,$$

and we can make this balance perfectly (for any value of $t$) with the choice $\lambda = -\alpha$. In other words, we have determined $\rho_\mathrm{new}$ as the homogeneous solution

$$\rho_\mathrm{h}(t) = e^{-\alpha t} \,.$$

The *general* solution Eq. (1.2.7) is the combination of $\rho_\mathrm{p}(t)$ and $\rho_\mathrm{h}(t)$ multiplied with an arbitrary constant $B$,

$$\rho(t) = c + B\,e^{-\alpha t} \,, \qquad (1.2.9)$$

where $B$ is still to be determined – remember $c$ and $\alpha$ are parameters considered known. We need the constant $B$ because we have yet to take into account how the concentration started, in other words, the initial condition. We assume that the dam is clear of contaminants initially, so at $t = 0$, $\rho(0) = 0$. This condition must be applied to Eq. (1.2.9) to determine $B$.

$$\rho(0) = c + B = 0 \,,$$

and $B = -c$ is the result. Finally, then, we have completed the construction of the *specific* solution, which we may write as

$$\rho(t) = c\left(1 - e^{-\alpha t}\right) . \qquad (1.2.10)$$

The claim, easily verified by direct substitution, is that Eq. (1.2.10) satisfies the differential equation Eq. (1.2.5) and the initial condition $\rho(0) = 0$.

What does the solution tell us? Let us look at a graph for a particular choice of the parameters. Suppose the dam has dimensions 7 km long, 2 km wide and 100 m deep: then $V = 1.4 \, \text{km}^3$. Let the flow of the river be $r = 7 \, \text{m}^3/\text{s}$. Consequently, Eq. (1.2.3) gives $\alpha = 10^{-9} \times 7/1.4 = 5 \times 10^{-9}$. If the concentration of the contaminant in the river is $c = 0.8 \, \text{kg/m}^3$, then Eq. (1.2.10) becomes

$$\rho = 0.8 \left( 1 - e^{-5 \times 10^{-9} t} \right), \tag{1.2.11}$$

where $t$ must be measured in seconds. But now we see that $t$ in seconds is unwise. Changes in the contaminant in the dam are not going to happen in seconds. For the argument of the exponential to be some reasonable value, $t$ would need to be about $10^9$ s, which is about $10^5$ days or 30 yr. So we should measure the time in years. The number of seconds in a year is approximately $3.15 \times 10^7$. Thus change the independent variable $t = 3.15 \times 10^7 T$ where $T$ is now in years. Rewrite Eq. (1.2.11) by replacing $t$,

$$\rho(t) = 0.8 \left( 1 - e^{-0.158T} \right).$$

This is a much more reasonable way to understand the changes in $\rho(t)$.

Incidentally, the change in scale of the time variable also changes the differential equation Eq. (1.2.5). We now regard the concentration as $\rho(T)$. By the change rule of differentiation,

$$\frac{d\rho}{dT} = \frac{dt}{dT} \frac{d\rho}{dt} = 3.15 \times 10^7 \frac{d\rho}{dt}.$$

The differential equation Eq. (1.2.5) becomes

$$\frac{d\rho}{dT}(T) + \beta \rho(T) = \beta c, \tag{1.2.12}$$

where $\beta = 3.15 \times 10^7 \alpha = 0.158 \, \text{yr}^{-1}$. In terms of $c$ and $\beta$,

$$\rho(T) = c \left( 1 - e^{-\beta T} \right) \tag{1.2.13}$$

is the solution to Eq. (1.2.12) with initial condition $\rho(0) = 0$.

The concentration Eq. (1.2.13) is shown as a function of time in Fig. 1.7. There are several observations worth making. The concentration $\rho(t)$ approaches a constant $0.8 \, \text{kg/m}^3$ quite rapidly (on the time scale of years). Already after 15 yr, the concentration is quite close to $c = 0.8 \, \text{kg/m}^3$. If we now look at Eq. (1.2.13) and consider $T \to \infty$, then $\rho(T) \to c = 0.8 \, \text{kg/m}^3$, which is what we see in Fig. 1.7. We call this an exponential approach to an equilibrium state. We may use this interpretation in understanding (1.2.13) for any choice of $c$ and $\beta$. The solution approaches an equilibrium $c$ on a time scale of $1/\beta$. In this particular example, it seems to take the dam a long time to be fully polluted.

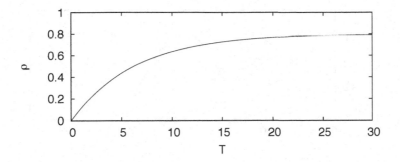

Fig. 1.7   Concentration of the contaminant in the dam over time.

---

**Reflection**

Looking back on what was done, there appears to be a few key ideas. The derivation of a differential equation that models a real-life situation depends very much on what is believed to control the process, in this case, the conservation of pollutant. When the differential equation is written in standard form, only terms with the unknown function on the left, a forcing term (function only of $t$) becomes apparent. A general solution can be constructed by finding an appropriate (particular) solution that balances the forcing term and adding to it a homogeneous solution (an exponential) multiplied with an unknown coefficient. The unknown coefficient is determined by the initial condition, leading to the final specific solution. Obvious questions arise. Does this strategy always work? How can we find the particular solution in general? The reader may have more. But let us consider a few more examples before drawing general conclusions.

---

## 1.2.2   *Periodic inflow as an input*

So far, we have considered a simple example of a forcing term in the differential equation Eq. (1.2.12) that is just a constant function, representing the inflow of a pollutant of constant concentration. Let us now consider other possible variations. Suppose the concentration in the river varies in time. For example, the pollutants are dumped into the river during the summer, but not as much in the winter. A simple model of this variation is

$$c(T) = 0.8\big[1 + \cos(2\pi T)\big]. \tag{1.2.14}$$

The variation is shown in Fig. 1.8. Note that the choice has a mean input of $0.8\,\mathrm{kg/m^3}$: This choice is deliberate so that a comparison with the result (1.2.13) can be made. In this way, we can assess the consequence of varying the inflow of pollutants subject to the same amount being dumped annually.

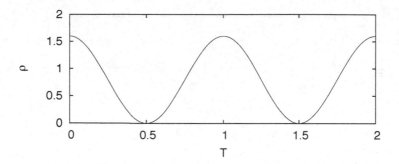

Fig. 1.8    Pollutant concentration in the river over a two-year cycle.

The differential equation that needs to be solved is simply Eq. (1.2.12) with $c$ replaced by Eq. (1.2.14).

$$\frac{d\rho}{dT}(T) + \beta\,\rho(T) = 0.8\,\beta\left[1 + \cos(2\pi T)\right].  \qquad (1.2.15)$$

We still have the parameter $\beta$, but the parameter $c$ has been replaced with a specified function of time $T$. The input in Eq. (1.2.15) contains two parts, a constant $0.8\beta$ and a term with a trigonometric function $0.8\beta\cos(2\pi T)$. The constant part we know how to treat, but what about the trigonometric part? An obvious guess, based on balancing $\beta\,\rho(T)$ with the right hand side, might be to add a cosine expression to the constant in the particular solution. Thus we might try

$$\rho_{\mathrm{p}}(T) = a_1 + a_2\,\cos(2\pi T).$$

Our hope is that we can find values for $a_1$ and $a_2$ that will ensure that our guess is a solution to Eq. (1.2.15). Upon substitution of our guess into Eq. (1.2.15) we obtain

$$-2\pi a_2\,\sin(2\pi T) + \beta\left[a_1 + a_2\,\cos(2\pi T)\right] = 0.8\beta + 0.8\beta\,\cos(2\pi T).$$

Clearly there is a constant term on the left and one on the right that we can balance. There is also a cosine on the left that can balance the one on the right, but there is also a sine that has no counterpart on the right to balance. We cannot add a sine term to the forcing term, but we can add one to our particular guess. Thus we try a particular solution of the form,

$$\rho_{\mathrm{p}}(T) = a_1 + a_2\,\cos(2\pi T) + a_3\,\sin(2\pi T),$$

and substitute it into Eq. (1.2.15). Since

$$\frac{d\rho_P}{dT}(T) = -2\pi a_2\,\sin(2\pi T) + 2\pi a_3\,\cos(2\pi T),$$

(1.2.15) becomes

$$-2\pi a_2\,\sin(2\pi T) + 2\pi a_3\,\cos(2\pi T) + \beta\left[a_1 + a_2\,\cos(2\pi T) + a_3\,\sin(2\pi T)\right]$$
$$= 0.8\beta + 0.8\beta\,\cos(2\pi T).$$

If our guess is to be successful, the behavior in time on the left must match perfectly the time behavior on the right. There are constants (no dependency on time), and trigonometric functions, $\sin(2\pi T)$ and $\cos(2\pi T)$. A perfect balance can be enforced by the choice,

balance constant terms: $\qquad\qquad \beta a_1 = 0.8\beta$,

balance sine terms: $\qquad\qquad -2\pi a_2 + \beta a_3 = 0$,

balance cosine terms: $\qquad\qquad 2\pi a_3 + \beta a_2 = 0.8\beta$.

This perfect balance is possible because:

- the forcing terms can be separated into different functions of time,
- the constant forcing term can be matched by a constant response,
- the trigonometric forcing term can be matched with a combination of trigonometric functions of the *same argument*,
- there are as many coefficients $a_n$ as there are functions to match.

The matching conditions provide equations for the unknown coefficients, hence *the method of undetermined coefficients*. The equation for the coefficient $a_1$ is simple. There are two equations for the coefficients $a_2$ and $a_3$; a review of systems of linear algebraic equations is provided in Appendix C. Once we determine the coefficients from the matching conditions, we have completed the construction of the particular solution,

$$\rho_{\mathrm{p}}(T) = 0.8 + \frac{0.8\beta^2}{4\pi^2 + \beta^2}\cos(2\pi T) + \frac{1.6\pi\beta}{4\pi^2 + \beta^2}\sin(2\pi T). \qquad (1.2.16)$$

The particular solution Eq. (1.2.16) does not satisfy the initial condition, $\rho(0) = 0$. We need to add another solution, the homogeneous solution $\rho_{\mathrm{h}}(T)$. Substitute the combination,

$$\rho(T) = \rho_{\mathrm{p}}(T) + \rho_{\mathrm{h}}(T)$$

into Eq. (1.2.15) to obtain

$$\left(\frac{\mathrm{d}\rho_{\mathrm{p}}}{\mathrm{d}T}(T) + \beta\,\rho_{\mathrm{p}}(T)\right) + \left(\frac{\mathrm{d}\rho_{\mathrm{h}}}{\mathrm{d}T}(T) + \beta\,\rho_{\mathrm{h}}(T)\right) = 0.8\,\beta\left[1 + \cos(2\pi T)\right].$$

The various terms have been grouped and placed in parentheses for emphasis. The point is that the first group of terms contain the particular solution and they balance the forcing terms by the matching conditions that we used to determine the unknown coefficients. Once the terms with the particular solution and the forcing terms are removed, we are left with the homogeneous equation,

$$\frac{\mathrm{d}\rho_{\mathrm{h}}}{\mathrm{d}T}(T) + \beta\,\rho_{\mathrm{h}}(T) = 0. \qquad (1.2.17)$$

From a strategic point of view, the particular solution "takes care of" the forcing terms, leaving a homogeneous equation which contains no forcing terms. The particular solution requires a guess based on the nature of the forcing terms, while the homogeneous equation can always be solved by guessing an exponential function as done in Sec. 1.1. Try $\rho(T) = \exp(\lambda T)$ and substitute into Eq. (1.2.17). A perfect match is possible with the choice $\lambda = -\beta$. The homogeneous solution is therefore

$$\rho_{\text{h}}(T) = A\,e^{-\beta T}\,,$$

where $A$ is an arbitrary constant.

The general solution becomes

$$\rho = a_1 + a_2 \cos(2\pi T) + a_3 \sin(2\pi T) + A\,e^{-\beta T}\,. \qquad (1.2.18)$$

Although we have determined the coefficients $a_n$ already, it is convenient to leave the coefficients in this expression as symbols to keep the remaining algebra as simple as possible. The point is that we know what they are; they are no longer unknown coefficients.

We are now in a position where the initial condition $\rho(0) = 0$ can be applied.

$$\rho(0) = a_1 + a_2 + A = 0\,. \qquad (1.2.19)$$

Bearing in mind that $a_1$ and $a_2$ have been determined – Eq. (1.2.16), Eq. (1.2.19) imposes a condition on $A$; $A = -a_1 - a_2$. Once $\beta$ is specified, the coefficients $a_n$ can be evaluated and then $A$ can be evaluated and the solution can be plotted. With the same previous choice $\beta = 0.158\,\text{yr}^{-1}$, the solution is shown in Fig. 1.9.

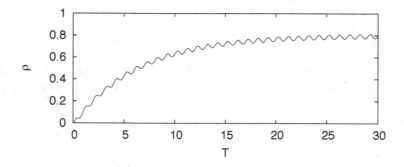

Fig. 1.9　Concentration of the contamination in the dam over time.

The behavior of the solution shown in Fig. 1.9 is that of an oscillatory pattern riding on top of what appears to be an exponential approach to the constant value $\rho = 0.8\,\text{kg/m}^3$. Indeed, the apparent exponential decay is very close to that described by Eq. (1.2.13) and shown in Fig. 1.7.[11] On the other hand, the long term behavior of the solution is different in that it is a steady oscillatory pattern around a mean value. As scientists and engineers, we are frequently interested in long term behavior and the two results, Eq. (1.2.13) and Eq. (1.2.18), show two

---

[11]An environmentalist might note that by taking the annual rate of deposit of contaminants into the river does not significantly affect the annual amount of contaminant in the dam in the long term. Such observations give credence to estimates of long term effects of pollution on the environment based on estimates of the annual rate of pollutant deposition.

possible steady patterns for the long term behavior, one a constant and the other a steady oscillation about a mean.

A new feature of the long term behavior is that there are now two time scales that affect the result. One is the exponential decay associated with the homogeneous solution, $T \sim 1/\beta$ (yr), while the other is a consequence of the time scale in the forcing term, specifically the period of oscillation in the cosine term $T \sim 1$ yr.

---

**Reflection**

We have seen two examples with different forcing terms, Eq. (1.2.5) and Eq. (1.2.15). The particular solutions have followed the behavior of the forcing term, and the general solution has been a combination of the particular solution and the homogeneous solution. Since the homogeneous equation is a separate equation without the forcing terms, its solution does not depend on the forcing term and is always the same. There are times when as scientists and engineers we wish to contemplate the consequences of different forcing effects in our mathematical model. We need to calculate the homogeneous solution just once, and be ready to add it to whatever particular solution that is constructed.

---

### 1.2.3 *Discontinuous inflow as an input*

The theme that is emerging is that certain forcing terms, those containing polynomials, sines and cosines, and exponentials, suggest obvious particular solutions. These classes of functions for the forcing terms may seem too restrictive, but scientists and engineers, realizing the value of working with functions whose properties are well known, create more complex forcing terms by selecting piecewise functions made up of polynomials, sines and cosines, and exponentials. A simple example is used here to illustrate how a piecewise solution can still be constructed.

Assume that the pollutants are dumped into the river at an ever increasing rate, until the concentration in the dam reaches a certain level $\rho_c$ (kg/m$^3$) that triggers a shutdown of the pollution (perhaps from some law enforcement). Suppose $T_c$ (yr) is the time that $\rho(T_c) = \rho_c$. The parameter $\rho_c$ is a control parameter but $T_c$ is a quantity that must be determined as part of the solution. A simple model for the forcing term is

$$c(T) = \begin{cases} sT, & \text{while } T < T_c, \\ 0, & \text{when } T > T_c. \end{cases} \tag{1.2.20}$$

Here $s$ (kg $\cdot$ m$^{-3}$ $\cdot$ yr$^{-1}$) is a constant, a parameter.

Replace $c$ in (1.2.12) by $c(T)$ as specified by (1.2.20).

$$\frac{\mathrm{d}\rho}{\mathrm{d}T}(T) + \beta\, \rho(T) = \beta\, c(T). \tag{1.2.21}$$

How do we now solve Eq. (1.2.21) when $c(T)$ is given by Eq. (1.2.20)? First observe that the right hand side is discontinuous at $T = T_c$. We do not want the solution to be discontinuous at $T = T_c$, because the derivative will not exist and the differential equation will have no meaning at $T = T_c$. By Principle 2 though, we must accept that the derivative can be discontinuous at $T = T_c$ to match the discontinuity in $c(T)$. What this observation suggests is that we seek separate solutions in each time range, $T < T_c$ and $T > T_c$, and then match them at $T = T_c$ so that the solution is continuous, while its derivative jumps in value.

During $T < T_c$, Eq. (1.2.21) is

$$\frac{d\rho}{dT}(T) + \beta\,\rho(T) = \beta\,sT\,. \tag{1.2.22}$$

The strategy to solve this equation is the same as before: find a particular solution that accounts for the presence of the forcing term, and add a homogeneous solution to Eq. (1.2.22) (with $sT = 0$). What should we select as a candidate for the particular solution $\rho_p$? An obvious candidate is $\rho_p(T) = aT$ since it is similar to $sT$, but its derivative is a constant which will not balance the right hand side. Instead, we should try $\rho_p(T) = a_1 + a_2 T$, where $a_1$ and $a_2$ are two undetermined coefficients. Upon substitution into Eq. (1.2.22) (remember Principle 1), we find

$$\frac{d}{dT}(a_1 + a_2 T) + \beta\,(a_1 + a_2 T) = \beta s T\,,$$

or

$$(a_2 + \beta a_1) + \beta a_2 T = \beta s T\,.$$

We can balance the behaviors in $T$ perfectly by setting $a_2 = s$ and $a_1 = -s/\beta$. Thus,

$$\rho_p(T) = -\frac{s}{\beta} + sT\,. \tag{1.2.23}$$

The response to a linear forcing term is a solution growing linearly. To this solution must be added the homogeneous solution, $A\exp(-\beta T)$, which we have calculated and used before in Eq. (1.2.9) and Eq. (1.2.18). The general solution is therefore

$$\rho(T) = -\frac{s}{\beta} + sT + A\,e^{-\beta T}\,.$$

The initial condition $\rho(0) = 0$ leads to $A = s/\beta$. Thus,

$$\rho(T) = \frac{s}{\beta}\left(e^{-\beta T} - 1\right) + sT\,, \tag{1.2.24}$$

is the solution valid for $T < T_c$.

Now it is time to determine $T_c$. It is defined to be the time when $\rho(T_c) = \rho_c$. Thus,

$$\rho_c = \frac{s}{\beta}\left(e^{-\beta T_c} - 1\right) + sT_c\,, \tag{1.2.25}$$

which provides an equation for $T_c$ when $\rho_c$ is given. Unfortunately this equation cannot be solved by algebraic means; instead it must be solved by numerical or graphical means.

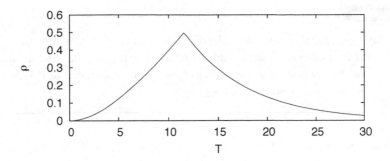

Fig. 1.10    Concentration of the contamination in the dam over time.

To complete the solution for all $T > 0$, we must construct the solution for $T > T_c$. The equation to be solved now is

$$\frac{d\rho}{dT}(T) + \beta\,\rho(T) = 0\,.$$

There is no forcing term. This is just the homogeneous problem with the solution $B \exp(-\beta T)$. But what is the initial condition? Since we want the solution to be continuous at $T = T_c$, we must require $\rho(T_c) = \rho_c,$[12] or

$$Be^{-\beta T_c} = \rho_c\,, \quad \text{so } B = \rho_c e^{\beta T_c}\,,$$

and the solution can be written as

$$\rho(T) = \rho_c e^{\beta(T_c - T)}\,. \tag{1.2.26}$$

This solution is valid for $T > T_c$.

To illustrate what this solution looks like, pick $s = 0.08\,\mathrm{kg/(m^3\,yr)}$. This choice means that after ten years the concentration of pollutants in the river will reach the constant value we took before in obtaining the solution (1.2.13). Further, set $\rho_c = 0.5\,\mathrm{kg/m^3}$. A solution for $T_c$ must be determined from Eq. (1.2.25). Numerical methods can be used and the result is $T_c = 11.56$ yr. The full solution is shown in Fig. 1.10. After an initial adjustment, called a transient, the solution approaches a straight line. If there were no restriction, this would be the long term behavior, linear growth. However, the solution terminates at $T_c = 11.56$ yr. After that, the concentration decays away as the pollutant now washes out of the dam. Note that there is a jump in the derivative of the solution at $T_c$; it matches the abrupt change in the forcing term (recall Principle 2).

Note again the different time scales that appear in the solution. The exponentially decaying term in (1.2.24) gives the transient adjustment on a time scale of $1/\beta \sim 6$ yr. After that the linear growth occurs with a time scale of $1/s \sim 12$ yr. Finally, the pollution decays away exponentially on a time scale $1/\beta \sim 6$ yr.

[12]This is an example where the initial condition does not occur at $T = 0$. Instead, a condition is applied at $T = T_c$. The moment in time when the condition is applied depends on the problem. Obviously, setting the clock to zero when the condition should be applied is convenient but not always possible as in this case because the clock has been set to zero when the pollution first began.

---

**Reflection**

The additional example just completed shows that it is possible to split a forcing term into piecewise parts and as long as each part contains a function that is one of those where a particular solution can be constructed, it is possible to link together the solutions in each piece.

---

### 1.2.4   *Abstract viewpoint*

We have seen a few examples of the presence of forcing terms in differential equations. Mathematically, they appear as known functions of the independent variable and can be placed as a term on the right hand side of a differential equation. In particular, Eq. (1.2.5), Eq. (1.2.15) and Eq. (1.2.25) are some examples of a general form for a first-order differential equation with forcing,

$$\frac{\mathrm{d}y}{\mathrm{d}x}(x) + a\,y(x) = r(x)\,. \tag{1.2.27}$$

Here $y(x)$ is the unknown function ($\rho$ in Eq. (1.2.5), (1.2.15) and Eq. (1.2.22)) and $x$ ($t$ in Eq. (1.2.5), and $T$ in Eq. (1.2.15) and Eq. (1.2.22)) is the independent variable. The forcing terms in Eq. (1.2.5), Eq. (1.2.15) and Eq. (1.2.22), written in terms of the independent variable $x$, are

$$
\begin{aligned}
r(x) &= a\,c, & &\text{in (1.2.5)}\,, \\
r(x) &= 0.8\,\beta\big[1 + \cos(2\pi x)\big], & &\text{in (1.2.15)}\,, \\
r(x) &= s\,x, & &\text{in (1.2.22)}\,.
\end{aligned}
$$

The strategy to solve Eq. (1.2.27) is to compose a general solution of two parts, a particular solution $y_{\mathrm{p}}(x)$ and a homogeneous part $y_{\mathrm{h}}(x)$. By invoking Principle 1, we substitute $y(x) = y_{\mathrm{p}}(x) + y_{\mathrm{h}}(x)$ into Eq. (1.2.27).

$$\frac{\mathrm{d}\big(y_{\mathrm{p}}(x) + y_{\mathrm{h}}(x)\big)}{\mathrm{d}x} + a\,\big(y_{\mathrm{p}}(x) + y_{\mathrm{h}}(x)\big) = r(x)\,,$$

or, by rearranging terms,

$$\left[\frac{\mathrm{d}y_{\mathrm{p}}}{\mathrm{d}x}(x) + a\,y_{\mathrm{p}}(x)\right] + \left[\frac{\mathrm{d}y_{\mathrm{h}}}{\mathrm{d}x}(x) + a\,y_{\mathrm{h}}(x)\right] = r(x)\,.$$

The brackets are inserted to emphasis the grouping of terms that let us ensure a balance with the choice,

$$\frac{\mathrm{d}y_{\mathrm{p}}}{\mathrm{d}x}(x) + a\,y_{\mathrm{p}}(x) = r(x)\,, \tag{1.2.28}$$

$$\frac{\mathrm{d}y_{\mathrm{h}}}{\mathrm{d}x}(x) + a\,y_{\mathrm{h}}(x) = 0\,. \tag{1.2.29}$$

The difference between Eq. (1.2.28) and Eq. (1.2.29) is what typifies the difference between a *particular* solution $y_p(x)$ and a *homogeneous* solution $y_h(x)$. The particular solution $y_p(x)$ satisfies the differential equation in the presence of the forcing term; with an appropriate guess, the method of undetermined coefficients determines the particular solution completely – see the examples Eq. (1.2.7) for Eq. (1.2.5), Eq. (1.2.16) for Eq. (1.2.15), and Eq. (1.2.23) for Eq. (1.2.22).

The homogeneous solution $y_h(x)$ satisfies the differential equation without the forcing term and can be found by guessing the solution has the form of an exponential $\exp(\lambda x)$. It can always by multiplied with an undetermined coefficient $A$; $y_h(x) = A \exp(\lambda x)$. We call the combination $y(x) = y_p(x) + y_h(x)$, the *general* solution.

**Principle 4**

*General solutions to linear equations can be constructed as a linear combination of separate solutions, usually a particular solution and a homogeneous solution.*

If we have been successful in calculating a general solution, then we anticipate that there will be an undetermined coefficient arising in the homogeneous solution. Principle 3 tells us we need some additional information to determine a unique solution. Consider the initial condition $y(0) = Y_0$ (a new parameter); it is called an initial condition even when the independent variable is $x$, because mathematics does not care what the variables mean. The observation is that the solution is specified at a specific choice for the independent variable. It distinguishes the type of additional information from other types, for example, global conditions such as $\int y(x)\, dx = 1$.

As a result of the initial condition,

$$y(0) = y_p(0) + y_h(0) = Y_0.$$

It is extremely important that the initial condition is applied to the general $y_p(x) + y_h(x)$ solution not just $y_h(x)$. Of course, if there is no $y_p(x)$, then $y_h(x)$ is the general solution – there should always be a $y_h(x)$.

Provided $r(x)$ is a smooth function, continuous with many derivatives, then we expect to have a solution which is continuous with many derivatives as indicated by Principle 2. But how do we know that there are no other solutions that also satisfy Eq. (1.2.27) and the initial condition? This question has already been settled for the homogeneous solution in Sec. 1.1.5. The strategy is the same as before: suppose there are two different solutions $y_1(x)$ and $y_2(x)$ and consider $Y(x) = y_1(x) - y_2(x)$.

Note that

$$\frac{\mathrm{d}Y}{\mathrm{d}x}(x) + a\,Y(x) = \left[\frac{\mathrm{d}y_1}{\mathrm{d}x}(x) + a\,y_1(x)\right] - \left[\frac{\mathrm{d}y_2}{\mathrm{d}x}(x) + a\,y_2(x)\right]$$
$$= r(x) - r(x) = 0\,.$$

So $Y(x)$ must satisfy the homogeneous equation Eq. (1.2.29). In addition, it must satisfy the initial condition,

$$Y(0) = y_1(0) - y_2(0) = Y_0 - Y_0 = 0\,.$$

As we saw in Sec. 1.1.5, the only solution is $Y(x) = 0$, which means $y_1(x) = y_2(x)$, and the two solutions must be the same. Thus we confirm Principle 3 even when forcing effects are present.

The challenging part of the construction of the general solution is to find the particular solution. Fortunately, there are several choices of $r(x)$ for which the form of the particular solution is known. By that it is meant that the particular solution has a specific dependency on the independent variable, and contains coefficients that must be determined by Principle 1. Table 1.4 gives some of these choices.

Table 1.4   Guesses for $y_\mathrm{p}(x)$.

| $r(x)$ | Form for $y_\mathrm{p}$ |
|---|---|
| $A$ | $c$ |
| $x$ | $c_0 + c_1 x$ |
| $x^2$ | $c_0 + c_1 x + c_2 x^2$ |
| $x^n$ | $c_0 + c_1 x + \cdots + c_n x^n$ |
| $\sin(\omega x)$ and/or $\cos(\omega x)$ | $c_1 \sin(\omega x) + c_2 \cos(\omega x)$ |
| $\exp(rx)$ | $c \exp(rx)$ |

This table may not be very large, but fortunately many more cases can be dealt with if $r(x)$ can be split into sums of the entries in the table, $r(x) = r_1(x) + r_2(x) + \cdots + r_n(x)$. The particular solutions for each $r_j(x)$ can be added to give the total particular solution. For example, suppose $r(x) = \alpha x + \beta \exp(-2x)$. Clearly, $r_1(x) = \alpha x$ and $r_2(x) = \beta \exp(-2x)$. By adding the corresponding forms from Table 1.4, the particular solution has the form, $y_\mathrm{p}(x) = c_0 + c_1 x + c_2 \exp(-2x)$. When this guess for the particular solution is substituted into Eq. (1.2.27), we find

$$c_1 - 2c_2\,\mathrm{e}^{-2x} + a\,c_0 + a\,c_1 x + a\,c_2\,\mathrm{e}^{-2x} = \alpha x + \beta\,\mathrm{e}^{-2x}\,.$$

Then we balance terms as follows:

| | |
|---|---|
| Terms that are constants; | $c_1 + a\,c_0 = 0\,,$ |
| Terms with $x$; | $a\,c_1 = \alpha\,,$ |
| Terms with $\exp(-2x)$; | $-2\,c_2 + a\,c_2 = \beta\,.$ |

We have three algebraic equations for the three unknown coefficients, $c_0$, $c_1$ and $c_2$, which are easy to solve.

**Principle 5**

*When a particular solution with several terms is substituted into the differential equation, Principle 1 demands that each term must be balanced exactly. This process will determine the unknown coefficients in the guess for the particular solution: The particular solution will be completely determined.*

As another example, consider $r(x) = \alpha + \beta x$. Since $r_1(x) = \alpha$, we should pick $y_{p1} = c_1$, and since $r_2(x) = \beta x$, we should pick $y_{p2} = c_2 + c_3\,x$. The combination will be

$$y_p(x) = c_1 + (c_2 + c_3\,x) = (c_1 + c_2) + c_3\,x\,.$$

There is something silly about this guess because there is a sum of two unknown coefficients. They should be combined into a single unknown coefficient. Thus the appropriate guess for the particular solution should be

$$y_p(x) = a_1 + a_2\,x\,. \tag{1.2.30}$$

The unknown coefficients are relabeled $a$ to avoid confusion.

**Principle 6**

*The sum or product of unknown constants can be replaced by a single unknown constant.*

The consequence of this principle is that we need to seek only the different types of functions that should be contained in the particular solution, and combine them with unknown coefficients. For the example we just considered where $r(x) = \alpha + \beta x$, we simply note that we need just two different functions in the particular solution, a constant and a term with $x$. Thus we multiply each term with an unknown coefficient and add the result together to produce Eq. (1.2.30).

This strategy is most effective when we are dealing with products of polynomials, exponentials and trigonometric functions in $r(x)$. They can be treated by the corresponding entries in Table 1.4. For example, if

$$r(x) = x\,\mathrm{e}^{\alpha x}\,,$$

then we should try the product of the entries in Table 1.4,

$$y_p(x) = \left[c_0 + c_1\,x\right]\left[c_3\,\mathrm{e}^{\alpha x}\right]\,.$$

The only different functions that appear in this expression are $\exp(\alpha x)$ and $x\exp(\alpha x)$. Thus the combination we need is simply

$$y_p(x) = a_0\,\mathrm{e}^{\alpha x} + a_1\,x\,\mathrm{e}^{\alpha x}\,.$$

A slightly more difficult case is the guess for $x \cos(x)$. The guess based on the entries in Table 1.4 is

$$y_{\mathrm{p}}(x) = \left[c_1 + c_2\, x\right] \left[d_1\, \cos(x) + d_2\, \sin(x)\right],$$

and the different functions present are

$$\cos(x), \quad x \cos(x), \quad \sin(x), \quad x \sin(x),$$

so the appropriate combination for the particular solution will be

$$y_{\mathrm{p}}(x) = a_1\, \cos(x) + a_2\, x\, \cos(x) + b_1\, \sin(x) + b_2\, x\, \sin(x).$$

**A word of advice:** If the guess for the particular solution leads to expressions that cannot be balanced, then note that the nature of the terms that are not balanced and consider the possibility of modifying the guess for the particular solution appropriately.

### 1.2.5   *Exercises*

**Some basic exercises to develop and improve skill.**

(1) Find particular solutions to:

   (a)

$$\frac{\mathrm{d}y}{\mathrm{d}x}(x) + a\, y(x) = x^3.$$

   **Answer:** $y_{\mathrm{p}}(1/a) = -2/a^4$.

   (b)

$$\frac{\mathrm{d}Q}{\mathrm{d}t}(t) - r\, Q(t) = \mathrm{e}^{-at}.$$

   **Answer:** For $r = a = 1$, $Q_{\mathrm{p}}(1) = -0.184$.

   (c)

$$\frac{\mathrm{d}Q}{\mathrm{d}t}(t) - r\, Q(t) = t\, \mathrm{e}^{-at}.$$

   **Answer:** For $r = a = 1$, $Q_{\mathrm{p}}(1) = -0.276$.

   (d)

$$\frac{\mathrm{d}f}{\mathrm{d}y}(y) + f(y) = \sin(2y).$$

   **Answer:** $f_{\mathrm{p}}(\pi) = -0.4$.

   (e)

$$\frac{\mathrm{d}x}{\mathrm{d}t}(t) + 3\, x(t) = t\, \sin(3t).$$

   **Answer:** $x_p(3) = 0.61$.

**More challenging exercises.**

(2)
$$\frac{dw}{dz}(z) + 2\,w(z) = \begin{cases} z\,, & \text{for } 0 \le z \le 1\,, \\ 1\,, & \text{for } z > 1\,, \end{cases}$$

subject to the initial condition $w(0) = 0$.

**Answer:** $w(2) = -0.462$.

(3) Suppose the forcing term is not of one of the possible choices in Table 1.4. For example,

$$\frac{dy}{dt}(t) + \beta y(t) = \frac{1}{1+t}\,.$$

Try to find a particular solution by making various guesses. What can you conclude?

**Some problems that arise in applications.**

(4) One view of the growth of fish is to measure its length $L$ (cm). A model developed by von Bertalanffy is a simple model for limited growth,

$$\frac{dL}{dt}(t) = A - k\,L(t)\,.$$

The idea is that some of the food is directed to sustaining the fish, described by $-k\,L(t)$, rather than increasing its size, described by $A$. Show that the fish will reach a terminal size. The typical adult size of one species of fish is 41 cm. Suppose at birth the fish is typically 1.2 cm and reaches length 33.5 cm after two years. Determine estimates for $A$ and $k$.

**Answer:** $A = 34.2$ cm/yr and $k = 0.834$ yr$^{-1}$.

(5) A simple chemical reaction that converts A to B with rate $k_1$ (min$^{-1}$) with a reverse reaction of B to A with rate $k_2$ (min$^{-1}$) can be modeled as

$$\frac{d[\mathbf{B}]}{dt}(t) = k_1\,[\mathbf{A}](t) - k_2\,[\mathbf{B}](t)\,,$$

where $[\mathbf{A}](t)$ (gm/cm$^3$) and $[\mathbf{B}](t)$ (gm/cm$^3$) are the concentrations of A and B, respectively. The usual diagram that describes the reaction is shown below.

$$A \;\;\underset{k_2}{\overset{k_1}{\rightleftharpoons}}\;\; B$$

Since the sum $[\mathbf{A}](t) + [\mathbf{B}](t)$ must be constant in time, it is determined by its initial value. Assume the initial concentration of A is $\alpha$ and of B is 0. Find the eventual concentration of B and the time it takes to reach half of this value.

**Answer:** For a particular case where the reaction rates are $k_1 = 0.037$ min$^{-1}$ and $k_2 = 0.051$ min$^{-1}$, the time takes 7.9 min.

(6) An RC circuit is described in Sec. 1.1.4. Here we consider the consequences of applying a direct voltage and an alternating voltage to the circuit. The applied voltage $E$ in Eq. (1.1.13) is replaced by

(a)

$$E(t) = V,$$

(b)

$$E(t) = V\sin(\omega t).$$

The differential equation Eq. (1.1.15) becomes

$$\frac{dQ}{dt}(t) + \frac{1}{RC}\,Q(t) = \frac{E(t)}{R}.$$

Assume that the charge on the capacitor is initially zero and determine the subsequent behavior of the charge in time for the two cases.

**Answer:** For the choice $R = 0.5\,\Omega$, $C = 2$ F and $\omega = \pi\,\mathrm{s}^{-1}$, the dimensionless charge $Q(t)/VC$ is shown as a function of time for the two cases in Fig. 1.11. Notice that the alternating voltage does not charge up the capacitor to a constant value.

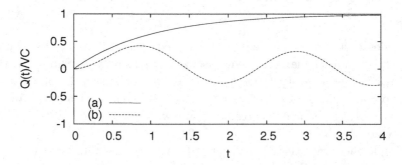

Fig. 1.11　The dimensionless charge as a function of time for the two cases.

(7) Let $v$ (ft/s) be the downward speed of a falling body. Gravity will continue to increase the velocity endlessly. However, air friction will counteract gravity. Assume that the friction force is proportional to the speed $v$. Then the equation of motion takes the form,

$$m\frac{dv}{dt}(t) = mg - kv(t)$$

where $m$ (lb) is its mass and $g$ is the gravitational constant. Determine the terminal velocity of the body when $m = 150$ lb and $k = 60$ lb/s.

**Answer:** 80 ft/s.

Suppose the body begins its fall at height $H$ (ft). How far does the body fall before it reaches 90% of its terminal velocity?

**Answer:** 280 ft.

(8) In Exercise 15 of Sec. 1.1.6, we considered a boat that coasts to a dock while subject to drag only. Now consider the additional influence of a reverse thrust $1000$ lb $\cdot$ ft/s$^2$ from the motor as the captain slows down the boat's speed. Assume that the reverse thrust is the same as the forward thrust and confirm that the steady speed of the boat would be 10 ft/s. How far does the boat go before it comes to a stop – at which point the captain turns off the motor and the boat has reached the dock?

(9) Newton's law of cooling (or heating) gives the rate of change of the temperature as being proportional to the temperature difference. As an example, suppose we are designing a building made with wood. If the outside (ambient) temperature is $T_a$ (°F), and the temperature inside the building is $T(t)$ (°F), then Newton's law suggests

$$\frac{\mathrm{d}T}{\mathrm{d}t}(t) = k\left(T_a - T(t)\right). \tag{1.2.31}$$

If $T(t) < T_a$ (it is colder inside than out), then the building naturally heats up and the temperature inside increases. The parameter $k$ (h$^{-1}$) reflects the properties of the building, for example, how wood influences the rate of adjustment to a temperature difference: $k = 2$ (h$^{-1}$) might be a reasonable guess. Note that we assume the time is measured as $t$ (h).

(a) Suppose the outside temperature is 60 °F and the building has a comfortable temperature of 75 °F. How long will it take for the temperature in the building to fall to 66 °F?
   **Answer:** 0.46 h.

(b) A more realistic view of the outside temperature is a daily variation. Assume now that $T_a(t) = 65 - 10\cos(\pi t/12)$ measured in degrees Fahrenheit. Assume that at $t = 0$ (midnight), the room is at 75 °F. How cold can it get inside?
   **Answer:** 55.1 °F.

(c) Now we consider the additional effects of heating/cooling systems. Suppose our design follows the principle of Newton's law. When it is colder/hotter than the desired temperature, the system heats/cools proportionally to the temperature difference. Thus we add a term $k_s\left(T_s - T(t)\right)$ to (1.2.31) with $T_a(t)$ as given in part (b). Consider the operating conditions for three different systems, $k_s = 0.1, 1, 10$ h$^{-1}$ and determine the consequences on the temperature inside. Pick a comfortable average temperature and determine what $T_s$ must be to ensure that choice ($T_s$ plays the role of a control parameter).

(10) The isotope 238 uranium radioactively decays into 234 thorium with a decay rate $k_1$ (yr$^{-1}$); it has a half-life of 4.47 billion years. If $y_1(t)$ (oz) is the amount of uranium at time $t$ (yr), then

$$\frac{\mathrm{d}y_1}{\mathrm{d}t}(t) = -k_1\, y_1(t),$$

describes its decay. Suppose the initial amount of uranium is $A_0$ (oz), determine the amount of uranium as a function of time.

Since 234 thorium decays into 234 protactinium with a decay rate of $k_2$ (yr$^{-1}$); it has a half-life of 24.1 days. The amount of thorium $y_2(t)$ (oz) changes in time according to the amount produced from the decay of the uranium and the loss due to the decay into protactinium. Thus,

$$\frac{dy_2}{dt}(t) = k_1\, y_1(t) - k_2\, y_2(t).$$

Solve this equation by using the known result for the decay of $y_1(t)$. What fraction of $A_0$ will there be of 234 thorium after a billion years?
**Answer**: $1.3 \times 10^{-11}$.

Although this is not the case, consider what happens to the solution when $k_2 \to k_1$.

(11) A pipe carries a concentration $c$ (lb/gal) of a chemical A dissolved in water into a reactor at rate $r$ (ft$^3$/min). The reactor holds $V$ (gal) of water. Inside the reactor, A is converted into another chemical B at a rate $k$ (s$^{-1}$). Another pipe carries the water containing both A and B out of the reactor at the same rate $r$ (ft$^3$/min). Assume the process is started instantaneously with no chemical B in the reactor and determine the time behavior of the production of B. Pick some values for the parameters and plot the time history of the production of B.

Under the following operating conditions, $r = 2\,\text{ft}^3/\text{min}$, $k = 0.03$ min$^{-1}$, $V = 120$ gal, $c = 0.1$ lb/gal, a chemical engineer records the measurements given in the table. Do your results agree with the measurements?

Table 1.5 Table of measurements.

| Time (min) | Concentration (lb/gal) |
|---|---|
| 10 | 0.0075 |
| 20 | 0.0144 |
| 30 | 0.0174 |
| 40 | 0.0185 |
| 50 | 0.0188 |
| 60 | 0.0190 |

**Some problems to extend the method of undetermined coefficients.**

(12) For the following problem,

$$\frac{dy}{dx}(x) + a\,y(x) = \alpha\,e^{-ax},$$

the guess for the particular solution $y_p(x) = C\exp(-ax)$ fails (Verify). Instead, determine the homogeneous solution and compare to the guess for $y_p(x)$. What interpretation can you give to your observations?

One way to deal with this difficulty is suggested by the result in Exercise 10 when $k_2 \rightarrow k_1$. Change the forcing term to be

$$\alpha\, e^{-(a+\varepsilon)\, x}$$

and solve the equation subject to the initial condition $y(0) = 0$. Finally, take the limit $\varepsilon \rightarrow 0$ of the solution and verify that the result is the solution to the equation with the original forcing function.

(13) The result of the previous exercise suggests that the obvious guess for the particular solution should be multiplied by $x$. Try this strategy on the following problem.

$$\frac{dy}{dx}(x) + a\, y(x) = \alpha\, x\, e^{-ax}\,.$$

## 1.2.6 *General forcing terms*

Surely there will be situations where the forcing term $r(x)$ in the differential equation

$$\frac{dy}{dx} + a\, y(x) = r(x) \tag{1.2.32}$$

will not be of the type composed from entries in Table 1.4. An example is given in Exercise 3 of Sec. 1.2.5 where

$$r(x) = \frac{1}{1+x}\,. \tag{1.2.33}$$

Hopefully, you tried several guesses for the particular solution and discovered none worked successfully. How then to treat this case?

The way forward is to exploit the properties of the homogeneous solution: any constant times the homogeneous solution is also a solution. Verification of this property is easy. First introduce the exponential for the homogeneous solution,

$$y_1(x) = e^{-ax}\,, \tag{1.2.34}$$

and then write the homogeneous solution as $y_h(x) = C\, y_1(x)$. The approach we will use to construct a particular solution allows a general argument to be valid whenever a homogeneous solution is known and will be useful in the next section.

Substitute the homogeneous solution into Eq. (1.2.32) with $r(x) = 0$.

$$C\frac{dy_1}{dx}(x) + a\, C\, y_1(x) = C\left(\frac{dy_1}{dx}(x) + a\, y_1(x)\right) = 0\,.$$

The last step uses the fact that $y_1(x)$ is a solution to the homogeneous equation. Note that the result is true for any value of $C$.

Now we make the guess that the particular solution to Eq. (1.2.32) can be obtained by replacing the coefficient $C$ by a function $C(x)$. In other words, we guess

$y_\mathrm{p}(x) = C(x)\,y_1(x)$ and substitute into Eq. (1.2.32).

$$\frac{\mathrm{d}C}{\mathrm{d}x}(x)\,y_1(x) + C(x)\,\frac{\mathrm{d}y_1}{\mathrm{d}x}(x) + a\,C(x)\,y_1(x)$$

$$= \frac{\mathrm{d}C}{\mathrm{d}x}(x) + C(x)\left(\frac{\mathrm{d}y_1}{\mathrm{d}x}(x) + a\,y_1(x)\right)$$

$$= \frac{\mathrm{d}C}{\mathrm{d}x}(x)\,y_1(x)\,. \tag{1.2.35}$$

Once again, the fact that $y_1(x)$ solves the homogeneous equation ensures that the expression in the parentheses must be zero and the result is just

$$\frac{\mathrm{d}C}{\mathrm{d}x}(x)\,y_1(x) = r(x)\,. \tag{1.2.36}$$

Since $y_1(x)$ is known and $r(x)$ is given, the result Eq. (1.2.36) states that the derivative of $C(x)$ is a known function of $x$.

$$\frac{\mathrm{d}C}{\mathrm{d}x}(x) = r(x)\,\mathrm{e}^{ax}\,. \tag{1.2.37}$$

### Principle 7

*General solutions to inhomogeneous, linear ordinary differential equations can be calculated by assuming that the coefficients in the homogeneous solution are functions of the independent variable. The method is called variation of parameters.*

There are two ways we can determine $C(x)$, depending on whether we wish to use the indefinite integral or the definite integral. One way is to find the anti-derivative of $\exp(ax)\,r(x)$, written here as the indefinite integral,

$$C(x) = \int^x \mathrm{e}^{a\xi}\,r(\xi)\,\mathrm{d}\xi + A\,,$$

where $A$ is a constant of integration. Consequently, by substituting $C(x)$ into

$$y(x) = C(x)\,\mathrm{e}^{-ax}$$

$$= \mathrm{e}^{-ax}\int^x \mathrm{e}^{a\xi}\,r(\xi)\,\mathrm{d}\xi + A\,\mathrm{e}^{-ax} \tag{1.2.38}$$

we obtain the general solution with two parts: the first part with the indefinite integral represents the particular solution, and the last term is simply the homogeneous solution now expressed with the arbitrary coefficient $A$.

To illustrate the procedure on a case that can also be solved by the method of undetermined coefficients, let us pick $r(x) = x^3$, the case chosen in Exercise 1(a) of Sec. 1.2.5. The equation to be solved is

$$\frac{\mathrm{d}y}{\mathrm{d}x}(x) + a\,y(x) = x^3\,.$$

By integrating Eq. (1.2.37),

$$C(x) = \int^x e^{a\xi}\, \xi^3 \, d\xi$$

$$= \frac{1}{a}\, e^{ax}\, x^3 - \frac{3}{a^2}\, e^{ax}\, x^2 + \frac{6}{a^3}\, e^{ax}\, x - \frac{6}{a^4}\, e^{ax} + A\,.$$

The integral is performed by using integration-by-parts several times. Finally, the general solution is

$$y(x) = \frac{x^3}{a} - \frac{3x^2}{a^2} + \frac{6x}{a^3} - \frac{6}{a^4} + A\, e^{-ax}\,. \tag{1.2.39}$$

This is the same result from applying the method of undetermined coefficients. I think you will agree that the method of undetermined coefficients is much easier to apply!

The other way $C(x)$ can be determined is to make use of the initial condition $y(0) = y_0$ and use the definite integral. Since $y(0) = C(0) = y_0$,

$$\int_0^x \frac{dC}{d\xi}(\xi)\, d\xi = C(x) - y_0\,. \tag{1.2.40}$$

Now by integrating the right hand side of Eq. (1.2.37),

$$\int_0^x e^{a\xi}\, r(\xi)\, d\xi = \left[ e^{a\xi} \left( \frac{\xi^3}{a} - \frac{3\xi^2}{a^2} + \frac{6\xi}{a^3} - \frac{6}{a^4} \right) \right]_0^x$$

$$= e^{ax} \left( \frac{x^3}{a} - \frac{3x^2}{a^2} + \frac{6x}{a^3} - \frac{6}{a^4} \right) + \frac{6}{a^4}\,. \tag{1.2.41}$$

By equating Eq. (1.2.40) with Eq. (1.2.41), we obtain $C(x)$ and hence the specific solution.

$$y(x) = \frac{x^3}{a} - \frac{3x^2}{a^2} + \frac{6x}{a^3} - \frac{6}{a^4} + \left( \frac{6}{a^4} + y_0 \right) e^{-ax}\,.$$

This result agrees with Eq. (1.2.39) once the initial has been applied to determine $A$.

Now let us return to the choice Eq. (1.2.33). The solution Eq. (1.2.38) is

$$y(x) = e^{-ax} \int^x \frac{e^{a\xi}}{1 + \xi}\, d\xi + A\, e^{-ax}\,. \tag{1.2.42}$$

There is no known anti-derivative, so the integral cannot be performed analytically but must be left as is. There is nothing wrong with the integral; it is just not possible to express the result in terms of simple functions. The result Eq. (1.2.42) is not entirely satisfactory because it is difficult to interpret the properties of the function that results from the integral, for example, its long term behavior. We ended up in this position because of the nature of the forcing term. The question arises just how important is the precise form of the forcing term. For example, if $1/(1 + x)$ is replaced by $\exp(-x)$ – plot both functions to see if the difference appears important – then the method of undetermined coefficients leads quickly to a simple solution. As engineers and scientists, we may often have the freedom to choose the nature of forcing, and, as this example shows, it is often useful to pick forms where the solution to the differential equation can be determined simply.

---
**Reflection**

Simply because there is a general formula Eq. (1.2.38) for the solution to (1.2.32) it does not make understanding the solution easy. We may prefer to simplify the forcing term and even the coefficient in the differential equation so that we can obtain a simple solution and understand its consequences.

---

## 1.3 Coefficients with time variation

Since coefficients have been used generically as unknown constants, for example, in particular and homogeneous solutions, the heading of this section may seem confusing. The coefficients referred to here are the ones that appear in the differential equation. It is best to introduce an example so that a clarification can be made.

### 1.3.1 *Variable river flow*

Let us return to the example of a river carrying pollution into a dam, as illustrated in Fig. 1.6. For a constant flow rate $r$ (m$^3$/s), the differential equation that models the situation is given by Eq. (1.2.12), rewritten here,

$$\frac{d\rho}{dT}(T) + \beta\,\rho(T) = \beta\,c,\qquad(1.3.1)$$

where $\beta = 3.15 \times 10^{-2} r/V$. The dam has volume $V$ (km$^3$) and the pollution carried by the river has concentration $c$ (kg/m$^3$). The concentration of the pollution in the dam is $\rho$ (kg/m$^3$), and $T$ is the time measured in years.

The derivation of the differential equation Eq. (1.3.1) does not change if the flow rate of the river depends on time and the outflow of the dam is the same as the river flow rate – thus no change in the volume of the dam. Suppose, then, that the river flow is seasonal, with an average annual rate $R$ (m$^3$/s) and a sinusoidal fluctuation of amplitude $\varepsilon$:

$$r(T) = R\left[1 - \varepsilon\,\cos(2\pi T)\right].$$

So the parameter $\beta$ is replaced by a function of time,

$$\beta(T) = b\left[1 - \varepsilon\,\cos(2\pi T)\right],$$

where $b = 3.15 \times 10^{-2} R/V$ gives the annual average of $\beta(T)$. The constant parameter $\beta$ in Eq. (1.3.1) has been replaced by a function, and this is the interpretation of a time-varying coefficient in a differential equation. So we need to solve

$$\frac{d\rho}{dT}(T) + \beta(T)\,\rho(T) = \beta(T)\,c,\qquad(1.3.2)$$

and we need to find a particular solution and a homogeneous solution.

Let us seek a homogeneous solution first. The usual approach, guessing an exponential form

$$\rho(T) = A\, e^{\lambda T},$$

does not work because, after we substitute the guess into the differential equation (1.3.2), we find

$$\lambda e^{\lambda T} + \beta(T) e^{\lambda T} = 0.$$

Since $\exp(\lambda T)$ is common, we are left with $\lambda + \beta(T) = 0$, or $\lambda = -\beta(T)$ which is not possible since $\lambda$ is assumed to be a constant.

So the remedy is to allow $\lambda$ to be a function of $T$! The new guess for the homogeneous solution is

$$\rho(T) = e^{\alpha(T)} \tag{1.3.3}$$

where $\alpha(T)$ replaces $\lambda$. Substitution into the differential equation Eq. (1.3.2) leads to

$$e^{\alpha(T)} \left( \frac{d\alpha}{dT}(T) + \beta(T) \right) = 0,$$

with a successful result by setting

$$\frac{d\alpha}{dT}(T) = -\beta(T) = -b\left(1 - \varepsilon\, \cos(2\pi T)\right). \tag{1.3.4}$$

Upon integration,

$$\alpha(T) = -b\left(T - \frac{\varepsilon}{2\pi}\, \sin(2\pi T)\right), \tag{1.3.5}$$

and the homogeneous solution is known!

$$\rho_{\mathrm{h}}(T) = A\, e^{\alpha(T)}. \tag{1.3.6}$$

A natural question is whether a constant should be added to Eq. (1.3.5). After all, we integrated Eq. (1.3.4) to obtain Eq. (1.3.5). If we use $\alpha(T) + C$ in place of Eq. (1.3.5), then the homogeneous solution will be

$$A\, e^{\alpha(T)+C} = A\, e^{C}\, e^{\alpha(t)},$$

and the net impact is that the constant $A$ has been modified by a factor $\exp(C)$. Since both $A$ and $C$ are unknown we may replace $A \exp(C)$ by $A$; this step is equivalent to setting $C = 0$, or ignoring the constant of integration in Eq. (1.3.5).

As a check, suppose $\varepsilon = 0$. Then $\beta(T) = b$, a constant and $\alpha(T) = -bT$. The homogeneous solution $A \exp(-bT)$ is exactly what we expect.[13]

The next step on the way to constructing the particular solution is to revisit the steps that led to Eq. (1.2.35). The key idea in these steps was to take the homogeneous solution and allow its coefficient to be an unknown function $F(T)$,

$$\rho_{\mathrm{p}}(T) = F(T)\, e^{\alpha(T)}. \tag{1.3.7}$$

---

[13] It is often very useful to check complicated results by making choices for parameters so that the results simplify and reduce to an expected result.

Upon substitution of Eq. (1.3.7) into Eq. (1.3.2), we obtain

$$\frac{\mathrm{d}\rho_\mathrm{p}}{\mathrm{d}T}(T) + \beta(T)\,\rho_\mathrm{p}(T) = \frac{\mathrm{d}F}{\mathrm{d}T}(T)\,\mathrm{e}^{\alpha(T)} + \frac{\mathrm{d}\alpha}{\mathrm{d}T}(T)\,F(T)\,\mathrm{e}^{\alpha(T)} + \beta(T)\,F(T)\,\mathrm{e}^{\alpha(T)}$$

$$= \frac{\mathrm{d}F}{\mathrm{d}T}(T)\,\mathrm{e}^{\alpha(T)} + \left(\frac{\mathrm{d}\alpha}{\mathrm{d}T}(T) + \beta(T)\right)F(T)\,\mathrm{e}^{\alpha(T)}\,. \qquad (1.3.8)$$

Because $\alpha(T)$ has been chosen to satisfy Eq. (1.3.4), the expression in parentheses in Eq. (1.3.8) is zero, and we are left with

$$\frac{\mathrm{d}\rho_\mathrm{p}}{\mathrm{d}T}(T) + \beta(T)\,\rho_\mathrm{p}(T) = \frac{\mathrm{d}F}{\mathrm{d}T}(T)\,\mathrm{e}^{\alpha(T)} = \beta(T)\,c\,. \qquad (1.3.9)$$

Bearing in mind that we have already determined $\alpha(T)$ by Eq. (1.3.5), Eq. (1.3.9) leads to

$$\frac{\mathrm{d}F}{\mathrm{d}T}(T) = \mathrm{e}^{-\alpha(T)}\beta(T)\,c\,.$$

Upon integration,

$$F(T) = \int^{T} \mathrm{e}^{-\alpha(s)}\,\beta(s)\,c\,\mathrm{d}s\,.$$

Fortunately, for this example it is possible to perform the integration since from Eq. (1.3.4),

$$\beta(s) = -\frac{\mathrm{d}\alpha}{\mathrm{d}s}(s)\,,$$

and the integral has the appearance of a change of variable.

$$F(T) = -c\int^{T} \mathrm{e}^{-\alpha(s)}\,\frac{\mathrm{d}\alpha}{\mathrm{d}s}(s)\,\mathrm{d}s = c\,\mathrm{e}^{-\alpha(T)}\,.$$

Finally, the particular solution is

$$\rho_\mathrm{p}(T) = F(T)\,\mathrm{e}^{\alpha(T)} = c\,,$$

and the general solution is

$$\rho(T) = c + A\,\mathrm{e}^{\alpha(T)}\,. \qquad (1.3.10)$$

The result shows the particular solution is just $c$. Bang our heads on the wall! We should have seen that $c$ is a particular solution directly from Eq. (1.3.1). Oh well, at least we know what we did must be correct, and it shows us how to proceed in the general case.

Once again, the question of the addition of a constant of integration to $F(T)$ arises. Suppose we add one, then the particular solution would be

$$\rho_\mathrm{p}(T) = \big(F(T) + C\big)\,\mathrm{e}^{\alpha(T)} = F(T)\,\mathrm{e}^{\alpha(T)} + C\,\mathrm{e}^{\alpha(T)}\,,$$

and the result is just the addition of the homogeneous solution which we will do in step Eq. (1.3.10) anyway. No need to have two copies of the homogeneous solution added, so ignore the constant of integration; in other words, set $C = 0$.

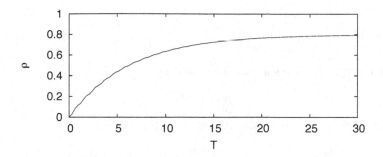

Fig. 1.12   Concentration of the contaminant in the dam over time.

Time to apply the initial condition. Since $\alpha(0) = 0$, the initial condition requires

$$\rho(0) = c + A = 0, \qquad \text{or} \quad A = -c,$$

and the specific solution is

$$\rho(T) = c\left(1 - e^{\alpha(T)}\right), \tag{1.3.11}$$

where $\alpha(T)$ is given by Eq. (1.3.5).

It is worth comparing the solution (1.3.11) to the previous ones we calculated, Eq. (1.2.13) and Eq. (1.2.18). To do so, pick $c = 0.8\,\text{kg/m}^3$ and $b = 0.158\,\text{yr}^{-1}$. In addition, select $\varepsilon = 0.5$. The result is shown in Fig. 1.12. Barely discernable is the annual variation in the concentration; it has been strongly reduced.

There seems to be little difference in all the models for the pollution in the dam, the main trends being captured very well by the simplest model, Eq. (1.2.13).

---

**Reflection**

We note the presence of a homogeneous solution, determined by assuming the form of an exponential with an unknown function as its argument, and the presence of a particular solution, determined by allowing the coefficient in the homogeneous solution to be a function. The general solution is still a combination of the particular solution and the homogeneous solution, and the initial condition must still be applied to obtain a specific solution.

---

We are ready to consider the general case for time-varying coefficients in a first-order differential equation.

### 1.3.2   *Abstract viewpoint*

**Integrating factor**

It is also possible to extend the ideas presented in Eq. (1.3.4) and Eq. (1.3.9) to establish existence and uniqueness for the general form of a first-order differential equation,

$$\frac{dy}{dx}(x) + r(x)\,y(x) = f(x)\,. \tag{1.3.12}$$

Start with the search for a solution expressed in the form,

$$y(x) = e^{R(x)}$$

and substitute into the homogeneous version of Eq. (1.3.12).

$$\frac{dy}{dx}(x) + r(x)\,y(x) = \left(\frac{dR}{dx}(x) + r(x)\right)e^{R(x)} = 0\,.$$

The solution is valid provided we set

$$\frac{dR}{dx}(x) = -r(x)\,. \tag{1.3.13}$$

Once we integrate Eq. (1.3.13), we know $R(x)$, and hence the homogeneous solution

$$y_{\mathrm{h}}(x) = A\,e^{R(x)}\,. \tag{1.3.14}$$

As noted before, there is no need to include a constant of integration when calculating $R(x)$.

Now we make the guess for the particular solution,

$$y_{\mathrm{p}}(x) = F(x)\,e^{R(x)}\,, \tag{1.3.15}$$

and substitute into Eq. (1.3.12). The result is

$$\frac{dy}{dx}(x) + r(x)\,y(x) = \frac{dF}{dx}(x)\,e^{R(x)} + \left(\frac{dR}{dx}(x) + r(x)\right)F(x)\,e^{R(x)}$$

$$= \frac{dF}{dx}(x)\,e^{R(x)}$$

$$= f(x)\,. \tag{1.3.16}$$

Notice how the choice Eq. (1.3.15) allows a tremendous simplification, leading to

$$\frac{dF}{dx}(x)\,e^{R(x)} = f(x)$$

or

$$\frac{dF}{dx}(x) = f(x)\,e^{-R(x)}\,. \tag{1.3.17}$$

Assuming that we have determined $R(x)$ by integrating Eq. (1.3.13), we know $\exp[-R(x)]$, which is called the integrating factor. Then we integrate Eq. (1.3.17) to obtain

$$F(x) = \int^{x} e^{-R(s)} f(s)\,ds\,,$$

and hence we have determined the particular solution by Eq. (1.3.15).

$$y_\mathrm{p}(x) = \mathrm{e}^{R(x)}\, F(x) = \mathrm{e}^{R(x)} \int^x \mathrm{e}^{-R(s)} f(s)\, \mathrm{d}s\,. \tag{1.3.18}$$

Finally, the general solution is

$$y(x) = \mathrm{e}^{R(x)} \int^x \mathrm{e}^{-R(s)} f(s)\, \mathrm{d}s + C\, \mathrm{e}^{R(x)}\,. \tag{1.3.19}$$

Provided the integrals exist, we have established the existence of a solution to the differential equation Eq. (1.3.12). What mitigates the value of this formal solution is the possibility that the integrals cannot be performed to give known simple functions; an example is given in Eq. (1.2.42). For most of the cases where the integrals can be done, the solution can also be determined by the method of undetermined coefficients which is a much easier process.

### Series solution

Is there no way we can obtain some understanding of the solution, even if just for early times? The answer is often yes! Return to the steps that led to the construction of a Taylor series representation for the solution Eq. (1.1.29). The steps involved differentiating the equation repeatedly and evaluating the derivatives at $t = 0$. Imagine applying these ideas to

$$\frac{\mathrm{d}y}{\mathrm{d}x}(x) + a\, y(x) = \frac{1}{1+x}\,.$$

The process is possible in this example but there is an adaption of the idea which is usually easier to implement. Expand the forcing term in a Taylor series – see Eq. (B.3.4).

$$\frac{\mathrm{d}y}{\mathrm{d}x}(x) + a\, y(x) = \sum_{n=0}^{\infty} (-1)^n\, x^n\,. \tag{1.3.20}$$

What should we pick for the particular solution? Well, if we follow the spirit of the entries in Table 1.4, then we should pick an infinite order polynomial.

$$y_\mathrm{p}(x) = \sum_{n=0}^{\infty} a_n\, x^n\,. \tag{1.3.21}$$

Of course this looks like the Taylor series representation Eq. (B.1.5) before the coefficients $a_n$ have been evaluated.

Substitute Eq. (1.3.21) into Eq. (1.3.20). To do so, we need the derivative of Eq. (1.3.21).

$$\frac{\mathrm{d}y_\mathrm{p}}{\mathrm{d}x}(x) = \sum_{n=1}^{\infty} n\, a_n\, x^{n-1} = \sum_{n=0}^{\infty} (n+1)\, a_{n+1}\, x^n\,.$$

In the last step, the counter $n$ is shifted by adding one. Thus Eq. (1.3.20) becomes

$$\sum_{n=0}^{\infty} (n+1)\, a_{n+1}\, x^n + a \sum_{n=0}^{\infty} a_n\, x^n = \sum_{n=0}^{\infty} (-1)^n\, x^n\,.$$

Now we must balance every power of $x$ separately,

$$(n+1)\, a_{n+1} + a\, a_n = (-1)^n\,.$$

This result constitutes a recursion relation. Apply the recursion for a few values of $n$:

$n = 0$:

$$a_1 + a\, a_0 = 1\,, \quad \text{or} \quad a_1 = 1 - a\, a_0\,.$$

$n = 1$:

$$2a_2 + a\, a_1 = -1\,, \quad \text{or} \quad a_2 = \frac{-1+a}{2} + \frac{a^2}{2}\, a_0\,.$$

$n = 2$:

$$3a_3 + a\, a_2 = 1\,, \quad \text{or} \quad a_3 = \frac{1}{3} + \frac{a}{6} - \frac{a^2}{6} - \frac{a^3}{6}\, a_0\,.$$

So far, the solution is

$$y_{\mathrm{p}}(x) = x + \frac{a-1}{2}\, x^2 + \frac{2+a-a^2}{6}\, x^3 + \cdots$$

$$+ a_0 \left( 1 - ax + \frac{a^2}{2}\, x^2 - \frac{a^3}{6}\, x^3 + \cdots \right). \quad (1.3.22)$$

There is something strange about this particular solution because it contains an unknown constant $a_0$. What has happened is that the recursion formula has included the contribution from the homogeneous solution. Thus the correct interpretation of (1.3.22) is that it is the general solution with contributions,

$$y_{\mathrm{p}}(x) = x + \frac{a-1}{2}\, x^2 + \frac{2+a-a^2}{6}\, x^3 + \cdots\,,$$

$$y_{\mathrm{h}}(x) = 1 - ax + \frac{a^2}{2}\, x^2 - \frac{a^3}{6}\, x^3 + \cdots\,.$$

The expression for $y_{\mathrm{h}}(x)$ looks like the first few terms of the Taylor series for $\exp(-ax)$ which is obviously the homogeneous solution. By applying the initial condition $y(0) = y_0$, it is easy to see that $a_0 = y_0$.

---

**Reflection**

If we want just to compute the Taylor series for the particular solution, we may use $a_0 = 0$ as the starting value in the recursion formula. This is particularly valuable when the homogeneous solution is known and so we do not want to include its contribution to the results of the recursion formula.

The general pattern for the coefficients in the Taylor series of $y_p(x)$ is difficult to determine. On the other hand, it is easy to code an algorithm on a computer to evaluate the recursion formula with $a_0 = 0$ which will construct the terms for $y_p(x)$. Simply specify $x$ and apply the recursion until $a_n$ is small enough that the Taylor series may be terminated to give a reasonably accurate evaluation of $y_p(x)$. Of course, the Taylor series may have a limit to the size of $x$ that ensures convergence, so the solution may be limited to small enough values of $x$.

### 1.3.3 *Exercises*

**Some basic exercises to develop and improve skill.**

(1) Construct the solution to the equation,

$$\frac{dy}{dx}(x) + \frac{2x}{1 + x^2}\, y(x) = 4x\,,$$

subject to the initial condition $y(0) = 1$.
**Answer:** It should be easy to verify your solution by direct substitution.

(2) Construct a solution to

$$\cos(\theta)\frac{df}{d\theta}(\theta) + \sin(\theta)\, f(\theta) = 1$$

subject to the initial condition $f(0) = 0$.
**Answer:** It is easy to verify by direct substitution.

(3) Consider the equation

$$\frac{dy}{dx}(x) + a\, y(x) = e^{-ax}$$

and construct the solution in two different ways:

(a) By the method of undetermined coefficients.
(b) By the use of an integrating factor.

**Answer:** The two solutions must agree.

(4) An example of a case that sometimes arises in applications is

$$\frac{d\phi}{dr}(r) - \frac{n}{r}\, \phi(r) = \frac{1}{r}\,.$$

Note that this equation seems to have a problem when $r = 0$. Ignore this difficulty and construct the solution as usual. Your solution should show how the difficulty at $r = 0$ is avoided. What initial condition must be supplied to select a specific solution?

**An exercise arising from an application.**

(5) Consider again the pollutant in a river flowing into a dam when the flow rate of the river varies seasonally as described by Eq. (1.3.4), but imagine the concentration also varies seasonally – see Eq. (1.2.14) –

$$c(T) = 0.8\big[1 + \cos(2\pi T)\big].$$

Construct the solution assuming $\rho(0) = 0$. Are you able to graph the solution?

### 1.3.4   *Summary*

With the knowledge gained in these first three sections, time has come to put together a checklist for constructing solutions to first-order linear differential equations:

$$\frac{dy}{dx}(x) + r(x)\,y(x) = f(x). \qquad (1.3.23)$$

The general outline is composed of several steps:

The *homogeneous solution.*  A homogeneous solution is always needed since its unknown coefficient is needed to handle the initial condition. Homogeneous solution can always be found by assuming an exponential form.

The *particular solution.*  Particular solutions take care of the forcing terms $f(x)$. There are essentially two methods for constructing a particular solution: the method of undetermined coefficients based on the forcing term being composed of polynomials, exponentials and cosine and/or sine functions; and the method of variation of parameters (1.3.15) – see also (1.2.36).

The *general solution.* The general solution results from combining the homogeneous and particular solutions.

A *specific solution.* After applying the initial condition to the general solution, the specific solution is completely determined and there are no remaining unknown coefficients.

What is particularly important about this general outline is that it expresses the general outline for the solution to all the linear differential equations that are considered in this text, including partial differential equations which will be tackled in a later chapter.

A detailed checklist is now provided which is presented also as a flowchart.

### Checklist

(1) Be sure that the differential equation has the correct form (1.3.23). It must be linear. The equation must contain a first derivative only. Higher-order differential equations are considered in the following sections.

(2) If the coefficient $r(x)$ in Eq. (1.3.23) is not constant but depends on $x$, then;

    (a) Find the homogeneous solution by determining $R(x)$ as in Eq. (1.3.13).

    (b) Find the particular solution by determining $F(x)$ as in Eq. (1.3.17).

    Continue with step 4.

(3) If $r(x) = r$ is a constant, then continue with the following steps;

    (a) Find the homogeneous solution by assuming the solution has the form of an exponential.

    (b) Examine the nature of the right hand side $f(x)$ in Eq. (1.3.23).

        i. If $f(x) = 0$, no particular solution is needed, so go to step 4.

        ii. If $f(x)$ has one of the forms in Table 1.4, then guess an appropriate particular solution. Remember combinations of the forms in Table 1.4 can also be treated with appropriate guesses for the particular solution.

        iii. If $f(x)$ contains an exponential that has the same argument as the exponential in the homogeneous solution, you may need to multiply the particular solution by $x$.

        iv. If $f(x)$ is none of the above, try the procedure that leads to Eq. (1.2.18).

(4) Compose the general solution by adding the homogeneous and particular solutions together. Obviously, if $f(x) = 0$ and there is no particular solution, then the general solution is just the homogeneous solution.

(5) Satisfy the initial condition. This will determine the constant multiplying the exponential in the homogeneous solution. The result should be a specific solution.

All the possibilities for a linear first-order differential equation have been considered, but there is no guarantee that the anti-derivatives for the integrals that determine $R(x)$ and $F(x)$ are known. By far the simplest case is the one with $r(x) = r$, a constant, and it is this case that also works for higher-order differential equations which we will consider next.

### 1.3.5 *Flowchart: Linear, first-order differential equations*

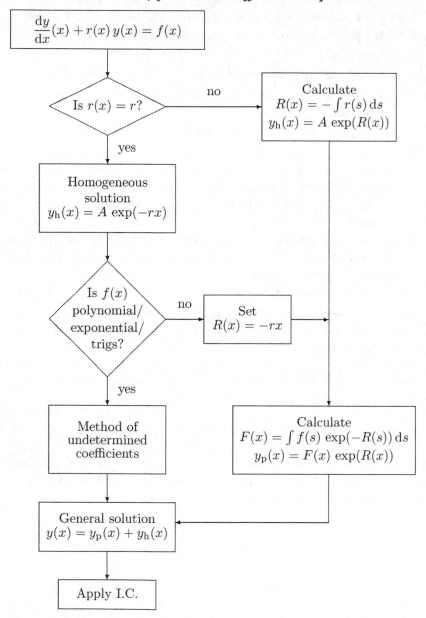

## 1.4 Second-order equations: Growth and decay

The first three sections of this chapter give a basic introduction to a first-order ordinary differential equation, that is, a single equation with only the first derivative of the unknown function. Of course, differential equations can arise in many more forms, for example, there may be several differential equations that contain only first derivatives but there are also several unknown functions. We call such a set a system of first-order differential equations and treat them fully in Chapter 5.

Another possibility is that a single equation may contain higher-order derivatives of the unknown function. At first sight, it may seem that we will have to develop an ever increasing list of methods to solve differential equations depending on the order of the highest derivative. Fortunately, this is not the case. The strategy for linear equations of all orders will be the same, only some small details will change. Restricting attention to second-order equations is then quite appropriate to learn a method of attack that is much easier to grasp than when higher-order derivatives are present.

Second-order differential equations arise in many ways and some of them will be explored in this section and later sections. One obvious way they arise is from Newton's law which states that mass times acceleration (a second derivative) must equal the total force. Another important way they arise is from the method of separation of variables for partial differential equations, a topic that will be explored in a later chapter. Also, second-order equations may arise from a combination of first-order equations. A simple example is the consecutive, first-order chemical reaction described next.

### 1.4.1   *A simple chain of chemical reactions*

Suppose that a chemical substance converts to a new substance which in turns converts to a final product. This two-step process illustrates what might happen, for example, in a chemical plant. Of course, in real chemical plants, the processes will be more complex and involve many steps, but this example introduces how mathematical models might arise in chemical engineering.

The process is illustrated with the diagram,

$$ A \xrightarrow{\quad k_1 \quad} B \xrightarrow{\quad k_2 \quad} C $$

The quantity A decomposes into a new quantity B at a rate proportional to the concentration $[A](t)$ $(kg/m^3)$. The constant of proportionality is $k_1$ $(s^{-1})$. The quantity B itself decomposes into another quantity C with a rate proportional to the concentration $[B](t)$ $(kg/m^3)$, the constant of proportionality being given by $k_2$ $(s^{-1})$.

We can also imagine this chemical reaction chain as a resource (oil) that goes through a refining step to produce an intermediate product (B), which is refined further to produce a final product (gasoline). An immediate question arises. How long does it take to produce the product once the resource A enters the production line? Once this question is resolved, new questions arise about the conditions to optimize the process. The situation is quite standard in engineering. Once a process is understood, subsequent questions arise about how to improve or optimize the process.

Obviously, the way to derive differential equations that describe the rate of change of the quantities is to appeal again to the conservation of matter. This time there is no amount entering or leaving the chemical reactor. Instead, the conservation of amount is expressed by the statement

> *The change in amount during a unit of time equals the amount gained minus the amount lost during a unit of time.*

For $[A](t)\,(\mathrm{kg/m^3})$, there is only loss due to conversion to B. For $[B](t)\,(\mathrm{kg/m^3})$, there is gain from the amount converted from A, and there is loss from conversion to C. Finally, there is only gain to C from conversion from B. Thus, the system of equations is:

$$\frac{\mathrm{d}[A]}{\mathrm{d}t}(t) = -k_1[A](t)\,, \tag{1.4.1}$$

$$\frac{\mathrm{d}[B]}{\mathrm{d}t}(t) = k_1[A](t) - k_2[B](t)\,, \tag{1.4.2}$$

$$\frac{\mathrm{d}[C]}{\mathrm{d}t}(t) = k_2[B](t)\,. \tag{1.4.3}$$

Note that the quantities $[A](t)\,(\mathrm{kg/m^3})$, $[B](t)\,(\mathrm{kg/m^3})$ and $[C](t)\,(\mathrm{kg/m^3})$ are unknown functions of time, while $k_1\,(\mathrm{s^{-1}})$ and $k_2\,(\mathrm{s^{-1}})$ are considered parameters (fixed quantities during the process).

There are three equations for three unknown functions – that is good since each equation should somehow determine one unknown. Unlike algebraic equations, however, differential equations need more information to determine solutions completely. We need to know how the process starts. Since there is a system of three differential equations, the expectation is that three initial conditions must be specified to determine the process completely. In other words, the starting value of each quantity must be known. Stated differently, the starting value of each quantity is a parameter that affects the process. Different starting values will cause the details of the process to be different. A simple picture assumes that there is only quantity A at the start and the process proceeds until it is all converted to product C and that is what will be assumed for this example.

The first equation (1.4.1) in this system of rate equations expresses how A becomes B. If $[A](t)\,(\mathrm{kg/m^3})$ starts with $[A](0) = [A_0]\,(\mathrm{kg/m^3})$ initially,[14] then Eq. (1.4.1) can be solved because it is a first-order differential equation with a specified initial condition. The next equation, Eq. (1.4.2), describes how B is created from A and produces C. The solution for $[A](t)\,(\mathrm{kg/m^3})$ can then be substituted into Eq. (1.4.2) and becomes a known forcing term. This allows Eq. (1.4.2) to be solved for $[B](t)\,(\mathrm{kg/m^3})$ assuming some initial value, and the choice made is $[B](0) = 0\,\mathrm{kg/m^3}$. Finally, $[B](t)$ can be substituted into Eq. (1.4.3) and $[C](t)\,(\mathrm{kg/m^3})$

---

[14]Standard practice sets the clock $t = 0$ s at the moment the process starts.

determined directly, assuming an initial condition such as $[C](0) = 0 \, \text{kg/m}^3$, the choice made for this example. The details of this solution strategy are left for an exercise.

Instead, a different approach will be followed. The equations are combined together to form a single equation for $[B](t)$ as follows. First, differentiate Eq. (1.4.2)

$$\frac{d^2[B]}{dt^2}(t) = k_1 \frac{d[A]}{dt}(t) - k_2 \frac{d[B]}{dt}(t) \,. \tag{1.4.4}$$

Now note that by adding Eq. (1.4.1), Eq. (1.4.2) and Eq. (1.4.3),

$$\frac{d[A]}{dt}(t) + \frac{d[B]}{dt}(t) + \frac{d[C]}{dt}(t) = 0 \,. \tag{1.4.5}$$

So, by using Eq. (1.4.3),

$$\frac{d[A]}{dt}(t) = -\frac{d[B]}{dt}(t) - k_2[B](t) \,,$$

and by substituting this result into Eq. (1.4.4), a single differential equation for $[B](t)$ is obtained

$$\frac{d^2[B]}{dt^2}(t) + \left(k_1 + k_2\right) \frac{d[B]}{dt}(t) + k_1 k_2[B](t) = 0 \,. \tag{1.4.6}$$

This equation has a second-order derivative, so it is called second-order. Each term involves $[B](t)$ or its derivatives by themselves: the equation is linear. The unknown function is $[B](t)$ and the independent variable is $t$. Only constant parameters $k_1$ and $k_2$ appear in the equation: the equation has constant coefficients.

The great news is that the strategy to solve Eq. (1.4.6) is the same as solving a first-order linear equation with constant coefficients. We simply substitute the guess

$$[B](t) = e^{\lambda t} \tag{1.4.7}$$

into Eq. (1.4.6). The result is

$$\lambda^2 \, e^{\lambda T} + \left(k_1 + k_2\right)\lambda \, e^{\lambda t} + k_1 k_2 \, e^{\lambda t} = 0 \,.$$

The exponential appears in every term and can be dropped. We are left with

$$\lambda^2 + \left(k_1 + k_2\right)\lambda + k_1 k_2 = 0 \,, \tag{1.4.8}$$

which is a quadratic equation for $\lambda$. It is a quadratic because there is a second derivative in the equation which causes a $\lambda^2$ factor to appear. In contrast, a first-order equation has a factor $\lambda$, leading to a simple linear equation for $\lambda$. This time we must solve the quadratic Eq. (1.4.8) for $\lambda$, but in this example it is easier to factor it:

$$\left(\lambda + k_1\right)\left(\lambda + k_2\right) = 0 \,. \tag{1.4.9}$$

The roots are clearly $-k_1$ and $-k_2$. What do we do with two roots? Obviously, we can write down two separate solutions

$$e^{-k_1 t} \quad \text{and} \quad e^{-k_2 t}$$

but which one should we pick? We take both by adding them together in a linear combination! So the full homogeneous solution is written as

$$[B](t) = c_1 e^{-k_1 t} + c_2 e^{-k_2 t} \tag{1.4.10}$$

and it is also the general solution to Eq. (1.4.6) since the equation is homogeneous – there is no forcing term on the right hand side of (1.4.6).

Initial conditions are needed to determine $c_1$ and $c_2$.[15] Recall we consider the situation where the resource A enters the production process with an initial concentration $[A](0) = [A_0]$ $(\text{kg/m}^3)$. At this moment there is no intermediate product $[B](0) = 0\,\text{kg/m}^3$ nor final product $[C](0) = 0\,\text{kg/m}^3$. Surely, starting with initial conditions for all three quantities, the differential equations determine their subsequent change in time completely. But how does these initial conditions help us determine the constants $c_1$ and $c_2$?

Since all the differential equations Eq. (1.4.1)–Eq. (1.4.3) are needed to derive the single equation Eq. (1.4.6), perhaps they are also needed to obtain appropriate initial conditions for the solution Eq. (1.4.10). The obvious choice is to evaluate Eq. (1.4.2) at $t = 0$. The consequence is a value for the rate of change of [B] at $t = 0$. Thus,

$$[B](0) = 0, \quad \frac{d[B]}{dt}(0) = k_1[A_0]. \tag{1.4.11}$$

The first condition in Eq. (1.4.11) can be applied immediately to the general solution Eq. (1.4.10). As a result,

$$c_1 + c_2 = 0. \tag{1.4.12}$$

To apply the second condition, we must first determine the derivative of the solution, which is

$$\frac{d[B]}{dt}(t) = -k_1 c_1 e^{-k_1 t} - k_2 c_2 e^{-k_2 t}. \tag{1.4.13}$$

We are now in a position to apply the second part of Eq. (1.4.11) to Eq. (1.4.13) at $t = 0$:

$$-k_1 c_1 - k_2 c_2 = k_1[A_0]. \tag{1.4.14}$$

We may view Eq. (1.4.12) and Eq. (1.4.14) as two equations for the two unknown constants $c_1$ and $c_2$.

---

[15] Since [B] has units kilograms per cubic meter, these constants must have the same units!

Appendix C provides a review of systems of equations and how to solve them. It also describes how such equations can by written as a matrix multiplication with a vector of unknowns. Thus, Eq. (1.4.12) and Eq. (1.4.14) may be expressed in the form,

$$\begin{bmatrix} 1 & 1 \\ -k_1 & -k_2 \end{bmatrix} \begin{pmatrix} c_1 \\ c_2 \end{pmatrix} = \begin{pmatrix} 0 \\ k_1[A_0] \end{pmatrix}. \tag{1.4.15}$$

As described in Appendix C, the solutions Eq. (C.1.7) and Eq. (C.1.9) to a system of equations can be calculated directly with little effort, leading to

$$c_1 = \frac{k_1}{k_2 - k_1}[A_0], \quad c_2 = -\frac{k_1}{k_2 - k_1}[A_0], \tag{1.4.16}$$

and the solution to the differential equation is

$$b(t) = \frac{[B](t)}{[A_0]} = \frac{k_1}{k_2 - k_1}\left(e^{-k_1 t} - e^{-k_2 t}\right). \tag{1.4.17}$$

We have cleverly written the results as a fraction of the initial input so that we can assess more readily how much the conversion from input to intermediate product has proceeded. There is another advantage. The quantity $b(t)$ has no units which means the result remains unchanged if we change the units of $k_1$, $k_2$ and time $t$.

There are several observations we can make about Eq. (1.4.17). First, it tells us how $b(t)$ behaves in time. There are obviously two time scales in the behavior of $b(t)$ associated with the two exponentials. Ultimately, $b(t)$ will just decay away. Next, let us find out how the other two quantities vary in time. From Eq. (1.4.2),

$$[A](t) = \frac{1}{k_1}\left(\frac{d[B]}{dt} + k_2[B]\right),$$

which leads to the result

$$a(t) = \frac{[A](t)}{[A_0]} = e^{-k_1 t}.$$

Obviously, $a(t)$ just decays away. From (1.4.5),

$$\frac{d}{dt}\Big([A](t)+]B](t) + [C](t)\Big) = 0,$$

which means $[A](t)+[B](t)+[C](t)$ is a constant in time. The initial values for these quantities tell us that the constant must be $[A_0]$. Incidentally, this is a statement of conservation: the total amount must remain unchanged in time as A converts to C. Thus

$$[C](t) = [A_0] - [B](t) - [A](t),$$

which leads to the result

$$c(t) = \frac{[C](t)}{[A_0]} = \left[1 - \frac{k_2}{k_2 - k_1}e^{-k_1 t} + \frac{k_1}{k_2 - k_1}e^{-k_2 t}\right]. \tag{1.4.18}$$

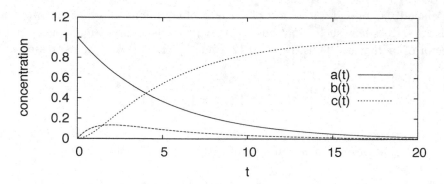

Fig. 1.13    Concentrations as functions of time: $k_1 = 0.2$ s$^{-1}$, $k_2 = 1.0$ s$^{-1}$.

Unlike $b(t)$, $c(t)$ continues to increase but eventually saturates at the value $c(t) \to 1$; all of $[A_0]$ has been converted into C.

It is often easier to understand the nature of the solutions by looking at their graphs as shown in Fig. 1.13 for the choice $k_1 = 0.2$ s$^{-1}$ and $k_2 = 1.0$ s$^{-1}$. Clearly visible is the conversion of A into B. As B grows in strength, its conversion into C becomes more prevalent.

While there are two time scales during this process, for $a(t)$ there is only one time scale $1/k_1$ (s) and the graph shows the exponential decay of $a(t)$ as it is converted into $b(t)$. Both $b(t)$ and $c(t)$ show an early rapid adjustment on the time scale $1/k_2$ (s), and then decay with the time scale of the exponential decay of $a(t)$. It is almost as though the intermediate product B is bypassed. Clearly, the longer time scale $1/k_1$ (s) dominates the process.

In contrast, the choice $k_1 = 1.0$ s$^{-1}$, $k_2 = 0.2$ s$^{-1}$, displayed in Fig. 1.14, shows a switch in the importance of the time scales. The concentration $a(t)$ decays rapidly on the time scale of $1/k_1$ (s). Most of it quickly becomes $b(t)$, which decays on the slower time scale $1/k_2$ (s) once $a(t)$ is mostly used up. After an initial adjustment, $c(t)$ is also produced over the slower time scale.

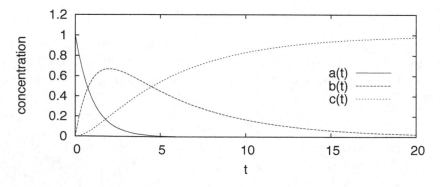

Fig. 1.14    Concentrations as functions of time: $k_1 = 1.0$ s$^{-1}$, $k_2 = 0.2$ s$^{-1}$.

> **Reflection**
>
> There are two transient parts to the solution. The one with the longer time scale dominates the approach of the solution to a steady state, $a(t) \to 0$, $b(t) \to 0$ and $c(t) \to 1$. The fact that there are two transient parts stems from the fact that the equation for $\lambda$, Eq. (1.4.8), is a quadratic with two roots, leading to two different exponentials in the homogeneous solution. Subsequently, we need two initial conditions to determine the two coefficients in the homogeneous solution. Most important in the example is that the roots to the quadratic are real and negative. Thus the exponentials decay in time.

So far, everything makes sense and we are satisfied, except we have glossed over an assumption. We have assumed $k_1 \neq k_2$. If $k_1 = k_2$, the solutions for $b(t)$ and $c(t)$, Eq. (1.4.17) and Eq. (1.4.18) fail because they require a division by $k_2 - k_1 = 0$. Now what? Let us look more closely at Eq. (1.4.17), and notice that the solution becomes indeterminate of the form $0/0$. Obviously we should take the limit as $k_2 \to k_1 = k$. So fix $k_1 = k$ and pick a moment $t$, then take the limit

$$\lim_{k_2 \to k} \frac{e^{-kt} - e^{-k_2 t}}{k_2 - k} = t\,e^{-kt} ,$$

and the solution for $b(t)$ becomes

$$b(t) = kt\,e^{-kt} . \tag{1.4.19}$$

Similarly, Eq. (1.4.18) is replaced with

$$c(t) = 1 + kt\,e^{-kt} - e^{-kt} .$$

How on earth could we know we have a problem like this in our general solution beforehand? Let us look at the equation for $\lambda$ again and note that with $k_2 = k_2 = k$, (1.4.8) becomes

$$\lambda^2 + 2k\lambda + k^2 = 0 ,$$

which is a perfect square. There is only one root $\lambda = -k$. We appear to have only one solution $b_h(t) = C \exp(-kt)$. Now we see that we must add another solution $b_h(t) = t \exp(-kt)$ and the homogeneous solution, and hence the general solution, has the form,

$$b(t) = c_1 e^{-kt} + c_2 t\, e^{-kt} . \tag{1.4.20}$$

Now we are ready to apply the initial conditions as before; see Exercise 6.

### 1.4.2 *Chained chemical reactions with forcing*

As chemical engineers, we might be more interested in supplying a constant rate $R\,(\mathrm{kg/(m^3 s)})$ of quantity A into the reactor. Consequently, the first equation Eq. (1.4.1) of the system is replaced by

$$\frac{d[A]}{dt}(t) = -k_1 [A](t) + R . \tag{1.4.21}$$

By following the steps that lead to Eq. (1.4.6), we obtain

$$\frac{d^2[B]}{dt^2}(t) + (k_1 + k_2)\frac{d[B]}{dt}(t) + k_1 k_2 [B](t) = k_1 R.$$

(1.4.22)

Now the equation is inhomogeneous because of the presence of $k_1 R$ in the right hand side of the equation. Fortunately, the forcing term is just a constant and a particular solution will be a constant as well,

$$[B]_p(t) = \frac{R}{k_2}.$$

The general solution combines the particular solution and the homogeneous solution Eq. (1.4.10).

$$[B](t) = \frac{R}{k_2} + c_1 e^{-k_1 t} + c_2 e^{-k_2 t}.$$

As before, the coefficients $c_1$ and $c_2$ are determined by initial conditions, and the initial conditions must reflect how we imagine the process is started. An obvious choice is that the process has a cold start; there is nothing inside the reacting unit until we start $(t = 0)$ adding the quantity A at the steady rate $R$. In other words,

$$[A](0) = 0, \quad [B](0) = 0, \quad [C](0) = 0.$$

This information must be converted to initial conditions on $[B](t)$. We have one already and the other comes from Eq. (1.4.2) evaluated at $t = 0$ – see Eq. (1.4.11).

$$\frac{d[B]}{dt}(0) = 0.$$

Consequently,

$$c_1 + c_2 = -\frac{R}{k_2},$$

$$-k_1 c_1 - k_2 c_2 = 0,$$

with the solution for the unknown coefficients,

$$c_1 = \frac{R}{k_1 - k_2}, \quad c_2 = \frac{k_1 R}{k_2(k_2 - k_1)}.$$

(1.4.23)

The algebraic steps to determine these solutions are presented in Appendix C.

As an aside, note again that the two equations for $c_1$ and $c_2$ can be written in matrix form as done in Eq. (1.4.15).

$$\begin{bmatrix} 1 & 1 \\ -k_1 & -k_2 \end{bmatrix} \begin{pmatrix} c_1 \\ c_2 \end{pmatrix} = \begin{pmatrix} -R/k_2 \\ 0 \end{pmatrix}.$$

(1.4.24)

The matrix is the same as in Eq. (1.4.15), but the right hand side vector is different. In Eq. (1.4.15), the contribution to the right hand side vector comes from the initial condition Eq. (1.4.11), whereas in Eq. (1.4.24) it comes from the contribution from the particular solution. In other words, the initial value of the particular solution will affect the initial conditions when applied to the general solution.

With the known values for the coefficients Eq. (1.4.23), the solution for [B](t) is complete. The other quantities are also easily found.

$$[A](t) = \frac{R}{k_1} - \frac{R}{k_1} e^{-k_1 t},$$

$$[B](t) = \frac{R}{k_2} + \frac{R}{k_1 - k_2} e^{-k_1 t} + \frac{k_1 R}{k_2(k_2 - k_1)} e^{-k_2 t},$$

$$[C](t) = Rt - \frac{R}{k_1} - \frac{R}{k_2} - \frac{k_2 R}{k_1(k_1 - k_2)} e^{-k_1 t} - \frac{k_1 R}{k_2(k_2 - k_1)} e^{-k_2 t}.$$

What is the eventual behavior of the forced system? In other words, what happens to the solutions as $t$ becomes large? Clearly the transient behavior, reflected by the behavior of the exponentials, dies away leaving

$$[A] = R/k_1, \quad [B] = R/k_2, \quad [C] = Rt - \frac{R}{k_1} - \frac{R}{k_2}.$$

By steadily supplying quantity A into the reactor, it reaches a steady state value $R/k_1$ (kg/m$^3$) with a steady rate of conversion to the quantity B. Similarly, B has reached a steady state $R/k_2$ (kg/m$^3$) with a steady conversion to C. But C just accumulates at a steady rate. Unless we withdraw some of C, it will eventually fill up the reactor and cause some kind of failure. Since the reaction is being used to create C, we would of course want to remove it from the reactor. Unlike the situation where we just added an initial amount of A and let the process convert it to C, we are now continually adding A and we should continually remove C. Some of the exercises explore this possibility.

### Reflection

The process of constructing the solution is the same; first, a *homogeneous* solution is found as a combination of all the exponential solutions, then a *particular* solution is constructed by the method of undetermined coefficients. The *general* solution combines the particular and homogeneous solutions, and only then are initial conditions applied to produce a *specific* solution. Since the homogeneous solution in this example contains two exponentially decaying contributions, the long term behavior is captured by the long term behavior of the particular solution.

### 1.4.3  *Abstract viewpoint*

#### 1.4.3.1  *Construction of a homogeneous solution*

A second-order linear differential equation with constant coefficients and without forcing effects takes the form,

$$\frac{d^2 y}{dx^2}(x) + a \frac{dy}{dx}(x) + b\, y(x) = 0. \tag{1.4.25}$$

To draw attention to the absence of forcing terms, this equation is referred to as *homogeneous*: it is simply the extension of the idea expressed in Sec. 1.1.7.

By invoking Principle 1, we seek a solution of the form

$$y(x) = e^{\lambda x} \tag{1.4.26}$$

and substitute into Eq. (1.4.25) to obtain

$$\lambda^2 C\, e^{\lambda x} + a\lambda\, C\, e^{\lambda x} + b\, C\, e^{\lambda x} = 0.$$

The exponential $\exp(\lambda x)$ is common to all terms and may be dropped, leading to a quadratic equation for $\lambda$.

$$\lambda^2 + a\lambda + b = 0. \tag{1.4.27}$$

---

**Principle 8**

*Exponential solutions of the form Eq. (1.4.26) to homogeneous, linear differential equations with constant coefficients will lead to a polynomial in $\lambda$, called the characteristic polynomial. The degree of the polynomial will match the order of the differential equation. The roots of the polynomial will determine the solution.*

---

The quadratic Eq. (1.4.27) may have

(1) two real roots, $\lambda_1$ and $\lambda_2$, then the homogeneous solution is

$$y(x) = c_1 e^{\lambda_1 x} + c_2 e^{\lambda_2 x}, \tag{1.4.28}$$

(2) one root (real), $\lambda$, then the homogeneous solution is

$$y(x) = c_1 e^{\lambda x} + c_2 x\, e^{\lambda x}, \tag{1.4.29}$$

(3) two complex roots, and we must find some way to express the solution as real functions, a topic to be covered in the next section.

There are several aspects of the procedure we should verify. First, can we add two solutions with different roots $\lambda_1$ and $\lambda_2$ as expressed in Eq. (1.4.28)? Let $y_1(x)$ and $y_2(x)$ be two solutions to (1.4.25) and consider the linear combination

$$y(x) = c_1 y_1(x) + c_2 y_2(x). \tag{1.4.30}$$

Principle 1 requires that we check whether $y(x)$ is a solution by direct substitution into Eq. (1.4.25). After a re-arrangement of the terms in the result of the substitution, we find

$$\frac{d^2 y}{dx^2}(x) + a\frac{dy}{dx}(x) + b\, y(x) = \left[\frac{d^2 y_1}{dx^2}(x) + a\frac{dy_1}{dx}(x) + b\, y_1(x)\right]$$

$$+ \left[\frac{d^2 y_2}{dx^2}(x) + a\frac{dy_2}{dx}(x) + b\, y_2(x)\right],$$

and each expression in brackets – added for emphasis – is zero because $y_1(x)$ and $y_2(x)$ are solutions to Eq. (1.4.25). Thus the right hand side is zero and the linear combination is a solution. Clearly, we can choose $y_1(x) = \exp(\lambda_1 x)$ and $y_2(x) = \exp(\lambda_2 x)$ since they are solutions. Thus the combination (1.4.28) is a solution. This conclusion is directly related to Principle 4.

### 1.4.3.2 *The importance of initial conditions*

If we can add two solutions together, why can we not add more, say three solutions? The issue is resolved by considering what initial conditions are needed. Recall that for a first-order differential equation of the form (1.1.19), we could construct the Taylor series for the solution provided we knew the initial condition $y(0) = y_0$. Then we showed this solution must be unique – there are no others. Our strategy then is to repeat this argument for (1.4.25).

Consider Eq. (1.4.25) evaluated at $t = 0$ with the hope that we might the able to start the construction of the Taylor series for the solution. Rewrite the result as

$$\frac{\mathrm{d}^2 y}{\mathrm{d}x^2}(0) = -a \frac{\mathrm{d}y}{\mathrm{d}x}(0) - b\,y(0)\,.$$

What this statement shows is if we want to know the second derivative at $t = 0$ uniquely, we had better know the first derivative and the function initially. These then become the required initial conditions, written generally as[16]

$$y(0) = \alpha \quad \text{and} \quad \frac{\mathrm{d}y}{\mathrm{d}x}(0) = \beta\,, \qquad (1.4.31)$$

then the second derivative of $y(x)$ can be determined initially as

$$\frac{\mathrm{d}^2 y}{\mathrm{d}x^2}(0) = -a\beta - b\alpha\,.$$

Repeated differentiation of Eq. (1.4.25) ensures all derivatives of $y$ can be determined at $t = 0$, and the solution is completely determined by its Taylor series. To illustrate, consider the derivative of Eq. (1.4.25) evaluated initially.

$$\frac{\mathrm{d}^3 y}{\mathrm{d}x^3}(0) = -a \frac{\mathrm{d}^2 y}{\mathrm{d}x^2}(0) - b \frac{\mathrm{d}y}{\mathrm{d}x}(0)$$

$$= (a^2 - b)\,\beta + ab\,\alpha\,.$$

As you can imagine, continuing the process for higher derivatives soon becomes complicated. However, we can imagine completing the task so that all derivatives at $t = 0$ can be determined.

---

[16]We have used the symbol $x$ as the independent variable. Mathematicians do not care whether this is a spatial or temporal variable. Many applications are for time evolving problems where $x$ represents time (and is often replaced by $t$). Thus Eq. (1.4.31) are referred to as initial conditions because they specify information at a "starting point."

### Principle 9: Extension of Principle 3

*Additional information is needed to select a unique solution to a differential equation. If the equation is second order, we need two more pieces of information, typically initial conditions for the solution and its derivative.*

#### 1.4.3.3    *The Wronskian*

The conclusion we reach is that the two initial conditions Eq. (1.4.31) should determine the solution completely. Since we have two solutions expressed as a linear combination Eq. (1.4.30), the two initial conditions should determine $c_1$ and $c_2$. We may verify this statement directly by applying Eq. (1.4.31) to Eq. (1.4.30) to obtain the two equations for $c_1$ and $c_2$:

$$c_1\, y_1(0) + c_2\, y_2(0) = \alpha\,, \tag{1.4.32}$$

$$c_1\, \frac{\mathrm{d}y_1}{\mathrm{d}x}(0) + c_2\, \frac{\mathrm{d}y_2}{\mathrm{d}x}(0) = \beta\,. \tag{1.4.33}$$

Note that in these equations, we consider $y_1(t)$ and $y_2(t)$ as known – we have presumably solved the differential equation and found two solutions – and the initial conditions provide $\alpha$ and $\beta$. As described in Appendix C, these equations may also be written as

$$\begin{bmatrix} y_1(0) & y_2(0) \\ \dfrac{\mathrm{d}y_1}{\mathrm{d}x}(0) & \dfrac{\mathrm{d}y_2}{\mathrm{d}x}(0) \end{bmatrix} \begin{pmatrix} c_1 \\ c_2 \end{pmatrix} = \begin{pmatrix} \alpha \\ \beta \end{pmatrix}\,. \tag{1.4.34}$$

It is clear now that the entries in the matrix play the role of parameters in the solution for $c_1$ and $c_2$.

By following the procedures given in Appendix C, the solution for $c_1$ and $c_2$ can be easily determined;

$$c_1 = \left( \alpha\, \frac{\mathrm{d}y_2}{\mathrm{d}x}(0) - \beta\, y_2(0) \right) \Big/ W(0)\,,$$

$$c_2 = \left( \beta\, y_1(0) - \alpha\, \frac{\mathrm{d}y_1}{\mathrm{d}x}(0) \right) \Big/ W(0)\,,$$

where the determinant of the matrix is

$$W(0) = y_1(0)\, \frac{\mathrm{d}y_2}{\mathrm{d}x}(0) - y_2(0)\, \frac{\mathrm{d}y_1}{\mathrm{d}x}(0) \tag{1.4.35}$$

and is called the *Wronskian*. Clearly, we require $W(0) \neq 0$, otherwise there are no solutions for $c_1$ and $c_2$. Obviously, we can define the Wronskian for any value of $x$ by

$$W(x) = y_1(x)\, \frac{\mathrm{d}y_2}{\mathrm{d}x}(x) - y_2(x)\, \frac{\mathrm{d}y_1}{\mathrm{d}x}(x) \tag{1.4.36}$$

which means that as long as $W(x) \neq 0$, it does not matter where we apply the "initial" conditions.

### 1.4.3.4 *Uniqueness*

In conclusion, provided we are given two initial conditions Eq. (1.4.31) and we have found two solutions that give $W(0) \neq 0$, we know we can construct a specific solution to Eq. (1.4.25). But can there be others? We can show that there cannot be others by a standard approach. Let $Y_1(x)$ and $Y_2(x)$ be two solutions that satisfy the differential equation Eq. (1.4.25) and the initial conditions Eq. (1.4.31). Now consider $Y(x) = Y_1(x) - Y_2(x)$ and substitute into the differential equation.

$$\frac{d^2 Y}{dx^2}(x) + a\frac{dY}{dx}(x) + b\,Y(x) = \frac{d^2 Y_1}{dx^2}(x) + a\frac{dY_1}{dx}(x) + b\,Y_1(x)$$

$$+ \frac{d^2 Y_2}{dx^2}(x) + a\frac{dY_2}{dx}(x) + b\,Y_2(x)$$

$$= 0,$$

because $Y_1(x)$ and $Y_2(x)$ separately satisfy the differential equation. Thus $Y(x)$ satisfies the same homogeneous differential equation.

Now apply the initial conditions to $Y(x)$.

$$Y(0) = Y_1(0) - Y_2(0) = \alpha - \alpha = 0, \tag{1.4.37}$$

$$\frac{dY}{dx}(0) = \frac{dY_1}{dx}(0) - \frac{dY_2}{dx}(0) = \beta - \beta = 0. \tag{1.4.38}$$

By repeated differentiation of the differential equation for $Y(x)$, we can establish the recursion formula,

$$\frac{d^{(n+2)} Y}{dx^{(n+2)}}(x) = -a\frac{d^{(n+1)} Y}{dx^{(n+1)}}(x) - b\frac{d^n Y}{dx^n}(x) = 0.$$

If we apply the initial conditions Eq. (1.4.37) and Eq. (1.4.38) to this recursion formula with $n = 0$, then we find that the second derivative of $Y(x)$ must be zero initially. Subsequently, we find that all the derivatives must be zero. In other words, the Taylor series is zero and $Y(x) = 0 \Rightarrow Y_1(x) = Y_2(x)$; there is only one unique solution.[17] However, it might be possible to write the solution in different, but equivalent, ways. Some of the exercises show how this might happen.

### Examples

Let us check the Wronskian for the solution given in Eq. (1.4.28) with the choice,

$$y_1(x) = e^{\lambda_1 x}, \quad y_2(x) = e^{\lambda_2 x}.$$

We find

$$W(x) = (\lambda_2 - \lambda_1)\,e^{\lambda_1 x} e^{\lambda_2 x}.$$

So for $\lambda_1 \neq \lambda_2$, $W(x) \neq 0$, specifically $W(0) \neq 0$. This means we will be able to find a unique solution, for example, Eq. (1.4.17).

---

[17]This argument depends on the fact that $Y(x)$ is an analytic function, that is, it has a Taylor series – a topic covered in studies of complex variables.

When $\lambda_1 = \lambda_2 = \lambda$, just one root to Eq. (1.4.27), then we should take

$$y_1(x) = e^{\lambda x}, \quad y_2(x) = x\,e^{\lambda x}.$$

The Wronskian Eq. (1.4.36) is

$$W(x) = e^{2\lambda x},$$

and these choices for the two solutions are acceptable.

What remains are the choices for the homogeneous solutions when the roots to the characteristic equation Eq. (1.4.27) are complex, the topic for the next section.

### 1.4.4   *Exercises*

**These are some basic exercises to help develop and improve skill.**

(1) Find a general solution to the following differential equations:

  (a)

$$\frac{d^2 y}{dx^2}(x) + \frac{dy}{dx}(x) - 6y(x) = 0.$$

  **Answer:** $W(x) = 5\exp(-x)$.

  (b)

$$\frac{d^2 w}{dy^2}(y) - (3+\pi)\frac{dw}{dy}(y) + 3\pi w(y) = 0.$$

  **Answer:** $W(y) = (\pi - 3)\exp\big((3+\pi)\,y\big).$

  (c)

$$2\frac{d^2 x}{dt^2}(t) + \frac{dx}{dt}(t) - 6x(t) = 0.$$

  **Answer:** $W(t) = -(7/2)\exp(-t/2).$

  Confirm your choice for the general solution by checking the Wronskian.

(2) Consider the differential equation,

$$\frac{d^2 y}{dt^2}(t) - y(t) = 0.$$

Show that $\exp(t)$ and $\exp(-t)$ are two solutions to this equation. Check the Wronskian to confirm that

$$y(t) = c_1 e^t + c_2 e^{-t}$$

is an acceptable general solution.

**Answer:** $W(t) = -2$.

  (a) Find the solution that satisfies the initial conditions,

$$y(0) = 1 \qquad \frac{dy}{dt}(0) = 0.$$

  This solution is called $\cosh(t)$, the *hyperbolic cosine* function. Its connection to cosine will be apparent in a later section.

(b) Find the solution that satisfies the initial conditions,

$$y(0) = 0 \quad \frac{dy}{dt}(0) = 1.$$

This solution is called sinh($t$), the *hyperbolic sine* function.

(c) Verify the following results for the derivatives of these functions:

$$\frac{d}{dt}\cosh(t) = \sinh(t), \quad \frac{d}{dt}\sinh(t) = \cosh(t).$$

Compare the pattern to the pattern for derivatives of cos($t$) and sin($t$).

(d) Now confirm that cosh($t$) and sinh($t$) are solutions to the differential equation. Check that the Wronskian for these solutions is not equal to zero. Thus we may write the general solution to the differential equation in an alternate form,

$$y(t) = a_1 \cosh(t) + a_2 \sinh(t).$$

**Answer:** $W(t) = 1$.

(e) Determine all the derivatives evaluated at $t = 0$ directly from the differential equation and from the initial condition given in part (a). This will allow you to determine the Taylor series for cosh($t$). Compare the Taylor series for cosh($t$) with that for cos($t$).

(3) Establish first that

$$y(t) = a\, e^t + b\, e^{-t} + c\, \cosh(t) + d\, \sinh(t)$$

solves the differential equation,

$$\frac{d^2 y}{dt^2}(t) - y(t) = 0.$$

This solution contains four different homogeneous solutions when we know that we need only two. To understand why two extra solutions are unnecessary, apply the initial conditions

$$y(0) = A, \quad \text{and} \quad \frac{dy}{dt}(0) = B.$$

There are two equations for four unknown coefficients. Determine $c$ and $d$ in terms of $A$, $B$, $a$ and $b$ and substitute the results into the solution. What is the consequence? What does it tell you about the number of independent solutions needed to satisfy the initial conditions?

(4) Let $y_1(x)$ and $y_2(x)$ be non-trivial solutions (that means neither $y_1(x) \neq 0$ nor $y_2(x) \neq 0$) to Eq. (1.4.25) and consider the derivative of their ratio $y_2(x)/y_1(x)$. What can you conclude if the Wronskian Eq. (1.4.36) $W(x) = 0$?

(5) Solve

$$\frac{d^2 b}{dt^2}(t) + (k_1 + k_2)\frac{db}{dt}(t) + k_1 k_2 b(t) = 0$$

with the initial conditions,

$$b(0) = 0 \quad \text{and} \quad \frac{db}{dt}(0) = k_1 \,.$$

At first assume, $k_1 \neq k_2$.

(a) Then take the limit as $k_2 \to k_1 = k$.
(b) Construct the solution by using the guess Eq. (1.4.20) and compare the result with the limit obtained in the previous part.
(c) Plot the solutions for $k_1 = 1$ and $k_2 = 0.9$, $k_2 = 1.0$, $k_2 = 1.1$. Is there anything strange about the solutions that might indicate a limiting process has occured?

**Answer:** The following table gives $b(t)$ at select times for various $k_2$. The parameter $k_1 = 1$.

Table 1.6   Some values
for $b(t)$.

| $k_2$ | $t = 1$ | $t = 2$ |
|---|---|---|
| 0.9 | 0.387 | 0.300 |
| 1.0 | 0.368 | 0.271 |
| 1.1 | 0.350 | 0.245 |

**Some problems that arise in applications.**

(6) Solve Eq. (1.4.1), Eq. (1.4.2) and Eq. (1.4.3) one at a time by using the solutions from the previous equations when needed.
  **Answer:** Your results should agree with those in Sec. 1.4.1, in particular with Eq. (1.4.17) and Eq. (1.4.18).
(7) With the definition $b(t) = [B](t)/[A_0]$, solve Eq. (1.4.6) subject to the initial conditions

$$b(0) = 1 \quad \text{and} \quad \frac{db}{dt}(0) = -k_1 \,.$$

What do you notice about the solution? Are there any difficulties with the choice $k_2 = k_1$?
(8) Suppose that quantity C is removed at some fixed rate from the reactor described in Sec. 1.4.2.

(a) At what rate should C be removed to maintain a steady operation of the reactor?
(b) How should the system of equations be changed with the inclusion of a steady rate of removal of C?
(c) Is it possible to remove C at this rate at the beginning of the process? If not, what would you change?
(d) Determine the solution for your mathematical model.

(e) Write a short report on what you did and provide justification for your choices.

(9) The apartments in a building have central heating/cooling systems that are controlled individually. The controls are designed so that heating or cooling will adjust the temperature according to

$$\frac{dT}{dt}(t) = -k\left(T(t) - T_a\right),$$

where $T_a$ (°F) is the desired temperature in the apartment. Assume that the temperature is initially at $T_1$ (°F) and show that the temperature will return back to $T_a$ (°F) in time.

Unfortunately, the electrician wired the controls incorrectly for two apartments adjacent to each other. The control in apartment A responds to the temperature $T_B(t)$ (°F) in the apartment next door, and vice versa. Thus,

$$\frac{dT_A}{dt}(t) = -k\left(T_B(t) - T_a\right),$$

$$\frac{dT_B}{dt}(t) = -k\left(T_A(t) - T_a\right).$$

Obtain a single differential equation for $T_A(t)$ (°F) and construct a general solution. Suppose the temperatures have the initial values $T_A(0) = T_1$ (°F) and $T_B(0) = T_2$ (°F). Determine the role of the initial values on the long term behavior of the system.

Suppose $k = 0.5$ h$^{-1}$ and the control temperature is $T_a = 72$ °F. Let $T_B$ be at the control temperature $T_a$ initially, but assume the temperature $T_A(0) = 71$ °F. Determine the temperature $T_A$ after ten hours.

**Answer:** $-2.2$ °F.

### 1.4.5 *Time-varying coefficients in the differential equation*

Hopefully, after completing the exercises of this section, you would have gained a good introduction to the solution of second-order ordinary differential equations with constant coefficients. You may have noticed that the procedure is constructive; it is based on determining the correct exponential behavior of the solutions. The process may seem too specialized, and based on guesswork rather than some special formula that describes all solutions. The truth of the matter is that unless the coefficients in the equation are constants we know few ways to find solutions except through numerical methods. There is no integrating factor here as presented in Sec. 1.3.2. However, it is possible to construct a series solution as done in Sec. 1.3.2 and it is this approach we will follow here.

It is certainly true that important engineering problems contain differential equations with variable coefficients. As a simple example, suppose there is a catalyst that can be added to speed up the chemical reaction rate $k_2$ (s$^{-1}$) in Eq. (1.4.2) and Eq. (1.4.3). Imagine adding the catalyst, possibly a fine powder, to the reacting unit at a steady rate until the maximum improvement is reached. A simple mathematical model for the consequence is that the reaction rate $k_2$ increases linearly in time as the powder is added until $t = T$ (s), when the maximum is reached. Thus, $k_2$ is replaced by

$$k_2(t) = \begin{cases} k_2 + (k_m - k_2)\,t/T = k_2 + \alpha\,t\,, & \text{for } t < T\,, \\ k_m\,, & \text{for } t > T\,, \end{cases}$$

where $\alpha = (k_m - k_2)/T$ (s$^{-2}$) is a new parameter.

The steps leading to Eq. (1.4.6) can be repeated to obtain a new equation containing $k_2(t)$. Note Eq. (1.4.5) is still true, but that Eq. (1.4.4) must be replaced with

$$\frac{d^2[\mathrm{B}]}{dt^2}(t) = k_1\frac{d[\mathrm{A}]}{dt}(t) - k_2(t)\frac{d[\mathrm{B}]}{dt}(t) - \frac{dk_2}{dt}(t)\,[\mathrm{B}](t)\,,$$

leading to

$$\frac{d^2[\mathrm{B}]}{dt^2}(t) + \big(k_1 + k_2(t)\big)\frac{d[\mathrm{B}]}{dt}(t) + \left(k_1\,k_2(t) + \frac{dk_2}{dt}(t)\right)[\mathrm{B}](t) = 0\,. \qquad (1.4.39)$$

When $t < T$, Eq. (1.4.39) becomes

$$\frac{d^2[\mathrm{B}]}{dt^2}(t) + \big(k_1 + k_2 + \alpha t\big)\frac{d[\mathrm{B}]}{dt}(t) + \big(k_1\,k_2 + k_1\,\alpha t + \alpha\big)[\mathrm{B}](t) = 0\,. \qquad (1.4.40)$$

Because the coefficients multiplying $[\mathrm{B}](t)$ and the rate of change of $[\mathrm{B}](t)$ contain functions of $t$, the guess of an exponential function does not work – try it! There is no simple way to construct a solution to Eq. (1.4.40). However, there are some things we can anticipate. We still expect there are two independent solutions $[\mathrm{B}]_1(t)$ and $[\mathrm{B}]_2(t)$ that lead to a general solution,

$$[\mathrm{B}](t) = c_1[\mathrm{B}]_1(t) + c_2[\mathrm{B}]_2(t) \qquad (1.4.41)$$

and that two initial conditions are needed to determine $c_1$ and $c_2$.

How can we construct solutions $[\mathrm{B}]_1(t)$ and $[\mathrm{B}]_2(t)$? We can use the idea we used before to define $\cosh(x)$ and $\sinh(x)$ in Exercise 2 by choosing special initial conditions! Let us require

$$[\mathrm{B}]_1(0) = 1\,, \qquad\qquad \frac{d[\mathrm{B}]_1}{dx}(0) = 0\,, \qquad (1.4.42)$$

$$[\mathrm{B}]_2(0) = 0\,, \qquad\qquad \frac{d[\mathrm{B}]_2}{dx}(0) = 1\,. \qquad (1.4.43)$$

How do these choices help? Well, general initial conditions become very simple! Assume the general initial conditions Eq. (1.4.31) and apply them to Eq. (1.4.41).

Since $[B]_1(t)$ and $[B]_2(t)$ must satisfy Eq. (1.4.42) and Eq. (1.4.43) respectively,

$$[\mathrm{B}](0) = c_1[\mathrm{B}]_1(0) + c_1[\mathrm{B}]_2(0) = c_1 = \alpha\,,$$

$$\frac{d[\mathrm{B}]_1}{dt}(0) = c_1\frac{d[\mathrm{B}]_1}{dt}(0) + c_2\frac{d[\mathrm{B}]_2}{dt}(0) = c_2 = \beta\,,$$

and the solutions for $c_1$ and $c_2$ are immediately known. Notice also that the Wronskian Eq. (1.4.35) becomes $W(0) = 1$. Because of these properties, solutions $[B]_1(t)$ and $[B]_2(t)$ that satisfy Eq. (1.4.42) and Eq. (1.4.43), respectively, are called *fundamental* solutions.

How do we know whether fundamental solutions exist? Well, let us see if we can construct the solutions by using a Taylor series representation. Recall we used a Taylor series representation in Eq. (1.1.29) to show how an initial condition determined a solution to a first-order differential equation completely. We also used a Taylor series representation Eq. (1.3.21) to construct the particular solution when the forcing terms is expressed as a Taylor series.

We start by writing

$$[\mathrm{B}](t) = \sum_{n=0}^{\infty} a_n t^n = a_0 + a_1 t + a_2 t^2 + \cdots + a_n t^n + \cdots. \tag{1.4.44}$$

To determine the coefficients $a_n$, we substitute the series into the differential equation and balance each of the terms with the same power of $t$.

The best way to organize the work is to find the expansions for each of the terms separately and then their addition is relatively straightforward, but first we need the derivatives of the series.

$$\frac{d[\mathrm{B}]}{dt}(t) = a_1 + 2a_2 t + \cdots + n a_n t^{n-1} + \cdots$$

$$= \sum_{n=1}^{\infty} n a_n t^{n-1}$$

$$= \sum_{n=0}^{\infty} (n+1)\, a_{n+1} t^n\,. \tag{1.4.45}$$

In the last step, the counter is shifted to make the balance of terms with $t^n$ easier when we solve the differential equation.[18] The second derivative of the series is

$$\frac{d^2[\mathrm{B}]}{dt^2}(t) = 2a_2 + 6a_3 t + 12a_4 t^2 + \cdots + n(n-1)a_n t^{n-2} + \cdots$$

$$= \sum_{n=2}^{\infty} n(n-1)a_n t^{n-2}$$

$$= \sum_{n=0}^{\infty} (n+1)(n+2)a_{n+2} t^n\,. \tag{1.4.46}$$

---

[18]It is always wise to write out a few terms of a sum to check that its counter is correctly set, especially when changing the counter.

In the last step, the counter is shifted by two.

The first term in the differential equation is given by Eq. (1.4.46). By using Eq. (1.4.45), the second term becomes

$$\left(k_1 + k_2 + \alpha t\right) \frac{\mathrm{d}[\mathrm{B}]}{\mathrm{d}t}(t) = \left(k_1 + k_2\right) \sum_{n=1}^{\infty} n\, a_n t^{n-1} + \alpha \sum_{n=1}^{\infty} n\, a_n t^n$$

$$= \left(k_1 + k_2\right) \sum_{n=0}^{\infty} (n+1)\, a_{n+1} t^n + \alpha \sum_{n=1}^{\infty} n\, a_n t^n . \qquad (1.4.47)$$

The third term becomes

$$\left(k_1 k_2 + k_1 \alpha t + \alpha\right) [\mathrm{B}](t) = \left(k_1 k_2 + \alpha\right) \sum_{n=0}^{\infty} a_n t^n + k_1 \alpha \sum_{n=1}^{\infty} a_{n-1} t^n . \qquad (1.4.48)$$

Substitute Eq. (1.4.46), Eq. (1.4.47) and Eq. (1.4.48) into Eq. (1.4.39) and balance the terms with the same powers of $t$.

**Balance terms with constants:**

$$2a_2 + \left(k_1 + k_2\right) a_1 + \left(k_1 k_2 + \alpha\right) a_0 = 0 . \qquad (1.4.49)$$

**Balance terms with $t$:**

$$6a_3 + \left(k_1 + k_2\right) 2a_2 + \alpha a_1 + \left(k_1 k_2 + \alpha\right) a_1 + k_1 \alpha a_0 = 0 . \qquad (1.4.50)$$

**Balance the general term $t^n$:**

$$(n+1)(n+2)a_{n+2} + \left(k_1 + k_2\right)(n+1)\, a_{n+1} + \alpha n a_n$$
$$+ \left(k_1 k_2 + \alpha\right) a_n + k_1 \alpha a_{n-1} = 0. \qquad (1.4.51)$$

We have obtained a recursion formula for the coefficients of the power series!

The recursion formula Eq. (1.4.51) requires three prior values before the next one can be computed. The initial conditions provide two starting values, $a_0$ and $a_1$. Then, Eq. (1.4.49) allows us to determine $a_2$. Next, Eq. (1.4.50) allows us to determine $a_3$. All the rest of the coefficients can be determined by applying Eq. (1.4.51) recursively with the result that the solution is completely determined. In particular, the initial conditions Eq. (1.4.42) give $a_0 = 1$ and $a_1 = 0$. Then Eq. (1.4.49) and Eq. (1.4.50) give $a_2$ and $a_3$, respectively:

$$a_0 = 1, \quad a_1 = 0, \quad a_2 = -\frac{1}{2}\left(k_1 k_2 + \alpha\right), \quad a_3 = \frac{1}{6}\left[\left(k_1 + k_2\right) k_1 k_2 + \alpha k_2\right].$$

An approximate solution for $[B]_1(t)$ is therefore

$$[\mathrm{B}]_1(t) = 1 - \frac{1}{2}\left(k_1 k_2 + \alpha\right) t^2 + \frac{1}{6}\left[\left(k_1 + k_2\right) k_1 k_2 + \alpha k_2\right] t^3 \ldots \qquad (1.4.52)$$

Similarly, the initial conditions Eq. (1.4.43) give $a_0 = 0$ and $a_1 = 1$. Then Eq. (1.4.49) and Eq. (1.4.50) give $a_2$ and $a_3$, respectively:

$$a_0 = 0, \quad a_1 = 1, \quad a_2 = -\frac{1}{2}\left(k_1 + k_2\right), \quad a_3 = \frac{1}{6}\left(2\alpha - k_1^2 - k_1 k_2 - k_2^2\right), \quad \ldots$$

and

$$[B]_2(t) = t - \frac{1}{2}\left(k_1 + k_2\right)t^2 + \frac{1}{6}\left(2\alpha - k_1^2 - k_1 k_2 - k_2^2\right)t^3 + \cdots . \qquad (1.4.53)$$

Unfortunately, in both cases the general pattern to the coefficients $a_n$ is difficult to guess, but we can notice that the coefficients depend on parameters $k_1$, $k_2$ and $\alpha$ and that there is a division by $(n+1)(n+2)$ in the calculation of $a_{n+2}$ which causes the magnitude of $a_n$ to decrease quickly as $n$ increases. Why is this important? Because we want to know whether the power series converges and that depends on how quickly the magnitude of $a_n$ decreases as $n$ increases! Also, the behavior of $a_n$ determines the interval in $t$ for which the series converges.[19]

So, although we do not have explicit formula for $[B]_1(t)$ or $[B]_2(t)$, we do know we can construct their power series and use that to determine them at least for some interval in $t$. There are better numerically methods to construct these solutions, but that must wait for an advanced course on numerical methods. The point now is simply that we can find two independent solutions to Eq. (1.4.39), that we can combine them in a linear combination to obtain a general solution Eq. (1.4.41), and that we can use the general solution to satisfy initial conditions.

Since Eq. (1.4.40) is only valid for $0 \leq t \leq T$, there is a good chance that the series in Eq. (1.4.52) and Eq. (1.4.53) converge in this time interval. The initial conditions for the example are the same as those given in Eq. (1.4.11), which means the specific solution is

$$[B](t) = k_1[A_0][B]_2(t), \qquad (1.4.54)$$

valid for $0 \leq t \leq T$.

So what about the solution for $t > T$? Well, the differential equation Eq. (1.4.39) simply reduces to Eq. (1.4.6) with $k_2$ replaced by $k_m$ and so the homogeneous solution is just

$$[B](t) = c_1 e^{-k_1 t} + c_2 e^{-k_m t} . \qquad (1.4.55)$$

The last step is to patch the two solutions Eq. (1.4.54) and Eq. (1.4.5) together at $t = T$.

How do we find the matching conditions? We used Principle 2 in Sec. 1.1.5 to answer this question when the forcing term is discontinuous, and the application to Eq. (1.4.39) is similar. Note that $k_2(t)$ has a jump discontinuity in its derivative at $t = T$. So by Principle 2, we require that $[B](t)$ and $d[B](t)/dt$ to be continuous at $t = T$ but $d^2[B](t)/dt^2$ to have a jump discontinuity to match the term $(dk_2/dt)[B]$ in the differential equation. Consequently, we apply matching conditions that require $[B](t)$ and $d[B](t)/dt$ to be continuous at $t = T$:

$$c_1 e^{-k_1 T} + c_2 e^{-k_m T} = k_1[A_0][B]_2(T), \qquad (1.4.56)$$

$$-k_1 c_1 e^{-k_1 T} - k_m c_2 e^{-k_m T} = k_1[A_0]\frac{d[B]_2}{dt}(T). \qquad (1.4.57)$$

---

[19]Perhaps now the value of convergent series may be appreciated.

These equations constitute two equations for $c_1$ and $c_2$. Once solved, $[B](t)$ is known for $t > T$ as well. The key consequence is that the time scale $1/k_2$ is replaced by $1/k_m$ which will speed up the chemical reaction.

---

**Reflection**

In general, solutions to second-order differential equations with coefficients that vary with the independent variable are difficult to solve. There is no special technique. Sometimes a clever guess can work, but mostly the Taylor series can be constructed which will determine the solution at least in the neighborhood of the initial condition. Fortunately, when the coefficients in the differential equations are constants, we are guaranteed that the solution is a combination of exponential functions.

---

## 1.5　Second-order equations: Oscillations

In the last section, we considered an example where the characteristic equation (1.4.27) is a quadratic with real roots. The general solution (1.4.28), then, is a linear combination of two exponential functions. What do we do if the roots turn out to be complex? The exponentials will have complex arguments. Can they be interpreted appropriately?

Obviously, this section will use complex variables, and a review of the necessary aspects of complex variables is given in Appendix D.

### 1.5.1　*Exponentials with complex arguments*

To explore the meaning behind complex roots to the characteristic equation, consider the simple example,

$$\frac{\mathrm{d}^2 y}{\mathrm{d}x^2}(x) + y(x) = 0 \,. \tag{1.5.1}$$

The standard procedure to find homogeneous solutions is to substitute a guess of the form $y(x) = \exp(\lambda x)$ into the differential equation and obtain the characteristic equation Eq. (1.4.27),

$$\lambda^2 + 1 = 0 \,. \tag{1.5.2}$$

The roots are complex $\lambda = \pm \mathrm{i}$.[20] We write the general solution as

$$y(x) = c_1 \, e^{\mathrm{i}x} + c_2 \, e^{-\mathrm{i}x} \,, \tag{1.5.3}$$

but what does it mean?

The first observation is that $\exp(-\mathrm{i}x)$ appears to be the complex conjugate of $\exp(\mathrm{i}x)$ since i has been replaced by $-\mathrm{i}$. What might be the significance of such an

---

[20]Obviously this is the reason the differential equation Eq. (1.5.1) is chosen – to ensure a purely imaginary argument to the exponential function.

observation? Well, note that if $z = a + ib$ is a complex number, and its complex conjugate is $\bar{z} = a - ib$, then $z$ is real if and only if $\bar{z} = z$ (because then $b = 0$). Now the solution Eq. (1.5.3) appears to be complex. How can this be? We expect the solution $y(x)$ to have only real values. It can only be real if $\bar{y}(x) = y(x)$! Let us take the complex conjugate of $y(x)$ as given in Eq. (1.5.3). To do so, we must use the following properties of complex conjugation: the complex conjugate of a sum is the sum of the complex conjugates Eq. (D.1.8); and the complex conjugate of a product is the product of the complex conjugates Eq. (D.1.9). Thus,

$$\bar{y}(x) = \bar{c}_1 \, e^{-ix} + \bar{c}_2 \, e^{ix} \, .$$

If this is to match Eq. (1.5.3), then

$$\bar{c}_1 = c_2, \quad \text{and} \quad \bar{c}_2 = c_1 \, . \tag{1.5.4}$$

In other words, if $c_1 = a + ib$, then $c_2 = a - ib$; $c_1$ and $c_2$ are complex conjugates of each other. Well, this is all very fine, but it has not helped us understand what $y(x)$ is as a real function. Instead, we take a different approach.

Fortunately, there are two other solutions to Eq. (1.5.1) that are easy to spot, $\sin(x)$ and $\cos(x)$, because their second derivatives just change their sign. For example,

$$\frac{d^2}{dx^2} \sin(x) = \frac{d}{dx} \cos(x) = -\sin(x) \, ,$$

which we may write as

$$\frac{d^2}{dx^2} \sin(x) + \sin(x) = 0 \, .$$

This equation is exactly the same as Eq. (1.5.1) with $y(x) = \sin(x)$. In other words, we have verified that $\sin(x)$ is a solution by Principle 1. In the same way, we can confirm $y(x) = \cos(x)$ is also a solution to Eq. (1.5.1). Thus another general solution is

$$y(x) = a_1 \cos(x) + a_2 \sin(x) \, . \tag{1.5.5}$$

Since the Wronskian Eq. (1.4.36) with $y_1(x) = \cos(x)$ and $y_2(x) = \sin(x)$ is $W(x) = 1$ we are sure that Eq. (1.5.5) is an acceptable solution.

Principle 9 tells us that we will have a specific solution if we add two initial conditions to Eq. (1.5.5). Let us isolate one of the solutions, $\cos(x)$, and ask what initial conditions must be supplied to make it the only solution to Eq. (1.5.1). They should be

$$y(0) = \cos(0) = 1 \, , \quad \text{and} \quad \frac{dy}{dx}(0) = -\sin(0) = 0 \, .$$

So now apply these initial conditions to the general solution Eq. (1.5.3) to determine $c_1$ and $c_2$.

$$c_1 + c_2 = 1 \, ,$$
$$c_1 i - c_2 i = 0 \, .$$

The solution is easily obtained by following the steps that give Eq. (C.1.7) and Eq. (C.1.9), except that complex arithmetic must be used: $c_1 = c_2 = 0.5$ – note that this result is consistent with Eq. (1.5.4) – and the solution becomes

$$y(x) = \frac{1}{2}\left(e^{ix} + e^{-ix}\right).$$

At the same time, this must be equivalent to the solution $\cos(x)$ because the solution is unique. In other words,

$$\cos(x) \equiv \frac{1}{2}\left(e^{ix} + e^{-ix}\right). \tag{1.5.6}$$

We have derived a connection formula between the trigonometric function $\cos(x)$ and exponentials with complex arguments.

The connection formula for $\sin(x)$ can be found by choosing the initial conditions,

$$y(0) = \sin(0) = 0, \quad \text{and} \quad \frac{dy}{dx}(0) = \cos(0) = 1.$$

Apply these initial conditions to the general solution Eq. (1.5.3) to determine new values for $c_1$ and $c_2$.

$$c_1 + c_2 = 0,$$
$$c_1 i - c_2 i = 1.$$

The solution is easily obtained: $c_1 = -c_2 = 0.5/i$,[21] and the solution becomes

$$y(x) = \frac{1}{2i}\left(e^{ix} - e^{-ix}\right).$$

Since the solution is unique, it must be the same as $\sin(x)$. In other words,

$$\sin(x) \equiv \frac{1}{2i}\left(e^{ix} - e^{-ix}\right). \tag{1.5.7}$$

How amazing! We take two solutions to a differential equation that satisfy the same initial conditions and find a connection between trigonometric functions and exponential functions with complex arguments. After multiplying Eq. (1.5.7) by i and adding to Eq. (1.5.6), we find the well-known Euler formula,

$$e^{ix} = \cos(x) + i\sin(x). \tag{1.5.8}$$

Either by replacing $x$ by $-x$ in Eq. (1.5.8), or by multiplying Eq. (1.5.7) with i and subtracting from Eq. (1.5.6), or even by just taking the complex conjugate of Eq. (1.5.8), the companion formula

$$e^{-ix} = \cos(x) - i\sin(x) \tag{1.5.9}$$

can be obtained. Both these formulas reveal a close connection between exponentials with complex arguments and trigonometric functions that describe oscillatory behavior.

---

[21] Note that this result is also consistent with Eq. (1.5.4).

The consequence of the Euler formula is a clear connection between the general solutions Eq. (1.5.3) and Eq. (1.5.5). Specifically, substitute Eq. (1.5.8) and Eq. (1.5.9) into Eq. (1.5.3):

$$
\begin{aligned}
y(x) &= c_1\, e^{ix} + c_2\, e^{-ix} \\
&= c_1\big[\cos(x) + i\sin(x)\big] + c_2\big[\cos(x) - i\sin(x)\big] \\
&= (c_1 + c_2)\cos(x) + (ic_1 - ic_2)\sin(x) \\
&= a_1\cos(x) + a_2\sin(x)\,.
\end{aligned}
\tag{1.5.10}
$$

We already know that $c_1$ and $c_2$ are complex conjugates of each other (1.5.4), so we can write instead $c_1 = a + ib$, $c_2 = a - ib$, where $a$ and $b$ are real numbers. Thus $a_1 = c_1 + c_2 = 2a$ and $a_2 = i(c_1 - c_2) = -2b$ are real numbers ensuring $y(x)$ is a real solution. Since the two different forms for the general solution are equivalent, we may choose the version that is best for our purposes. Often the trigonometric form Eq. (1.5.5) is best since it reveals the oscillatory nature of the solution.

**Reflection**

Properties of functions can be deduced from their behavior as solutions to differential equations. This idea leads to understanding the exponential function, but it also leads to understanding many of the special functions of science and engineering, such as Bessel functions, Chebyshev polynomials, etc.

### 1.5.2 *Application to differential equations*

The value of Euler's formula becomes more apparent when we consider the solution to a typical ordinary differential equation, for example,

$$
\frac{d^2y}{dt^2}(t) + 2\frac{dy}{dt}(t) + 5\,y(t) = 0\,.
\tag{1.5.11}
$$

The characteristic equation is

$$
\lambda^2 + 2\lambda + 5 = 0\,,
$$

with the roots,

$$
\lambda = -1 \pm 2i\,.
\tag{1.5.12}
$$

The general solution can be written as

$$
\begin{aligned}
y(t) &= c_1\, e^{-t+2it} + c_2\, e^{-t-2it} \\
&= c_1\, e^{-t}\, e^{2it} + c_2\, e^{-t}\, e^{-2it} \\
&= e^{-t}\big(c_1\, e^{2it} + c_2\, e^{-2it}\big)\,.
\end{aligned}
\tag{1.5.13}
$$

The exponentials with real arguments and purely complex arguments have been separated and factored by property Eq. (A.1.3) of exponential functions. The final step is to use Eq. (1.5.10) in Eq. (1.5.13) to obtain

$$y(t) = e^{-t} \left[ a_1 \cos(2t) + a_2 \sin(2t) \right].$$
(1.5.14)

Now we have a purely real solution with two parts, and initial conditions would be used to obtain specific values for $a_1$ and $a_2$, but we pause first to address the following question.

What is the nature of the two solutions,

$$y_1(t) = e^{-t} \cos(2t),$$
(1.5.15)

$$y_2(t) = e^{-t} \sin(2t),$$
(1.5.16)

that appear in the general solution Eq. (1.5.14)? They are exponentially decaying oscillations! To understand this statement, look at Fig. 1.15. It shows these two functions bounded by the exponentials $\pm \exp(-t)$; the exponential $\exp(-t)$ plays the role of a time dependent amplitude of the trigonometric functions in Eq. (1.5.15) and Eq. (1.5.16).

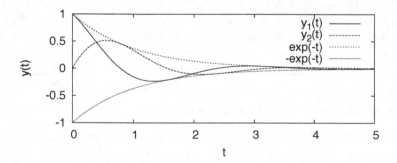

Fig. 1.15   Exponentially decaying oscillations.

Certainly we observe decaying oscillations in Fig. 1.15 for each contribution to the general solution Eq. (1.5.14) but how can we understand what the nature is of the addition of two decaying oscillations? Fortunately, there is a way to rewrite the result in a more transparent form. Whenever there is a linear combination of cosine and sine terms with the same argument $\omega t$, it can be rewritten in terms of $\cos(\omega t - \phi)$.[22] Of course, we could also choose $\sin(\omega t - \phi)$; it is a matter of taste. To see how this can be done, note that the addition angle formula Eq. (D.2.3) gives

$$a \cos(\omega t - \phi) = a \cos(\phi) \cos(\omega t) + a \sin(\phi) \sin(\omega t)$$

$$= a_1 \cos(\omega t) + a_2 \sin(\omega t).$$
(1.5.17)

---

[22]Why write the argument with $a - \phi$? For the simple reason the cosine profile is shifted to the right by $\phi$, and that is easy to imagine.

Now identify

$$a \cos(\phi) = a_1, \tag{1.5.18}$$

$$a \sin(\phi) = a_2. \tag{1.5.19}$$

How do we solve these two equations for $a$ and $\phi$? Observe first that these equations are similar to those when constructing the polar representation of a vector or a complex number, see Eq. (D.1.3) and Eq. (D.1.4). Start with

$$a^2 \cos^2(\phi) + a^2 \sin^2(\phi) = a^2 = a_1^2 + a_2^2, \tag{1.5.20}$$

and then

$$\frac{a \sin(\phi)}{a \cos(\phi)} = \tan(\phi) = \frac{a_2}{a_1}. \tag{1.5.21}$$

Clearly $a$ is just the square root of a sum of squares, but care must be taken to determine the correct quadrant for $\phi$, guided by the signs for $a_2$ and $a_1$ in Eq. (1.5.18) and Eq. (1.5.19).

As a result of the transformation of a sum of cosine and sine to a single cosine, there is an alternate form for the general solution Eq. (1.5.14),

$$y(t) = a \, e^{-t} \cos(2t - \phi). \tag{1.5.22}$$

The coefficients $a_1$ and $a_2$ have been traded for two new coefficients $a$ and $\phi$. The advantage of the form Eq. (1.5.22) is that it is easy to read what it looks like. First, it is a cosine shifted to the right by the phase $\phi$ and the amplitude of the cosine is decaying as an exponential $\exp(-t)$.

### Reflection

There are three different ways the homogeneous solution can be expressed when the roots to the characteristic equation are complex, for example, Eq. (1.5.2) and Eq. (1.5.12): as a combination of exponentials with complex arguments as in Eq. (1.5.3), as a combination of trigonometric functions multiplied by an exponential function Eq. (1.5.14), or as a single trigonometric function with an amplitude that changes exponentially Eq. (1.5.22) and shifted in phase. Having several choices gives us the flexibility to decide the form most useful for satisfying additional conditions such as initial conditions and for interpreting the results.

### 1.5.3  *The LCR circuit*

The value in trading exponential functions with complex arguments for trigonometric functions with real arguments is that the behavior of the solution is clearly oscillatory, and oscillatory behavior is very common in science and engineering. A great example is the behavior of the solutions to the mathematical model for the

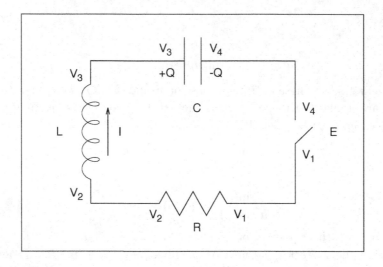

Fig. 1.16   The LCR circuit.

electric circuit shown in Fig. 1.16. This circuit is similar to the one in Fig. 1.5 with
a resistor of resistance $R$ ($\Omega$) and a capacitor with capacitance $C$ (F), but includes
an inductor with inductance $L$ (H). As before, an external voltage $E(t)$ (V) can be
applied to drive the circuit.

The way to derive a mathematical equation is to use Kirchoff's laws. They were
used in deriving a differential equation in Eq. (1.1.15). Here we restate the laws for
the circuit in Fig. 1.16:

- The voltage drop across each component must add up to balance the applied
  voltage. Let $V_R = V_2 - V_1$, $V_L = V_3 - V_2$ and $V_C = V_4 - V_3$ be the voltage drops
  across the resistor, the inductor and the capacitor, respectively. Then,

$$V_L + V_R + V_C = E .$$

- The current $I$ (A) must be the same in all parts of the circuit.

To proceed, we must determine the voltage drop across the various components.
The voltage drop across the resistor is $V_R = I(t)R$. The voltage drop across the
capacitor is $V_C = Q(t)/C$, where $Q(t)$ (C) is the charge on the capacitor – it can
change in time. The voltage drop across the inductor is $V_L = L\,dI(t)/dt$. The first
law becomes

$$L\frac{dI}{dt}(t) + R\,I(t) + \frac{1}{C}\,Q(t) = E(t) . \qquad (1.5.23)$$

As it stands, there are two unknown functions, $I(t)$ and $Q(t)$. To complete the
formulation of an equation, we must add the connection between the current $I(t)$
through the circuit and the charge $Q(t)$ on the capacitor,

$$I(t) = \frac{dQ}{dt}(t) . \qquad (1.5.24)$$

Thus (1.5.23) becomes

$$\frac{d^2Q}{dt^2}(t) + \frac{R}{L}\frac{dQ}{dt}(t) + \frac{1}{LC}Q(t) = \frac{E(t)}{L}. \tag{1.5.25}$$

This is a second-order differential equation for the unknown charge $Q(t)$ across the capacitor. It is linear and has constant coefficients that involve the parameters $R$, $L$ and $C$. It has a forcing term $E(t)$ which must still be specified. The solution will contain two parts, a particular solution which will depend on the nature of $E(t)$ and a homogeneous solution. We will need two initial conditions to determine a unique solution, and once the solution for $Q(t)$ is known, we can determine the current in the circuit from Eq. (1.5.24).

We anticipate that the homogeneous solution will describe the transient adjustment in response to the initial conditions, and that is where we will start. For example, suppose that the capacitor has been charged fully with $Q_0$. No external voltage will be applied, $E(t) = 0$. We simply close the circuit to see what happens. So initially,

$$Q(0) = Q_0, \quad \text{and there is no current,} \quad \frac{dQ}{dt}(0) = 0. \tag{1.5.26}$$

We start with the trial solution

$$Q(t) = e^{\lambda t},$$

and substitute into Eq. (1.5.25) to obtain

$$\lambda^2 e^{\lambda t} + \frac{R}{L}\lambda e^{\lambda t} + \frac{1}{LC}e^{\lambda t} = 0.$$

As always, $e^{\lambda t}$ may be dropped from each term, leaving us with the characteristic equation,

$$\lambda^2 + \frac{R}{L}\lambda + \frac{1}{LC} = 0.$$

The roots to this quadratic are

$$\lambda_\pm = -\frac{R}{2L} \pm \frac{1}{\sqrt{LC}}\sqrt{\frac{R^2C}{4L} - 1}. \tag{1.5.27}$$

Whether the roots are real or imaginary depends on the parameter $R^2C/(4L)$.

**Real roots**

If $R^2C/(4L) > 1$, then the roots are real and the solution takes the form

$$Q(t) = c_1 e^{\lambda_+ t} + c_2 e^{\lambda_- t}.$$

By applying the first initial condition $Q(0) = Q_0$ in Eq. (1.5.26), we have

$$c_1 + c_2 = Q_0.$$

By applying the other initial condition in Eq. (1.5.26), we have

$$\lambda_+ c_1 + \lambda_- c_2 = 0 \, ,$$

leading to the solution

$$Q(t) = -\frac{\lambda_- Q_0}{\lambda_+ - \lambda_-} e^{\lambda_+ t} + \frac{\lambda_+ Q_0}{\lambda_+ - \lambda_-} e^{\lambda_- t} \, . \tag{1.5.28}$$

To illustrate this solution, let us make the choice $L = 1$ H, $C = 4$ F and $R = 2\,\Omega$. Then $\lambda_+ = -1 + \sqrt{3}/2$ s$^{-1}$, $\lambda_- = -1 - \sqrt{3}/2$ s$^{-1}$. The evolution of $Q(t)/Q_0$ is shown in Fig. 1.17. Clearly, the charge just decays away. The two time scales are $1/|\lambda_+| \approx 7.7$ s and $1/|\lambda_-| \approx 0.5$ s. The decay is dominated by the slower decay rate $|\lambda_+| \approx 0.13$ s$^{-1}$.

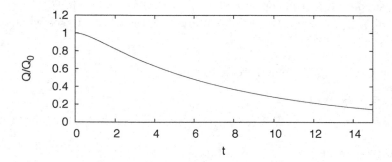

Fig. 1.17   The charge density on the capacitor as a function of time.

**Complex roots**

The case $R^2 C/(4L) < 1$ shows completely different behavior. First we rewrite (1.5.27) in the form

$$\lambda_\pm = \lambda_r \pm i\lambda_i = -\frac{R}{2L} \pm i\frac{1}{\sqrt{LC}}\sqrt{1 - \frac{R^2 C}{4L}} \, . \tag{1.5.29}$$

Scientists and engineers have introduced some special parameters to express the result in a more readable form and to identify important behavior in the solution.[23]

The *natural frequency* $\omega_0$ (rad/s) and the *damping factor* $\alpha$ (s$^{-1}$) are defined by

$$\omega_0 = \frac{1}{\sqrt{LC}} \qquad \text{and} \qquad \alpha = \frac{R}{2L} \, . \tag{1.5.30}$$

As a consequence, the roots to the characteristic equation can be written as

$$\lambda = -\alpha \pm i\sqrt{\omega_0^2 - \alpha^2} \, . \tag{1.5.31}$$

---

[23]It is often convenient in science and engineering to introduce new parameters that consolidate expressions into more readable forms.

One more parameter proves helpful, the *damped frequency*,

$$\omega_d = \sqrt{\omega_0^2 - \alpha^2}\,, \tag{1.5.32}$$

because we can write the three versions of the homogeneous solution in the simple forms,

$$Q(t) = c_1\, e^{-\alpha t} e^{i\omega_d t} + c_2\, e^{-\alpha t} e^{-i\omega_d t} \tag{1.5.33}$$

$$= e^{-\alpha t} \left[a_1 \cos(\omega_d t) + a_2 \sin(\omega_d t)\right] \tag{1.5.34}$$

$$= a e^{-\alpha t} \cos(\omega_d t - \phi)\,. \tag{1.5.35}$$

Now it becomes clear why the new parameters are renamed as they are: $\alpha$ control the damping rate in the decaying exponential; $\omega_0$ is the frequency of the oscillation in the absence of any damping; while $\omega_d$ gives the frequency when damping is present.

In each of these forms for the general solution, there are two unknown coefficients; $c_1$ and $c_2$ in Eq. (1.5.33), $a_1$ and $a_2$ in Eq. (1.5.34), and $a$ and $\phi$ in Eq. (1.5.35). The forms are equivalent; the coefficients are connected through Eq. (1.5.10) and Eq. (1.5.20) and Eq. (1.5.21). Which form is chosen often depends on other factors such as the nature of the initial conditions, or the form of the solution needed to understand the result.

Let us now illustrate how initial conditions determine the unknown coefficients in the various general forms. The application of Eq. (1.5.26) to Eq. (1.5.33) produces two equations,

$$c_1 + c_2 = Q_0\,,$$

$$(\alpha - i\omega_d)\, c_1 + (\alpha + i\omega_d)\, c_2 = 0\,,$$

which, after some algebra, leads to

$$\frac{Q(t)}{Q_0} = i e^{-\alpha t} \left[-\frac{1}{2\omega_d} (\alpha + i\omega_d)\, e^{i\omega_d t} + \frac{1}{2\omega_d} (\alpha - i\omega_d)\, e^{-i\omega_d t}\right].$$

The easiest part of this process is the differentiation of Eq. (1.5.30) in preparation for the application of the derivative initial condition. Otherwise the algebra is tedious and the result not very informative.

The application of the initial conditions to the form of the general solution Eq. (1.5.34) leads to

$$a_1 = Q_0\,,$$

$$-\alpha\, a_1 + \omega_d a_2 = 0\,,$$

which gives the solution as

$$\frac{Q(t)}{Q_0} = e^{-\alpha t} \left(\cos(\omega_d t) + \frac{\alpha}{\omega_d} \sin(\omega_d t)\right). \tag{1.5.36}$$

While the derivative requires the product rule and so is a little more difficult to obtain, the coefficients are easily determined from the initial conditions and the form of the solution is more readable.

Finally, the initial conditions applied to the form Eq. (1.5.35) lead to

$$a\cos(\phi) = Q_0\,,$$

$$-a\,\alpha\cos(\phi) + a\,\omega_d\sin(\phi) = 0\,,$$

and the solution follows in two steps; first, replace $a\cos(\phi)$ in the second equation by $Q_0$.

$$a\,\sin(\phi) = \frac{\alpha\,Q_0}{\omega_d}\,.$$

The resulting equations for $a$ and $\phi$ are exactly like Eq. (1.5.20) and Eq. (1.5.21) and can be solved in the same way. The algebra is a little more difficult than the steps leading to Eq. (1.5.36), but if the solution is required in the form of a decaying cosine, then it is the best one.

To compare with the behavior in Fig. 1.17, let us keep $L = 1$ H, $C = 4$ F but change the resistance to $R = 1/5$ Ω. Then $\alpha = 1/10$ s$^{-1}$ and $\omega_d = \sqrt{6}/5$ s$^{-1}$, and the behavior of the solution

$$\frac{Q(t)}{Q_0} = 1.02\,e^{-0.1t}\cos(0.49t - 0.2)$$

is shown in Fig. 1.18. What we see is an oscillation that dies away. The oscillation occurs on a time scale $2\pi/\lambda_i \approx 12$ s and its amplitude dies away on a scale of $1/|\lambda_r| \approx 10$ s.

## One (double) root

Note that there is a value of $R$ given by $2\sqrt{L/C}$ where there is only one root $\lambda = -R/(2L)$ to the quadratic Eq. (1.5.13). The general solution is given by Eq. (1.4.29),

$$Q(t) = \left(c_1 + c_2 t\right)e^{-Rt/(2L)}$$

and the application of the initial conditions Eq. (1.5.26) leads to the solution

$$\frac{Q(t)}{Q_0} = \left(1 + \frac{Rt}{2L}\right)e^{-Rt/(2L)}\,. \tag{1.5.37}$$

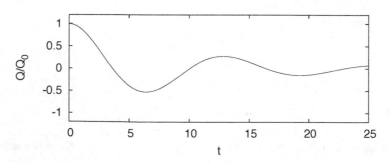

Fig. 1.18    The charge density on the capacitor as a function of time.

Make the same choice as before, $L = 1$ H and $C = 4$ F, then the resistance must be $R = 1\ \Omega$. If a graph of this function is made, it will look remarkably similar to Fig. 1.17 with the exception that the time scale of the decay now being 1 s. But this case, called critically damped, marks a transition point in the values of $R$ between over-damped Eq. (1.5.28) ($R > 2\sqrt{L/C}$) and under-damped Eq. (1.5.36) behavior ($R < 2\sqrt{L/C}$). When the resistance is large, the charge quickly damps away. When the resistance is small, the charge oscillates with a slow decay. When the resistance is just below the critical value $R = 2\sqrt{L/C}$, then $\omega_d$ is small, or the period of the oscillation is long and maybe difficult to observe. Further decreases in the resistance shorten the period and the oscillation is noticeable.

> ### Reflection
>
> For a second-order equation we expect two different contributions to the homogeneous solutions. The nature of these contributions depend on whether the roots to the characteristic equation are real or complex. Real roots give two exponentials with different time scales, while complex roots give oscillatory behavior with growth or decay depending on the sign of the real part of the roots. Thus there is a period and a growth or decay time. Which of the cases arise depends strongly on the parameters in the differential equation. For the engineer, the ability to select desirable behavior is granted by understanding the behavior of the solutions and their dependency on the choice for the parameters.

### 1.5.4   *Abstract viewpoint*

When the roots of the characteristic equation Eq. (1.4.27) are complex,

$$\lambda = \lambda_r \pm i\lambda_i\,,$$

the solution to the differential equation

$$\frac{d^2 y}{dx^2}(x) + a\,\frac{dy}{dx}(x) + b\,y(x) = 0 \tag{1.5.38}$$

can take several different forms:

(1) The complex form,

$$y(x) = c_1\,e^{(\lambda_r + i\lambda_i)\,x} + c_2\,e^{(\lambda_r - i\lambda_i)\,x}\,. \tag{1.5.39}$$

The coefficients $c_1$ and $c_2$ will be complex in general and satisfy $\bar{c}_2 = c_1$ in order to ensure $y(x)$ is real-valued.

(2) The real form most convenient for initial conditions,

$$y(x) = e^{\lambda_r x}\left(a_1\,\cos(\lambda_i x) + a_2\,\sin(\lambda_i x)\right)\,. \tag{1.5.40}$$

The coefficients $a_1$ and $a_2$ are real-valued.

(3) The real form most convenient for interpreting the behavior,
$$y(x) = a\, e^{\lambda_r x} \cos(\lambda_i\, x - \phi)\,. \tag{1.5.41}$$
The coefficient $a$ gives the initial amplitude which either grows ($\lambda_r > 0$) or decays ($\lambda_r < 0$) in $x$. The oscillation has a wavenumber (frequency) $\lambda_i$.

In Sec. 1.4.3, the general solution was expressed in a form with two solutions,
$$y(x) = c_1\, y_1(x) + c_2\, y_2(x)\,,$$
with the coefficients $c_1$ and $c_2$ to be determined by two initial conditions Eq. (1.4.31);
$$y(0) = \alpha\,, \qquad \frac{dy}{dx}(0) = \beta\,. \tag{1.5.42}$$
Previously, the two solutions $y_1(x)$ and $y_2(x)$ were exponentials with different real arguments, and we were able to verify that initial conditions Eq. (1.5.42) can be used to determine $c_1$ and $c_2$ uniquely. Indeed, we showed that the Wronskian
$$W(0) = y_1(0)\,\frac{dy_2}{dx}(0) - y_2(0)\,\frac{dy_1}{dx}(0) \neq 0$$
was sufficient to ensure unique solutions to $c_1$ and $c_2$. The choice for $y_1(x)$ and $y_2(x)$ is obvious here too.

The form Eq. (1.5.39) suggests the choice
$$y_1(x) = e^{(\lambda_r + i\lambda_i)\, x} \quad \text{and} \quad y_2(x) = e^{(\lambda_r - i\lambda_i)\, x}\,.$$
The Wronskian is non-zero and there is always a solution to initial conditions (1.5.42). Specifically,
$$c_1 + c_2 = \alpha\,,$$
$$\left(\lambda_r + i\,\lambda_i\right) c_1 + \left(\lambda_r - i\,\lambda_i\right) c_2 = \beta\,,$$
which can always be solved to give $c_1$ and $c_2$. They will be complex in general, but $y(x)$ will be real.

The form Eq. (1.5.40) suggests
$$y_1(x) = e^{\lambda_r x} \cos(\lambda_i x) \quad \text{and} \quad y_2(x) = e^{\lambda_r x} \sin(\lambda_i x)\,.$$
The Wronskian for these choices is also non-zero. The application of the initial conditions (1.5.42) leads to
$$a_1 = \alpha\,,$$
$$\lambda_r a_1 + \lambda_i\, a_2 = \beta\,,$$
and the solutions for $a_1$ and $a_2$ are easily obtained.

Form Eq. (1.5.41) is different. There is a single function with two unknown constants $a$ and $\phi$, although it is equivalent to the form Eq. (1.5.40) with two functions each with an unknown constant. Instead of checking a Wronskian, a solution for $a$ and $\phi$ can be derived directly from the initial conditions Eq. (1.5.42).
$$a \cos(\phi) = \alpha\,,$$
$$\lambda_r\, a \cos(\phi) + \lambda_i\, a \sin(\phi) = \beta\,.$$
Use the first condition to remove $a \cos(\phi)$ from the second condition. The result is
$$a \sin(\phi) = \frac{\beta - \lambda_r \alpha}{\lambda_i}\,,$$
and $a$ and $\phi$ can be determined in the same way as in Eq. (1.5.20) and Eq. (1.5.21).

The bottom line is that a solution can always be found for Eq. (1.5.38) subject to Eq. (1.5.42) when the roots to the characteristic equation are complex.

## 1.5.5 *Exercises*

**These are some basic exercises to help develop and improve skill.**

(1) Verify that Eq. (1.5.39), Eq. (1.5.40) and Eq. (1.5.41) are solutions to Eq. (1.5.38) by direct substitution – see Principle 1. Do not forget that $\lambda_r \pm i\lambda_i$ must be solutions to $\lambda^2 + a\lambda + b = 0$.

(2) Rewrite the solution Eq. (1.5.41) in terms of exponentials with complex arguments to establish a connection between $a$ and $\phi$ and $c_1$ and $c_2$.

(3) Here are some basic problems: Find the solution to the differential equations that satisfy the given initial conditions.

(a)
$$\frac{d^2 f}{dx^2}(x) + 4\frac{df}{dx}(x) + 8\,f(x) = 0\,,$$

subject to

$$f(0) = 2\,, \quad \text{and} \quad \frac{df}{dx}(0) = -2\,.$$

**Answer:** $f(\pi/2) = -0.0864$.

(b)
$$\frac{d^2 A}{dt^2}(t) - 2\frac{dA}{dt}(t) + 2\,A(t) = 0\,,$$

subject to

$$A(0) = 0\,, \quad \text{and} \quad \frac{dA}{dt}(0) = 2\,.$$

**Answer:** $A(\pi/2) = 9.621$.

(c)
$$\frac{d^2 y}{dt^2}(t) + x^2\,y(t) = 0\,,$$

subject to

$$y(0) = 1\,, \quad \text{and} \quad \frac{dy}{dt}(0) = 0\,.$$

**Answer:** for $x = \pi$, $y(1) = -1$.

(4) The hyperbolic cosine and sine functions – see Exercise 2 in Sec. 1.4.4 – are defined by

$$\cosh(x) = \frac{e^x + e^{-x}}{2}\,, \quad \sinh(x) = \frac{e^x - e^{-x}}{2}\,.$$

(a) Show that

$$\cos(ix) \equiv \cosh(x)\,, \quad \sin(ix) \equiv i\sinh(x)\,.$$

These relations inspire the term hyperbolic in the names of $\cosh(x)$ and $\sinh(x)$.

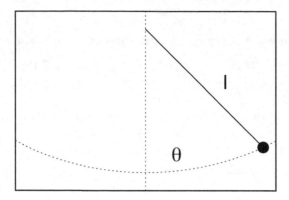

Fig. 1.19   Schematic of a pendulum.

(b) Use the relations in (a) to show that they have the derivatives

$$\frac{d}{dx}\cosh(x) = \sinh(x), \qquad \frac{d}{dx}\sinh(x) = \cosh(x).$$

(c) Use the relations in (a) to prove the identity

$$\cosh^2(x) - \sinh^2(x) \equiv 1.$$

**Some problems that arise in applications.**

(5) The equation for the motion of a pendulum, shown in Fig. 1.19, is derived by considering Newton's law of motion. Since the body of mass $m$ (gm) is restricted to move on a circle with radius $l$ (cm), the forces on the body in the radial direction must balance. If $T$ (gm $\cdot$ cm/s$^2$) is the tension in the rod supporting the weight, then $T = m\,g\,\cos(\theta)$ where $\theta$ is the angle in radians. The acceleration in the angular direction is driven by the tangential force.

$$m\,l\,\frac{d^2\theta}{dt^2}(t) = -m\,g\,\sin\big(\theta(t)\big).$$

This is an example of a nonlinear differential equation because $\theta$ appears in the argument of a function $\sin(\theta)$. Special methods must be used to solve this equation – they are discussed in Chapter 5. Instead we will restrict the pendulum to cases where the deflection is small. In other words $\theta(t)$ is always small, and we approximate $\sin\big(\theta(t)\big) \approx \theta(t)$. Thus the *linearized* version of the differential equation is

$$\frac{d^2\theta}{dt^2}(t) + \frac{g}{l}\,\theta(t) = 0.$$

Suppose we pull the pendulum to the side by $\theta = 4°$ and let it go. Find an expression for the subsequent motion.

In the derivation of the pendulum equation, we have neglected any resistance to the motion. Assume that a resistive force proportional to the speed of the pendulum is added – perhaps as a consequence of air resistance. Now the equation of motion becomes

$$\frac{d^2\theta}{dt^2}(t) + k\frac{d\theta}{dt}(t) + \frac{g}{l}\theta(t) = 0,$$

where $k$ ($s^{-1}$) is a coefficient of friction. Notice the similarity of this equation with (1.5.23).
Obtain the solution for the three cases where the characteristic equation has two real roots, one real root, and two complex roots with the initial conditions given above. Take $l = 10$ cm, and plot the solutions for the choices $k = 20$, $19.8$, $19.6\,s^{-1}$. What do you notice about the differences in the solutions? Evaluate the solutions at $t = 0.5$ s.
**Answer:** $\theta/\theta_0 = 0.045, 0.042, 0.039$.

(6) Suppose you are a clockmaker and have been commissioned to make a grandfather clock with an old fashion pendulum to keep time. The length of the rod must be 3 ft. Assume no resistance. What should be the mass of the body attached to the end of the rod so that the pendulum swings with a one-second interval?
**Answer:** ha! ha!

(7) Resistance in the circuit shown in Fig. 1.16 acts as a damping agent. Set $R = 0\,\Omega$ in Eq. (1.5.23) and simply close the circuit that has a charge $Q_0$ (C) across the capacitor initially, but no current. The result should indicate a steady oscillation for all time. Plot your result along with the results in Fig. 1.18 for the choice $L = 1$ H and $C = 4$ F. Often resistance is very small in circuits and provided we consider the solution for a moderate time, the oscillatory solution is accurate, but it is not valid for very long times! Eventually the presence of resistance, no matter how small, will cause the oscillations to decay away.
**Answer:** The two graphs are shown in Fig. 1.20.

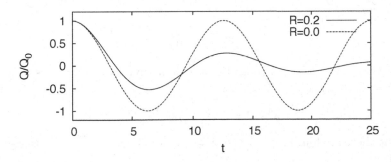

Fig. 1.20   The temporal behavior of the charge on the capacitor for two choices of the resistance.

(8) For the electric circuit shown in Fig. 1.16, suppose that a surge has caused an initial charge $Q_0$ (C) to be deposited on the capacitor. The circuit is closed, but otherwise dead. Suppose $C = 0.1$ F, $L = 2$ H, and $R = 0.01$ $\Omega$. How long

will it take for the charge $Q$ to be always less than $0.01Q_0$?

**Answer**: 30.7 min.

(9) Consider a cylindrical piece of wood of density 0.6 gm/cm$^3$. Its radius is 3 cm, and its length is 2 cm. It is floating undisturbed on a water surface. It is suddenly pushed downwards by 0.5 cm and then let go. Describe its subsequent motion by invoking Archimedes' principle: An object that is completely or partially immersed in a fluid is acted on by an upward force equal to the weight of fluid displaced. The piece of wood bobs up and down. What is the period of the motion?

**Answer**: 0.22 s.

### 1.5.6 *Higher-order differential equations*

Higher-order homogeneous differential equation with constant coefficients are solved in a standard way, by guessing the solution in an exponential form. Thus the equation,

$$\frac{\mathrm{d}^n y}{\mathrm{d}x^n}(x) + a_0 \frac{\mathrm{d}^{n-1}y}{\mathrm{d}x^{n-1}}(x) + \cdots + a_{n-1}y(x) = 0,$$

has the characteristic polynomial

$$\lambda^n + a_0\lambda^{n-1} + \cdots + a_{n-1} = 0. \tag{1.5.43}$$

The challenge now is to find all the roots to Eq. (1.5.43). While we are guaranteed that the roots are either real or occur in complex conjugate pairs, there is no general formula for polynomials with $n > 4$. So use whatever techniques you know!

A specific example may help to show the possible ways to find roots and to use them to compose a general solution. Consider the equation

$$\frac{\mathrm{d}^4 y}{\mathrm{d}x^4}(x) - y(x) = 0. \tag{1.5.44}$$

The characteristic equation is

$$\lambda^4 - 1 = 0.$$

Obviously $\lambda = (1)^{1/4} = 1$ is a root. Perhaps less obviously, $\lambda = -1$ is a root. Are there more? If we write

$$\lambda^4 - 1 = \left(\lambda^2 - 1\right)\left(\lambda^2 + 1\right)$$

we see that there are four roots; $\lambda = 1$, $\lambda = -1$, $\lambda = \mathrm{i}$ and $\lambda = -\mathrm{i}$. Thus the general solution is

$$y(x) = c_1 \mathrm{e}^x + c_2 \mathrm{e}^{-x} + c_3 \mathrm{e}^{\mathrm{i}x} + c_4 \mathrm{e}^{-\mathrm{i}x}$$

or by converting to trigonometric functions,

$$y(x) = c_1 \mathrm{e}^x + c_2 \mathrm{e}^{-x} + a_1 \cos(x) + a_2 \sin(x).$$

The importance of finding four roots should be clear. We expect four initial conditions to construct unique solutions to Eq. (1.5.41) and that means we need four independent solutions and four coefficients.

How about a slightly more difficult example? Suppose

$$\frac{\mathrm{d}^4 y}{\mathrm{d}x^4}(x) + y(x) = 0 \qquad (1.5.45)$$

with the characteristic equation

$$\lambda^4 + 1 = 0\,.$$

The answer appears to be $\lambda = (-1)^{1/4}$, but we need four separate results! Fortunately, exponentials come to our rescue again! The details appear in Appendix D: the roots are

$$\frac{1+\mathrm{i}}{\sqrt{2}}, \quad \frac{-1+\mathrm{i}}{\sqrt{2}}, \quad \frac{-1-\mathrm{i}}{\sqrt{2}}, \quad \frac{1-\mathrm{i}}{\sqrt{2}}$$

and the general solution to (1.5.45) is

$$
\begin{aligned}
y(x) &= c_1 e^{(1+\mathrm{i})x/\sqrt{2}} + c_2 e^{(-1+\mathrm{i})x/\sqrt{2}} + c_3 e^{(-1-\mathrm{i})x/\sqrt{2}} + c_4 e^{(1-\mathrm{i})x/\sqrt{2}} \\
&= e^{x/\sqrt{2}} \left( c_1 e^{\mathrm{i}x/\sqrt{2}} + c_4 e^{-\mathrm{i}x/\sqrt{2}} \right) + e^{-x/\sqrt{2}} \left( c_2 e^{\mathrm{i}x/\sqrt{2}} + c_3 e^{-\mathrm{i}x/\sqrt{2}} \right) \\
&= e^{x/\sqrt{2}} \left[ a_1 \cos\left(x/\sqrt{2}\right) + a_2 \sin\left(x/\sqrt{2}\right) \right] \\
&\quad + e^{-x/\sqrt{2}} \left[ a_3 \cos\left(x/\sqrt{2}\right) + a_4 \sin\left(x/\sqrt{2}\right) \right].
\end{aligned}
$$

## 1.6 Forcing terms: Resonances

In Sec. 1.5.3, we considered a simple electric circuit – see Fig. 1.16 – and studied its natural transient behavior in the absence of any applied voltage. Now we would like to understand how the circuit responds to an applied voltage. There are two obvious cases to consider, a constant applied voltage $E(t) = V_0$ (V) and an alternating voltage $E(t) = A \sin(\omega t)$ with amplitude $A$ (V) and frequency $\omega/(2\pi)$ (rad/s).

The determination of the response to these forcing effects follows the strategy in Sec. 1.2. A form for a particular solution with unknown coefficients is chosen based on the form of the forcing. By invoking Principle 1, these coefficients will be determined by requiring the guess for the particular solution to satisfy the differential equation. Furthermore, we anticipate that the particular solution will be the long term response of the circuit. The transient adjustment to the initial state of the circuit will be captured by the homogeneous solution, and together with the particular solution will compose a general solution by Principle 4.

### 1.6.1 *LCR circuit with constant applied voltage*

For a constant applied voltage, the equation to be solved is

$$\frac{d^2Q}{dt^2}(t) + \frac{R}{L}\frac{dQ}{dt}(t) + \frac{1}{LC}Q(t) = \frac{V_0}{L}. \tag{1.6.1}$$

Since the forcing term is just a constant function of time, the obvious choice for a particular solution is a constant $Q_p(t) = a$ – see Table 1.4. Upon substitution into Eq. (1.6.1), the constant must be

$$Q_p(t) = CV_0. \tag{1.6.2}$$

To complete the study, we must choose an initial condition. Suppose the circuit was dead initially – no charge nor current;

$$Q(0) = 0, \quad \text{and} \quad I(0) = \frac{dQ}{dt}(0) = 0. \tag{1.6.3}$$

Obviously the particular Eq. (1.6.2) does not satisfy the first initial condition. A homogeneous solution must be added to construct the general solution, and the homogeneous solution was determined in the previous section. The characteristic equation may have real roots or complex roots Eq. (1.5.27) depending on the parameter $R^2C/(4L)$. Let us take the case where the homogeneous solutions are decaying oscillations. Specifically, one of the choices made in Sec. 1.5.3 is $L = 1$ H, $C = 4$ F, and $R = 1/5$ Ω, and the most useful choice for the homogeneous solution Eq. (1.5.35) is

$$Q_h(t) = a\,e^{-\alpha t}\cos(\omega_d t - \phi), \tag{1.6.4}$$

where the roots of the characteristic equation Eq. (1.5.27) are complex,

$$\lambda = \alpha \pm i\,\omega_d = -0.1 + i\frac{\sqrt{6}}{5} \tag{1.6.5}$$

and the parameters $\alpha$ and $\omega_d$ are defined in Eq. (1.5.30) and Eq. (1.5.32).

The initial conditions Eq. (1.6.3) must be applied to the general solution

$$Q(t) = Q_p(t) + Q_h(t)$$
$$= CV_0 + a\,e^{-\alpha t}\cos(\omega_d t - \phi).$$

As a result, there will be two equations for $a$ and $\phi$:

$$CV_0 + a\cos(\phi) = 0,$$
$$-\alpha\,a\cos(\phi) + \omega_d\,a\sin(\phi) = 0$$

which lead to

$$a\cos(\phi) = -CV_0, \tag{1.6.6}$$

$$a\sin(\phi) = -\frac{\alpha}{\omega_d}CV_0. \tag{1.6.7}$$

Then, by following the steps in Eq. (1.5.20) and Eq. (1.5.21),

$$a^2 = \left(1 + \frac{\alpha^2}{\omega_d^2}\right)(CV_0)^2 \approx 1.042\,(CV_0)^2,$$

and

$$\tan(\phi) = -\frac{\alpha}{\omega_d} \approx -0.204\,.$$

From Eq. (1.6.6) and Eq. (1.6.7), $\phi$ must lie in the second quadrant; $\phi \approx 0.294$. So the solution is

$$\frac{Q(t)}{CV_0} \approx 1 + 1.02\,e^{-0.1\,t}\,\cos(0.49t - 2.94)\,.$$

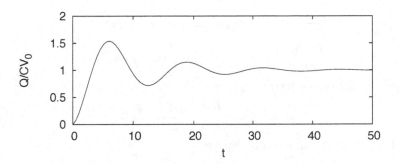

Fig. 1.21   The evolution of the charge density.

The result is shown in Fig. 1.21. As expected, the transient oscillation, which is quite large initially, eventually dies away and the charge settles to the constant value predicted by the particular solution. From a design perspective, the transient behavior might not be the desired behavior. For example, suppose the circuit is a simple model for a battery charger – the battery is represented by the capacitor. We would rather have the charge reach the constant value quickly and without oscillation. We should increase the resistance so that the transients decay as quickly as possible and that occurs for the critically damped case.

**Reflection**

The homogeneous solution which captures the transient adjustment of the initial condition decays away provided the real part of the complex roots is negative. The consequence is that the homogeneous solution tends to zero as time advances, leaving the particular solution as the long term behavior.

### 1.6.2   *LCR circuit with alternating applied voltage*

Now let us consider an alternating applied voltage $E(t) = A\sin(\omega t)$ with an amplitude $A$ (V) and an angular frequency $\omega$ (rad/s). The differential equation for the charge density now reads

$$\frac{d^2Q}{dt^2}(t) + \frac{R}{L}\frac{dQ}{dt}(t) + \frac{1}{LC}Q(t) = \frac{A}{L}\sin(\omega t)\,. \tag{1.6.8}$$

The first step is to construct a particular solution. Fortunately, the choice of trial solution is given in Table 1.4:

$$Q_{\mathrm{p}}(t) = a_1 \cos(\omega t) + a_2 \sin(\omega t). \tag{1.6.9}$$

Note that the argument of the cosine and sine must match that of the forcing voltage. The coefficients $a_1$ and $a_2$ are as yet undetermined. By substitution of Eq. (1.6.9) into Eq. (1.6.8), we will have terms with $\cos(\omega t)$ and $\sin(\omega t)$ and they must balance to ensure that $Q_{\mathrm{p}}$ is a solution – Principle 1. Since

$$\frac{\mathrm{d}Q_{\mathrm{p}}}{\mathrm{d}t}(t) = -\omega\, a_1 \sin(\omega t) + \omega a_2 \cos(\omega t),$$

$$\frac{\mathrm{d}^2 Q_{\mathrm{p}}}{\mathrm{d}t^2}(t) = -\omega^2 a_1 \cos(\omega t) - \omega^2 a_2 \sin(\omega t),$$

the balance of the cosine terms requires

$$-\omega^2 a_1 + \omega \frac{R}{L} a_2 + \frac{1}{LC} a_1 = 0, \tag{1.6.10}$$

and the balance of the sine terms requires

$$-\omega^2 a_2 - \omega \frac{R}{L} a_1 + \frac{1}{LC} a_2 = \frac{A}{L}. \tag{1.6.11}$$

The solution to Eq. (1.6.10) and Eq. (1.6.11) is given in Eq. (C.1.7) and Eq. (C.1.9), but now is the time to use the parameters introduced in Eq. (1.5.30) and Eq. (1.5.32) to keep the algebra as simple as possible. The system of equations becomes

$$-\left(\omega^2 - \omega_0^2\right) a_1 + 2\,\omega\,\alpha\, a_2 = 0,$$

$$-2\,\omega\,\alpha\, a_1 - \left(\omega^2 - \omega_0^2\right) a_2 = \frac{A}{L}$$

or in matrix form

$$\begin{bmatrix} -\left(\omega^2 - \omega_0^2\right) & 2\,\omega\,\alpha \\ -2\,\omega\,\alpha & -\left(\omega^2 - \omega_0^2\right) \end{bmatrix} \begin{pmatrix} a_1 \\ a_2 \end{pmatrix} = \begin{pmatrix} 0 \\ A/L \end{pmatrix}.$$

From Eq. (C.1.7) and Eq. (C.1.9),

$$a_1 = -2\frac{\alpha\,\omega}{D}\frac{A}{L},$$

and

$$a_2 = -\frac{\left(\omega^2 - \omega_0^2\right)}{D}\frac{A}{L},$$

where

$$D = \left(\omega^2 - \omega_0^2\right)^2 + 4\omega^2\alpha^2 \tag{1.6.12}$$

is the determinant of the matrix. Since $D$ is a sum of squares, $D \neq 0$ and the solution is always valid.

Thus the particular solution is

$$Q_\mathrm{p}(t) = -\frac{A}{LD}\left[2\,\omega\,\alpha\cos(\omega t) + \left(\omega^2 - \omega_0^2\right)\sin(\omega t)\right]. \tag{1.6.13}$$

The form of $Q_\mathrm{p}(t)$ is not so informative. Instead use Eq. (1.5.17) to cast it into the alternate form,

$$Q_\mathrm{p}(t) = a\cos(\omega t - \phi), \tag{1.6.14}$$

where

$$a\cos(\phi) = -\frac{2\,\omega\,\alpha}{D}\frac{A}{L}, \tag{1.6.15}$$

$$a\sin(\phi) = -\frac{\left(\omega^2 - \omega_0^2\right)}{D}\frac{A}{L}. \tag{1.6.16}$$

Now square Eq. (1.6.15) and Eq. (1.6.16) and add the results together.

$$a^2 = \frac{A^2}{L^2 D^2}\left[4\omega^2\alpha^2 + \left(\omega^2 - \omega_0^2\right)^2\right]$$
$$= \frac{A^2}{L^2 D} \tag{1.6.17}$$

because the expression in the square brackets is just $D$. Obviously $a = A/(L\sqrt{D})$. The phase is determined by

$$\tan(\phi) = \frac{\left(\omega^2 - \omega_0^2\right)}{2\,\omega\,\alpha}. \tag{1.6.18}$$

If $\omega < \omega_0$, the phase lies in the second quadrant, but if $\omega > \omega_0$, then the phase lies in the third quadrant. If $\omega = \omega_0$, then $\phi = \pi$ which means the response of the particular solution is completely out of phase with the input signal.

As a consequence, the particular solution can be written as

$$Q_\mathrm{p}(t) = \frac{A}{L\sqrt{D}}\cos(\omega t - \phi). \tag{1.6.19}$$

It is preferable to leave $\phi$ as is; substitution of an arctan from Eq. (1.6.18) simply makes the resulting expression far too difficult to read. As it stands, Eq. (1.6.19) tells us that the amplitude of the particular solution is $A/(L\sqrt{D})$, where $D$ is given by Eq. (1.6.12), and the phase is $\phi$, where $\phi$ is determined by Eq. (1.6.18).

To the particular solution Eq. (1.6.19) we must add the homogeneous solution Eq. (1.6.4), written as

$$Q_\mathrm{h}(t) = B\,\mathrm{e}^{-\alpha t}\cos(\omega_\mathrm{d}t - \psi), \tag{1.6.20}$$

to compose the general solution,

$$Q(t) = \frac{A}{L\sqrt{D}}\cos(\omega t - \phi) + B\,\mathrm{e}^{-\alpha t}\cos(\omega_\mathrm{d}t - \psi). \tag{1.6.21}$$

We apply the initial conditions by substituting Eq. (1.6.3) into Eq. (1.6.21);

$$\frac{A}{L\sqrt{D}}\cos(\phi) + B\cos(\psi) = 0,$$

$$\frac{A\omega}{L\sqrt{D}}\sin(\phi) - \alpha B\cos(\psi) + \omega_\mathrm{d}B\sin(\psi) = 0,$$

which leads to

$$B\cos(\psi) = -\frac{A}{L\sqrt{D}}\cos(\phi)\,, \tag{1.6.22}$$

$$B\sin(\psi) = -\frac{A}{L\sqrt{D}}\frac{\omega\sin(\phi) + \alpha\cos(\phi)}{\omega_d}\,. \tag{1.6.23}$$

These two equations for $B$ and $\psi$ are solved in exactly the same manner as we determined $\alpha$ and $\phi$ in Eq. (1.6.15) and Eq. (1.6.16).

Rather than complete the algebraic steps, let the computer do the work of evaluating all the necessary expressions, and then simply graph the results. Often, it is more convenient to use intermediate variables such as $\alpha$, $\omega_d$, $D$, $a$, $B$, $\phi$ and $\psi$ as part of a series of steps of evaluation, rather than trying to express the final result in terms of the original parameters. For example, start with the same choice as before for the circuit parameters: $L = 1$ H, $C = 4$ F and $R = 1/5$ $\Omega$. Now evaluate the necessary quantities one at a time:

(1) First determine $\alpha$ and $\omega_d$ from Eq. (1.5.30) and Eq. (1.5.32) – see also Eq. (1.6.5);

$$\alpha = \frac{R}{2L} = 0.1\,, \quad \omega_0 = 0.5\,, \quad \omega_d = \sqrt{\omega_0^2 - \alpha^2} = 0.499\,.$$

(2) To continue we need two more choices, $A = 1$ V for convenience since we should plot $Q/(C\,A)$ anyway to assess the relative output,[24] and $\omega = 1/4$ rad/s as a first example. Note that this choice of $\omega$ corresponds to a oscillatory forcing with period $2\pi/\omega = 8\pi \approx 25$ s.

(3) Now $D$ can be determined from Eq. (1.6.12),

$$D = \left(\omega^2 - \omega_0^2\right)^2 + 4\omega^2\alpha^2 = 0.0377\,.$$

(4) Then $a$ and $\phi$ follow from Eq. (1.6.15) and Eq. (1.6.16),

$$a\cos(\phi) = -\frac{2\,\omega\,\alpha}{D}\frac{A}{L} = -1.33\,, \quad a\sin(\phi) = -\frac{\left(\omega^2 - \omega_0^2\right)}{D}\frac{A}{L} = 4.97\,,$$

which give $a = 5.15$ and $\phi = 1.83$.

(5) Finally, solve Eq. (1.6.22) and Eq. (1.6.23) to obtain $B$ and $\psi$.

$$B\cos(\psi) = -\frac{A}{L\sqrt{D}}\cos(\phi) = 1.34\,,$$

$$B\sin(\psi) = -\frac{A}{L\sqrt{D}}\frac{\omega\sin(\phi) + \alpha\cos(\phi)}{\omega_d} = -2.22\,.$$

So, $B = 2.25$ and $\psi = -1.03$.

---

[24]The input amplitude $A$ is measured as a voltage. We must multiply by $C$ to obtain an effective input charge. Then $Q/(C\,A)$ measures the ratio of the response to the forcing, or the ratio of the output to the input.

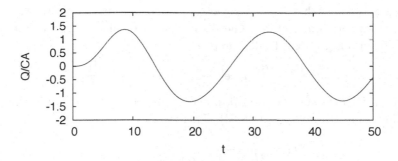

Fig. 1.22 The evolution of the charge density with $\omega = 0.25$ s$^{-1}$.

Now it is easy to graph $Q(t)/(CA)$ as shown in Fig. 1.22. There is a rapid adjustment on the time scale of $1/\alpha = 10$ s, to a steady oscillation of period $2\pi/\omega \approx 25$ s. This steady oscillation is of course described by the particular solution. The amplitude of the particular solution is $A/(L\sqrt{D})$ but we want to compare this with an effective input amplitude which must be measured in coulombs. An appropriate choice is $CA$ so the ratio of the steady amplitude to the input amplitude is given by $Q/(CA) = \omega_0^2/\sqrt{D} \approx 1.29$.

---

**Reflection**

The homogeneous solution contains two time scales, one associated with exponential decay $1/\alpha$, and the other with the frequency of the oscillation $2\pi/\omega_{\mathrm{d}}$. The forcing term describing the alternating applied voltage also has a time scale which induces a time scale in the particular solution $2\pi/\omega$. Once the homogeneous solution has decayed to small values, the long term behavior is given by the particular solution Eq. (1.6.19) which is a steady oscillation.

---

### 1.6.3 *Complex solutions*

The complicated process, involving much tedious algebra, in constructing the particular solution Eq. (1.6.19) raises the question of whether there is a simpler way to proceed. There is and it is based on using exponentials with complex arguments. Of course, the approach depends entirely on the forcing being cosines and/or sines and starts with the conversion of the forcing term into exponentials. To illustrate the approach, consider again Eq. (1.6.8). Rewrite the forcing term as

$$\frac{A}{L}\sin(\omega t) = -\frac{Ai}{2L}\,e^{i\omega t} + \frac{Ai}{2L}\,e^{-i\omega t}. \qquad (1.6.24)$$

Obviously the particular solution must be a combination of the two exponentials in Eq. (1.6.24).

$$Q_p(t) = c_1\,e^{i\omega t} + c_2\,e^{-i\omega t}, \qquad (1.6.25)$$

but we already know that $\bar{c}_2 = c_1$ since $Q_p(t)$ must be real – see Eq. (1.5.4). We need only to determine $c_1$, and we need to substitute only the first exponential in Eq. (1.6.25) into Eq. (1.6.8) to determine $c_1$.

$$-\omega^2 c_1 \, e^{i\omega t} + \frac{R\omega}{L} \, i c_1 \, e^{i\omega t} + \frac{1}{LC} c_1 \, e^{i\omega t} = -\frac{Ai}{2L} \, e^{i\omega t}.$$

The solution for $c_1$ is easily obtained, but it is convenient to introduce the parameters $\alpha$ and $\omega_0$ as before – Eq. (1.5.30). Then,

$$
\begin{aligned}
c_1 &= \frac{Ai}{2L} \frac{1}{\omega^2 - 2\omega\,\alpha\,i - \omega_0^2} \\
&= \frac{Ai}{2LD} \left[ (\omega^2 - \omega_0^2) + 2\omega\,\alpha\,i \right] \\
&= -\frac{A\omega\,\alpha}{LD} + \frac{Ai}{2LD} (\omega^2 - \omega_0^2).
\end{aligned}
\tag{1.6.26}
$$

The quantity $D$ is given by Eq. (1.6.12). The last step to obtain the solution Eq. (1.6.14) is to write the result in Eq. (1.6.26) in polar form.

$$c_1 = \frac{a}{2} \, e^{-i\phi} = \frac{a}{2} \cos(\phi) - i\frac{a}{2} \sin(\phi),$$

with the consequence that

$$a \cos(\phi) = -\frac{2\omega\,\alpha}{D} \frac{A}{L},$$

$$a \sin(\phi) = -\frac{(\omega^2 - \omega_0^2)}{D} \frac{A}{L}.$$

These equations are the same as Eq. (1.6.15) and Eq. (1.6.16).

To return the solution to real form, we simply add the second term in Eq. (1.6.25) using the result $c_2 = \bar{c}_1$. Now the advantage of expressing $c_1$ in polar form becomes apparent.

$$
\begin{aligned}
Q_p(t) &= \frac{a}{2} \, e^{i(\omega t - \phi)} + \frac{a}{2} \, e^{-i(\omega t - \phi)} \\
&= a \, \cos(\omega t - \phi).
\end{aligned}
$$

The solution agrees with the previous result Eq. (1.6.14) and is arguably easier to obtain. Instead of solving the two equations, Eq. (1.6.10) and Eq. (1.6.11), for $a_1$ and $a_2$ and then converting to a single trigonometric function, there is just one equation, Eq. (1.6.26), for $c_1$ but the solution requires complex arithmetic.

Finally, the homogeneous solution must be added as before Eq. (1.6.21) and the initial conditions satisfied as before – see Eq. (1.6.22) and Eq. (1.6.23).

### 1.6.4   *Resonance*

In Fig. 1.22, we show the result for the choice of forcing frequency $\omega = 0.25 \text{ s}^{-1}$. For our second choice $\omega = 0.5 \text{ s}^{-1}$, the period of the steady oscillation is $2\pi/\omega \approx 12$ s. The result is shown in Fig. 1.23. Again there is a rapid adjustment to the particular solution but now the steady amplitude is a factor 2.5 larger than the input

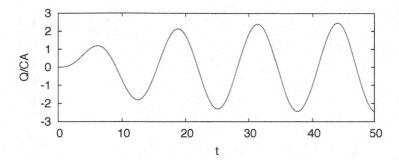

Fig. 1.23   The evolution of the charge density with $\omega = 0.5$ s$^{-1}$.

amplitude. How can this be? The amplitude of the particular solution is $A(L/\sqrt{D}\,)$ but to measure its strength compared to the effective input charge $CA$, we must consider the ratio $Q/(CA) = \omega_0^2/(\sqrt{D}\,)$. This factor changes as $\omega$ changes because $D$ depends on $\omega$ – see Eq. (1.6.12). Because the factor depends on the inverse of the square root of $D$, the smaller the value of $D$ the bigger the responding amplitude.

To understand better the way $D$ varies with $\omega$, we rewrite it as

$$D(\omega) = \left(\omega^2 - \omega_0^2\right)^2 + 4\omega^2\alpha^2$$

$$= \omega^4 - 2\left(\omega_0^2 - 2\,\alpha^2\right)\omega^2 + \omega_0^4$$

$$= \left[\omega^2 - \left(\omega_0^2 - 2\,\alpha^2\right)\right]^2 + 4\,\alpha^2\left(\omega_0^2 - \alpha^2\right) \qquad (1.6.27)$$

by using the standard procedure to complete the square. Recall that $\alpha < \omega_0$ gives complex roots for the characteristic equation, and hence oscillatory homogeneous solutions. When this is the case we are also guaranteed that $D > 0$. The largest the multiplication factor can be is when $D$ is at a minimum, and that obviously occurs with the choice

$$\omega_R^2 = \omega_0^2 - 2\,\alpha^2\,. \qquad (1.6.28)$$

At this value, the minimum is

$$D_R = 4\,\alpha^2\left(\omega_0^2 - \alpha^2\right). \qquad (1.6.29)$$

There is an important approximation when the resistance is very small, that is, $\alpha$ is very small. First, the frequency $\omega_R \approx \omega_0$ and $D_R \approx 4\alpha^2\omega_0^2$ becomes very small with the consequence that the ratio of output to input amplitudes becomes very large $Q_R \approx \omega_0/(2\alpha)$. This phenomenon is called *resonance*. By tuning $\omega$ to $\omega_R \approx \omega_0$, we obtain a huge amplification of the input signal. To make the point clear, the increase in the amplitude of the steady response as a function of $\omega$ is shown in Fig. 1.24 for two choices of $R$ while keeping $L = 1$ H, $C = 4$ F fixed. Notice how the peak response doubles with the halving of $R$. The peak occurs at $\omega_R \approx \omega_0 = 1/\sqrt{LC} \approx 0.5$ s$^{-1}$.

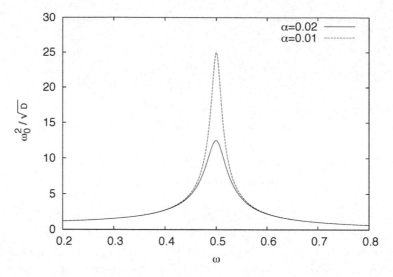

Fig. 1.24  The ratio of the output to input amplitudes of the steady response as a function of applied frequency for two choices of the damping factor.

What is so special about $\omega_0$? If we simply look at the natural frequency, we see that it is the frequency of the homogeneous solution Eq. (1.5.32) in the absence of any resistance. In other words, when we force a system at its natural frequency, we get resonance if the resistance is small enough!

> **Reflection**
>
> Looking for important consequences often rests on the ability to manage the algebra that arises in constructing the solution. The introduction of new parameters can keep the algebra tractable and can help focus on what is important in the results. For the example of resonance, it is the ratio of output to input amplitudes which depends on $1/\alpha$.

### 1.6.5  *Abstract view*

Because of the complexity of the algebra in constructing the solution Eq. (1.6.21), it is easy to miss the strategy and the underlying principles. Indeed, the strategy follows that expressed in Sec. 1.2 to construct solutions to forced, first-order differential equations. Nevertheless, this is a good opportunity to review the strategy and reinforce the principles underlying the construction of solutions.

The solution to the differential equation

$$\frac{d^2 y}{dx^2}(x) + a\,\frac{dy}{dx}(x) + b\,y(x) = r(x)\,, \tag{1.6.30}$$

is composed of two parts, a particular solution $y_p(x)$ designed to account for the presence of $r(x)$ and a homogeneous solution $y_h(x)$ with $r(x) = 0$ – Principle 4.

### 1.6.5.1 *Particular solution*

When $r(x)$ is a function composed of polynomials, exponentials, and/or sines and cosines, the particular solution $y_p(x)$ can be guessed in a similar form. Table 1.4 gives some specific examples. The guess for the particular solution will contain some undetermined coefficients – Principle 5. As an illustration, suppose $r(x) = 1 + x + x^2$. Then Table 1.4 suggests a guess of the form

$$y_p(x) = \alpha_0 + \alpha_1 x + \alpha_2 x^2. \tag{1.6.31}$$

The coefficients $\alpha_0$, $\alpha_1$, $\alpha_2$ are undetermined. By substitution of Eq. (1.6.31) into Eq. (1.6.30) – Principle 1, terms will emerge that are either constants, contain $x$, or contain $x^2$. Each type must balance: The ones on the left contain the undetermined coefficients and they must balance the known coefficients on the right – Principle 5. Specifically for Eq. (1.6.31),

$$\left(b\,\alpha_0 + a\,\alpha_1 + 2\,\alpha_2\right) + \left(b\,\alpha_1 + 2\,a\,\alpha_2\right) x + \left(b\,\alpha_2\right) x^2 = 1 + x + x^2.$$

Balancing the constants,

$$b\,\alpha_0 + a\,\alpha_1 + 2\,\alpha_2 = 1.$$

Balancing the terms with $x$,

$$b\,\alpha_1 + 2\,a\,\alpha_2 = 1.$$

Balancing the terms with $x^2$,

$$b\,\alpha_2 = 1.$$

It is now straightforward to find the solutions for $\alpha_0$, $\alpha_1$ and $\alpha_2$. At the end of this procedure, we have no unknown coefficients in the solution.

### 1.6.5.2 *Homogeneous solution*

The particular solution is unlikely to satisfy the initial conditions. Hence the need for the homogeneous solution, a solution that must not upset the solution that already takes $r(x)$ into account.

We construct the homogeneous solution to

$$\frac{d^2 y}{dx^2}(x) + a\frac{dy}{dx}(x) + b\,y(x) = 0, \tag{1.6.32}$$

by seeking solutions of the form – Principle 8,

$$y_h(x) = e^{\lambda x}.$$

After substitution into Eq. (1.6.32), we obtain the characteristic equation,

$$\lambda^2 + a\lambda + b = 0. \tag{1.6.33}$$

In general, we expect to find two solutions, associated with the roots of Eq. (1.6.33), and we express the homogeneous solution as a linear combination of the two solutions.

$$y_h(x) = C_1 Y_1(x) + C_2 Y_2(x).  \quad (1.6.34)$$

The nature of these solutions depend on whether the roots to Eq. (1.6.33) are real or complex:

(1) If the roots are real, $\lambda_1$ and $\lambda_2$ say, then

$$Y_1(x) = e^{\lambda_1 x}, \quad Y_2(x) = e^{\lambda_2 x}.$$

(2) If the roots are complex, $\lambda_r \pm i\lambda_i$, then

$$Y_1(x) = e^{(\lambda_r + i\lambda_i)x}, \quad Y_2(x) = e^{(\lambda_r + i\lambda_i)x},$$

or

$$Y_1(x) = e^{\lambda_r x}\cos(\lambda_i x), \quad Y_2(x) = e^{\lambda_r x}\sin(\lambda_i x).$$

The second form can also be written as

$$Y_h(x) = A e^{\lambda_r x}\cos(\lambda_i x - \phi),$$

where the two unknown coefficients $C_1$ and $C_2$ in Eq. (1.6.34) are effectively replaced by $A$ and $\phi$.

(3) If there is only one root $\lambda$ – called a double root – then

$$Y_1(x) = e^{\lambda x}, \quad Y_2(x) = x e^{\lambda x}.$$

### 1.6.5.3 *General solution*

The general solution – Principle 4 – will be

$$y(x) = y_p(x) + y_h(x) = y_p(x) + C_1 Y_1(x) + C_2 Y_2(x).  \quad (1.6.35)$$

### 1.6.5.4 *Specific solution*

The final step is to apply the initial conditions to the general solution – Principle 9:

$$y(0) = \alpha, \quad \text{and} \quad \frac{dy}{dx}(0) = \beta.  \quad (1.6.36)$$

As a consequence, we obtain two equations for the two unknown coefficients $C_1$ and $C_2$.

$$Y_1(0)C_1 + Y_2(0)C_2 = \alpha - y_p(0),  \quad (1.6.37)$$

$$\frac{dY_1}{dx}(0)C_1 + \frac{dY_2}{dx}(0)C_2 = \beta - \frac{dy_p}{dx}(0).  \quad (1.6.38)$$

The important point here is that the particular solution is known completely, so the right hand sides of Eq. (1.6.37) and Eq. (1.6.38) are fully known.

By comparing Eq. (1.6.37), Eq. (1.6.38) to Eq. (1.4.32), Eq. (1.4.33), we find the same system of equations except for a modified right hand side that includes contributions from the particular solution. The solution to Eq. (1.6.37) and Eq. (1.6.38) is

$$W(0)\, C_1 = \frac{dY_2}{dx}(0)\,(\alpha - y_p(0)) - Y_2(0)\left(\beta - \frac{dy_p}{dx}(0)\right),$$

$$W(0)\, C_2 = -\frac{dY_1}{dx}(0)\,(\alpha - y_p(0)) + Y_1(0)\left(\beta - \frac{dy_p}{dx}(0)\right),$$

where the Wronskian is

$$W(0) = Y_1(0)\frac{dY_2}{dx}(0) - Y_2(0)\frac{dY_1}{dx}(0).$$

Provided we have found two solutions $Y_1(x)$ and $Y_2(x)$ that ensure $W(0) \neq 0$, we have completed the construction of the specific solution to the differential equation and the initial conditions.

### 1.6.6    *Exercises*

**These are some basic exercises to help develop and improve skill.**

(1) Find a solution to the following differential equations that satisfy the given initial conditions.

(a)

$$\frac{d^2 f}{dx^2}(x) + 4\frac{df}{dx}(x) + 8\, f(x) = x + 2x^2,$$

subject to

$$f(0) = 2 \quad \text{and} \quad \frac{df}{dx}(0) = -2.$$

**Answer:** $f(1) = 0.143$.

(b)

$$\frac{d^2 A}{dt^2}(t) - 2\frac{dA}{dt}(t) + 2\, A(t) = 1 - e^{-3t},$$

subject to

$$A(0) = 0 \quad \text{and} \quad \frac{dA}{dt}(0) = 2.$$

**Answer:** $A(0.5) = 1.35$.

(c)

$$\frac{d^2 y}{dt^2}(t) + x^2\, y(t) = \frac{t}{x^2},$$

subject to

$$y(0) = 1 \quad \text{and} \quad \frac{dy}{dt}(0) = 0.$$

**Answer:** With $x = 0.5$, $y(\pi/2) = 3.21$.

(d)

$$\frac{d^2 y}{dx^2}(x) + \frac{dy}{dx}(x) - 6y(x) = \sin(2x),$$

subject to

$$y(0) = \pi \quad \text{and} \quad \frac{dy}{dx}(0) = 0.$$

**Answer:** $y(0.5) = 5.442$.

(e)

$$\frac{d^2 w}{dy^2}(y) - (3 + \pi)\frac{dw}{dy}(y) + 3\pi w(y) = \sin(\pi t),$$

subject to

$$w(0) = 0 \quad \text{and} \quad \frac{dw}{dy}(0) = 0.$$

**Answer:** With $t = 0.5$, $w(1) = 4.84$.

(f)

$$2\frac{d^2 x}{dt^2}(t) + \frac{dx}{dt}(t) - 6x(t) = t\, e^{-t},$$

subject to

$$x(0) = 0 \quad \text{and} \quad \frac{dx}{dt}(0) = 0.$$

**Answer:** $x(0.5) = 0.29$.

(g)

$$L\frac{d^2 Q}{dt^2}(t) + R\frac{dQ}{dt}(t) = V,$$

subject to

$$Q(0) = 0 \quad \text{and} \quad \frac{dQ}{dt}(0) = 0.$$

**Answer:** With the choice $L = 1$, $V = 2$ and $R = 0.2$, $Q(1) = 0.94$.

**Some problems that arise in applications.**

(2) Suppose we apply the following voltages to the LCR circuit:

(a)

$$E(t) = V_0\left(1 - e^{-rt}\right).$$

(b)

$$E(t) = V_0 \begin{cases} rt & \text{for } 0 < rt < 1, \\ 1 & \text{for } rt > 1. \end{cases}$$

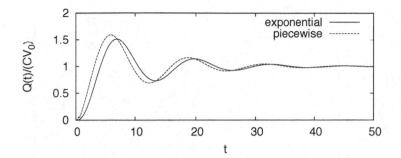

Fig. 1.25 The dimensionless charge as a function of time for two different ramps in the applied voltage.

In both cases the voltage is ramped up until it reaches $V_0$ (V). Make the choice $L = 1$ H, $C = 4$ F, and $R = 0.2$ $\Omega$ and compare the results to Fig. 1.21 by plotting $Q(t)/(CV_0)$ when $r = 2$ s$^{-1}$. Is there any significant difference? Which case is easier to solve?

**Answer:** The two solutions are shown in Fig. 1.25.

(3) Let us explore the role of the resistance $R$ in an LCR circuit by setting $R = 0$ in Eq. (1.6.1) and then comparing the result with the solution found in Fig. 1.21. The particular solution Eq. (1.6.2) remains unchanged. Determine the homogeneous solution and satisfy the initial conditions Eq. (1.6.3). Plot the result and compare it with Fig. 1.21.

**Answer:** The results show a persistent oscillation around the mean value $Q = CV_0$. There is no decay in the oscillation as shown in Fig. 1.21.

(4) Continue the investigation of the role of $R$ by considering Eq. (1.6.8) with $R = 0$. Find the new particular solution and homogeneous solution to replace Eq. (1.6.19) and Eq. (1.6.20) respectively.

(a) Apply the initial conditions Eq. (1.6.3) to the new general solution.
**Answer:** $Q(1)/(CA) = 0.01$.

(b) Study the conditions for resonance as done in Sec. 1.6.4.
**Answer:** Resonance occurs at $\omega = 0.5$ s$^{-1}$.

(c) Add the result as an additional curve in Fig. 1.24.
**Answer:** See Fig. 1.26.

(d) Study the time dependent behavior of the solution at resonance.
**Answer:** $Q(20)/(CA) = 3.92$.

(5) A ball of mass $m$ (gm) and radius $r$ (cm) is released from rest in a liquid of viscosity $\mu$ (gm/(cm $\cdot$ s)). The ball falls under the influence of gravity but the liquid exerts a drag proportional to its velocity $v$ (cm/s). If $y$ (cm) measures the vertical distance the ball has fallen, then its equation of motion is

$$m \frac{d^2 y}{dt^2} + 6\pi\mu r \frac{dy}{dt} = mg.$$

Determine $y(t)$. What can you say about the velocity $v(t)$?

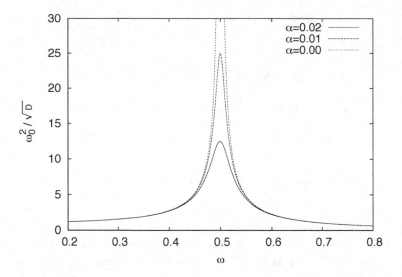

Fig. 1.26 The ratio of output to input amplitudes, $\omega_0^2/\sqrt{D}$, as a function of the forcing frequency $\omega$ for several choices of the damping factor $\alpha$.

Suppose a steel ball of density $\rho = 8$ gm/cm$^3$ is dropped in water ($\mu = 0.01$ gm/(cm $\cdot$ s)). Determine the radius of the ball that will fall steadily at 5.05 mph (the fastest speed of a swimmer).

**Answer:** 0.35 mm.

**Some exercises to extend an abstract view.**

(6) The Wronskian of two solutions $Y_1(x)$ and $Y_2(x)$ to the homogeneous equation (1.6.34) is

$$W(x) = Y_1(x)\frac{dY_2}{dx}(x) - Y_2(x)\frac{dY_1}{dx}(x).$$

Show that it satisfies the equation,

$$\frac{dW}{dx}(x) = -a\,W(x).$$

Solve this equation with $W(0) = W_0$, and show that $W(x)$ is never zero if $W_0 \neq 0$. What conclusion can you draw from this result?

(7) Suppose the homogeneous equation has only a single root to its characteristic equation. The equation may be written as

$$\frac{d^2 y}{dx^2}(x) + a\frac{dy}{dx}(x) + \frac{a^2}{4}\,y(x) = 0.$$

There is one exponential solution $y_1(x) = \exp(-ax/2)$. Find the other solution by using the Wronskian as determined in the previous exercise.

(8) Show that if the Wronskian $W(x) = 0$, then

$$\frac{d}{dx}\left(\frac{Y_1(x)}{Y_2(x)}\right) = 0.$$

What does this say about the relationship between $Y_1(x)$ and $Y_2(x)$?

### 1.6.7 *General forcing terms*

Both examples given in Secs. 1.6.1 and 1.6.2 contain forcing terms that appear in Table 1.4 and so a particular solution is readily available. What can we do if the forcing term is not of a type that suggests a guess for the particular solution? We faced this challenge in Sec. 1.2.6 and we adapt the method used there to constructing a particular solution to (1.6.34), restated here,

$$\frac{d^2y}{dx^2}(x) + a\frac{dy}{dx}(x) + b\,y(x) = r(x). \tag{1.6.39}$$

#### 1.6.7.1 *Variation of parameters*

The first step is to construct the homogeneous solutions Eq. (1.6.34), and then replace the unknown coefficients by unknown functions – the procedure is called *the variation of parameters.*

$$y_p(x) = C_1(x)\,Y_1(x) + C_2(x)\,Y_2(x). \tag{1.6.40}$$

Notice the similarity of this guess with that in Eq. (1.3.15) for first-order differential equation. The choices for $Y_1(x)$ and $Y_2(x)$ depend on the roots to the characteristic equation and are listed after Eq. (1.6.34). The important point is that they are both solutions to the homogeneous equation and that they are known functions.

Now substitute Eq. (1.6.40) into Eq. (1.6.39). We proceed in several steps. First, determine

$$\frac{dy_p}{dx}(x) = \frac{dC_1}{dx}(x)\,Y_1(x) + C_1(x)\frac{dY_1(x)}{dx}(x)$$
$$+ \frac{dC_2}{dx}(x)\,Y_2(x) + C_2(x)\frac{dY_2(x)}{dx}(x). \tag{1.6.41}$$

Before we plunge into computing the second derivative, let us pause and contemplate the ramifications. We will certainly obtain second derivatives of both $C_1(x)$ and $C_2(x)$. Can we avoid that? It seems that we need $C_1(x)$ and $C_2(x)$ to help balance the forcing term $R(x)$. But why do we need two unknown functions to balance a single function? We do not! We are free to impose a single condition on $C_1(x)$ and $C_2(x)$ so that we will end up with two equations for $C_1(x)$ and $C_2(x)$, and the clever choice is

$$\frac{dC_1}{dx}(x)\,Y_1(x) + \frac{dC_2}{dx}(x)\,Y_2(x) = 0, \tag{1.6.42}$$

because now the second derivative

$$\frac{d^2 y_p}{dx^2}(x) = \frac{dC_1}{dx}(x)\frac{dY_1}{dx}(x) + C_1(x)\frac{d^2 Y_1}{dx^2}(x)$$

$$+ \frac{dC_2}{dx}(x)\frac{dY_2}{dx}(x) + C_2(x)\frac{d^2 Y_2}{dx^2}(x) \quad (1.6.43)$$

contains no second derivatives of $C_1(x)$ or $C_2(x)$.

Substitute Eq. (1.6.40), Eq. (1.6.41), bearing in mind Eq. (1.6.42), and Eq. (1.6.43) into Eq. (1.6.39) to obtain

$$C_1(x)\left[\frac{d^2 Y_1}{dx^2}(x) + a\frac{dY_1}{dx}(x) + b\, Y_1(x)\right] + C_2(x)\left[\frac{d^2 Y_2}{dx^2}(x) + a\frac{dY_2}{dx}(x) + b\, Y_2(x)\right]$$

$$+ \frac{dC_1}{dx}(x)\frac{dY_1}{dx}(x) + \frac{dC_2}{dx}(x)\frac{dY_2}{dx}(x) = r(x).$$

The assumption that $Y_1(x)$ and $Y_2(x)$ are solutions to the homogeneous equation means that the expressions within the square brackets must both be zero! Wonderful, because we are simply left with

$$\frac{dC_1}{dx}(x)\frac{dY_1}{dx}(x) + \frac{dC_2}{dx}(x)\frac{dY_2}{dx}(x) = r(x). \quad (1.6.44)$$

The two equations, Eq. (1.6.42) and Eq. (1.6.44), will determine the derivatives of $C_1(x)$ and $C_2(x)$ – remember we assume we have already determined $Y_1(x)$ and $Y_2(x)$.

After some algebra – see Appendix C,

$$\frac{dC_1}{dx}(x) = -\frac{r(x)\,Y_2(x)}{W(x)}, \quad (1.6.45)$$

$$\frac{dC_2}{dx}(x) = +\frac{r(x)\,Y_1(x)}{W(x)}, \quad (1.6.46)$$

where $W(x)$ is the Wronskian,

$$W(x) = Y_1(x)\frac{dY_2}{dx}(x) - Y_2(x)\frac{dY_1}{dx}(x).$$

Obviously, to complete the construction of the solution we need the anti-derivatives for Eq. (1.6.45) and Eq. (1.6.46). Good luck! We can at least though write the solutions in terms of integrals.

$$C_1(x) = -\int^x \frac{r(\xi)}{W(\xi)}\, Y_2(\xi)\, d\xi, \quad (1.6.47)$$

$$C_2(x) = +\int^x \frac{r(\xi)}{W(\xi)}\, Y_1(\xi)\, d\xi, \quad (1.6.48)$$

and the integrals can be substituted into Eq. (1.6.40) to complete the determination of the particular solution.

### 1.6.7.2  *Example*

To illustrate the procedure, let us pick Exercise 1(d) in Sec. 1.6.6.

$$\frac{d^2y}{dx^2}(x) + \frac{dy}{dx}(x) - 6y(x) = \sin(2x).$$

The characteristic equation for the homogeneous solution is

$$\lambda^2 + \lambda - 6 = 0,$$

which has roots $\lambda = -3, 2$. The homogeneous solution is

$$y_h(x) = A_1\,e^{-3x} + A_2\,e^{2x}.$$

Now substitute the guess

$$y_p(x) = C_1(x)\,e^{-3x} + C_2(x)\,e^{2x}$$

into the differential equation. First, calculate the derivative,

$$\frac{dy_p}{dx}(x) = \frac{dC_1}{dx}(x)\,e^{-3x} + \frac{dC_2}{dx}(x)\,e^{2x} - 3C_1(x)\,e^{-3x} + 2C_2\,e^{2x},$$

and set

$$\frac{dC_1}{dx}(x)\,e^{-3x} + \frac{dC_2}{dx}(x)\,e^{2x} = 0.$$

Next, calculate the second derivative,

$$\frac{d^2y_p}{dx^2}(x) = -3\frac{dC_1}{dx}(x)\,e^{-3x} + 2\frac{dC_2}{dx}(x)\,e^{2x} + 9C_1\,e^{-3x} + 4C_2\,e^{2x}.$$

Substitute into the differential equation,

$$-3\frac{dC_1}{dx}(x)\,e^{-3x} + 2\frac{dC_2}{dx}(x) = \sin(2x).$$

You should find that all terms containing $C_1(x)$ and $C_2(x)$ (but not their derivatives) should cancel. The solution for the derivatives of $C_1(x)$ and $C_2(x)$ is

$$\frac{dC_1}{dx}(x) = -\frac{1}{5}\,e^{3x}\sin(2x),\qquad \frac{dC_2(x)}{dx}(x) = \frac{1}{5}\,e^{-2x}\sin(2x).$$

The anti-derivatives are known,

$$C_1(x) = -\frac{1}{65}\,e^{3x}\big[3\,\sin(2x) - 2\,\cos(2x)\big],$$

$$C_2(x) = -\frac{1}{20}\,e^{-2x}\big[\sin(2x) + \cos(2x)\big].$$

Finally, the particular solution is

$$y_p(x) = -\frac{5}{52}\,\sin(2x) - \frac{1}{52}\,\cos(2x).$$

After all this work, we end up with the particular solution we would have obtained by the method of undetermined coefficients, and the effort would have been much less.

If $r(x) = 1/(1+x)$ in Eq. (1.6.39), we would end up with the need to find the anti-derivatives of

$$\frac{\mathrm{d}C_1}{\mathrm{d}x}(x) = -\frac{1}{3}\frac{e^{3x}}{1+x}, \qquad \frac{\mathrm{d}C_2(x)}{\mathrm{d}x}(x) = \frac{1}{3}\frac{e^{-2x}}{1+x}$$

and we would be stuck.

---

**Reflection**

While it is possible to solve linear differential equations with constant coefficients that have any forcing term, in general the solution is expressed in terms of integrals Eq. (1.6.47) and Eq. (1.6.48), but they are unlikely to have anti-derivatives unless $r(x)$ has a special form, for example, as in Table 1.4.

---

### 1.6.7.3 *Series solutions*

Just as we did in Sec. 1.3.2, we can at least gain some understanding of the solution for early times by calculating the solution as a Taylor series. The starting point is to calculate the Taylor series for $r(x)$ written as a general series expansion – note the similarity with (1.6.20).

$$r(x) = \sum_{n=0}^{\infty} R_n x^n. \tag{1.6.49}$$

The point here is that $R_n$ is known. This step of course may not be easily done, but normally it is possible to calculate the first few coefficients.

This is where the power of the method of undetermined coefficients is notable, because according to Table 1.4 we should take

$$y_{\mathrm{p}}(x) = \sum_{n=0}^{\infty} a_n x^n \tag{1.6.50}$$

as the form of the particular solution – again note the similarity with Eq. (1.3.21). The limitation of this approach is clear; the series expansions for $r(x)$ and $y_{\mathrm{p}}(x)$ may have a finite radius of convergence. This means the solution is valid only in a specific range in $x$. In other words, we know something about the solution only for small enough values of $x$. Nevertheless, this may prove useful.

To proceed we substitute Eq. (1.6.49) and Eq. (1.6.50) into Eq. (1.6.39). To do so, we need the first and second derivatives of Eq. (1.6.50):

$$\frac{\mathrm{d}y_{\mathrm{p}}}{\mathrm{d}x}(x) = \sum_{n=1}^{\infty} n\,a_n x^{n-1} = \sum_{n=0}^{\infty} (n+1)\,a_{n+1} x^n,$$

$$\frac{\mathrm{d}^2 y_{\mathrm{p}}}{\mathrm{d}x^2}(x) = \sum_{n=2}^{\infty} n(n-1)\,a_n x^n = \sum_{n=0}^{\infty} (n+1)\,(n+2)\,a_{n+2} x^n,$$

where the counters in the sums have been shifted to align the series representation with $x^n$. Thus the differential equation Eq. (1.6.39) becomes

$$\sum_{n=0}^{\infty} (n+1)(n+2) a_{n+2} x^n + a \sum_{n=0}^{\infty} (n+1) a_n x^n + b \sum_{n=0}^{\infty} a_n x^n = \sum_{n=0}^{\infty} R_n x^n.$$

The balance of each term in $x^n$ leads to the recursion relation,

$$(n+1)(n+2) a_{n+2} + a(n+1) a_{n+1} + b a_n = R_n, \tag{1.6.51}$$

valid for $n = 0, 1, \ldots, \infty$.

Start the recursion with $n = 0$,

$$2 a_2 + a a_1 + b a_0 = R_0.$$

To make progress, we will need to know $a_0$ and $a_1$ to determine $a_2$. The coefficients $a_0$ and $a_1$ play the role of the unknown coefficients that appear in the homogeneous solution. Since we know the homogeneous solutions, we are only interested in calculating the particular solution. Note that the standard initial conditions would determine $a_0$ and $a_1$. To calculate the particular solution, we simply set $a_0 = a_1 = 0$. Thus the first few applications of the recursion formula give

$$2 a_2 = R_0 \quad \Rightarrow \quad a_2 = \frac{R_0}{2},$$

$$6 a_3 + 2a\, a_2 = R_1 \quad \Rightarrow \quad a_3 = \frac{R_1 - aR_0}{6},$$

$$12 a_4 + 6a\, a_3 + b\, a_2 = R_2 \quad \Rightarrow \quad a_4 = \frac{2R_2 - 2aR_1 + (2a - b) R_0}{24}.$$

Clearly, the procedure becomes increasingly complicated to perform, but we can always use the computer to evaluate the recursion for as many terms as we need.

The finally steps require the addition of the homogeneous solution as expressed in Eq. (1.6.34) and the application of the initial conditions.

### 1.6.8 *Summary*

Time to create a checklist for constructing solutions to linear, second-order differential equations,

$$\frac{d^2 y}{dx^2}(x) + a \frac{dy}{dx}(x) + b\, y(x) = r(x).$$

Notice that the coefficients in the differential equation are assumed to be constants. There is a simple reason for this assumption: there are no known methods to produce a general form of the homogeneous solution when $a$ and/or $b$ are functions. This stands in contrast to the case of a first-order differential equation Eq. (1.3.12) where the homogeneous solution can be constructed as an exponential function Eq. (1.3.14).

For constant coefficients in the differential equation, we know the homogeneous solutions are combinations of exponentials, and if we know the homogeneous solutions, then we can try to construct a particular solution for any $r(x)$.

## Checklist

(1) Be sure that the second-order differential equation is linear and has constant coefficients $a$ and $b$.

(2) Find the roots to the quadratic characteristic equation.

(3) There are three possible cases;

**Real roots:** The solutions are

$$Y_1(x) = e^{\lambda_1 x} \quad \text{and} \quad Y_2(x) = e^{\lambda_2 x}.$$

**Double roots:** The solutions are

$$Y_1(x) = e^{\lambda x} \quad \text{and} \quad Y_2(x) = x\,e^{\lambda x}.$$

**Complex roots:** The solutions are

$$Y_1(x) = e^{\lambda_r x} \cos(\lambda_i x) \quad \text{and} \quad Y_2(x) = e^{\lambda_r x} \sin(\lambda_i x).$$

(4) The homogeneous solution is

$$y_h(x) = A_1\,Y_1(x) + A_2\,Y_2(x)$$

where $A_1$ and $A_2$ are unknown coefficients to be determined by the application of the initial conditions, but only after the general solution is known.

(5) Is $r(x)$ of the form that allows a guess for the particular solution $y_p(x)$? If yes, determine the particular solution by the method of undetermined coefficients. If no, then follow the method of variation of parameters.

(6) Compose the general solution $y(x) = y_p(x) + y_h(x)$.

(7) Apply the initial conditions to determine the specific solution.

### 1.6.9 *Flowchart: Linear, second-order differential equations*

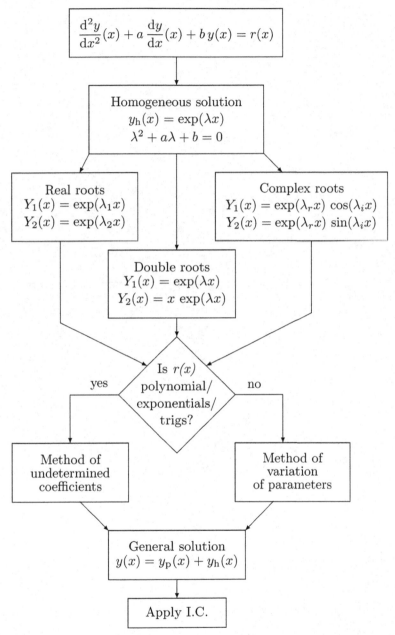

# Periodic Behavior

**2**

## 2.1 Periodic functions

In Sec. 1.6.2, an alternating external voltage is applied to an LCR electric circuit, and its response, the particular solution, is also an alternating charge with the same frequency. These alternating quantities exhibit what is called periodicity. They are examples of temporal periodicity, patterns that repeat in time like the ticking a grandfather clock, rotating blades of a fan, the rotating light in a lighthouse, and so on. There are also spatially periodic patterns, such as tiles on a floor, patterns on wallpaper, crystalline structure, etc. Sometimes patterns do not repeat perfectly, such as waves crashing on a beach, but we assume they do as part of a mathematical model. A good starting point, then, is to make more precise what is meant by periodicity before treating it as a useful tool.

The choice for the periodic applied voltage in the previous section was a trigonometric function $V_0 \sin(\omega t)$. But this choice is far too simple for general purposes. For example, inputs might correspond to digital signals that represent sound or data in computers. All they have in common is that the signals contain some periodicity, some repeated pattern. Our first challenge is to express the idea of periodicity in mathematical terms.

### 2.1.1 *Mathematical expression*

To start, suppose we have an electronic signal that has a periodic pattern. Let $f(t)$ be the function that describes this pattern. How can we express the periodicity of this function? Said simply, the pattern (function) must repeat exactly after the duration of a period $T$. In other words,

$$f(t+T) = f(t). \tag{2.1.1}$$

In Fig. 2.1, a periodic function is displayed with $T = 2\pi$.[1] It allows us to see the consequence of the definition Eq. (2.1.1). Just pick a point inside the interval $[0, 2\pi]$

---

[1]The choice $T = 2\pi$ is common because it is the period in radians when moving around the boundary of a circle.

and notice that the function has the same value as that located at a point shifted $2\pi$ to the right. Obviously, we can shift as many times as we like either to the left or the right. In practice, then, it is the nature of the periodic function in the interval $[0, 2\pi]$ that determines it for all time.[2]

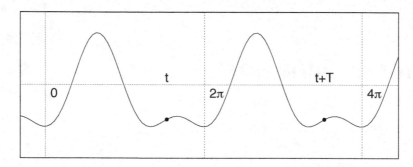

Fig. 2.1   An example of a $2\pi$-periodic function.

The only basic functions that exhibit periodicity are the sines and cosines. But fortunately there are many of them. To understand this apparently strange statement, consider the choices

$$\sin(\omega t), \quad \sin(2\omega t), \quad \sin(3\omega t), \dots . \qquad (2.1.2)$$

The first choice in this sequence $\sin(\omega t)$ has the period $T = 2\pi/\omega$. How do we know? By checking Eq. (2.1.1),

$$\sin\big(\omega(t + T)\big) = \sin(\omega t + \omega T) = \sin(\omega t + 2\pi) = \sin(\omega t).$$

The second choice in Eq. (2.1.2) $\sin(2\omega t)$ has period $\pi/\omega$, which is half that of $\sin(\omega t)$. Similarly, the period of the third choice in Eq. (2.1.2) is one third that of $\sin(\omega t)$. These statements become quite obvious when we look at these functions displayed in Fig. 2.2 with the choice $T = 2\pi$. Notice for example that $\sin(3\omega t)$ has a pattern that fits three times inside of the interval $[0, 2\pi]$.

Suppose now we add the sine functions in Eq. (2.1.2) to form

$$f(t) = b_1 \sin(\omega t) + b_2 \sin(2\omega t) + b_3 \sin(3\omega t). \qquad (2.1.3)$$

This function is composed of three periodic functions each with a different period. However, $\sin(2\omega t)$ is also periodic with $T = 2\pi/\omega$ because

$$\sin\big(2\omega t + 2\omega(2\pi/\omega)\big) = \sin(2\omega t + 4\pi) = \sin(2\omega t).$$

Hmmm. A $T$-periodic function is also $nT$ periodic. So we must be careful what we mean by $T$. Of all the choices for $T$, we should pick the smallest![3] So $\sin(\omega t)$ has

---

[2]Of course no pattern persists for all time. Normally, we select a range of time that contains the pattern and ignore what happens outside this range. We just let the pattern repeat forever as a mathematical convenience.

[3]Of course, we exclude the choice $T = 0$. Also, there is an oddity in that the constant function $f(t) = C$ has any period and there is no smallest. We turn a blind eye to this oddity, but exploit the property that any constant added to a periodic function does not disrupt the periodicity.

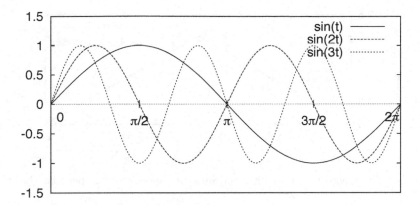

Fig. 2.2 The first three in the family of $2\pi$-periodic sine functions.

period $2\pi/\omega$, $\sin(2\omega t)$ has period $\pi/\omega$ and $\sin(n\omega t)$ has period $2\pi/(n\omega)$. When we consider a function such as $f(t)$ in Eq. (2.1.3) which is composed of several periodic functions, we must pick the largest common period of all the components which is $T = 2\pi/\omega$ in this case. All the other periods can fit an exact multiple of times in this longest period so the periodicity of $f(t)$ is $T$.

Not surprisingly, we see the same pattern of behavior for the cosine functions,

$$\cos(\omega t), \quad \cos(2\omega t), \quad \cos(3\omega t), \ldots$$

as shown in Fig. 2.3. Together with the sine functions, the set of trigonometric functions $\{\cos(k\omega t), \sin(k\omega t)\}$ cover an extraordinary range of oscillations from "slow" $(T = 2\pi/\omega)$ to very "rapid" $(T = 2\pi/(k\omega)$ with $k$ as large as we like). How can this pattern be exploited?

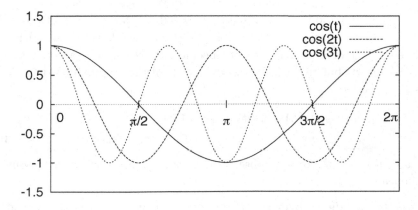

Fig. 2.3 Some $2\pi$-periodic cosine functions.

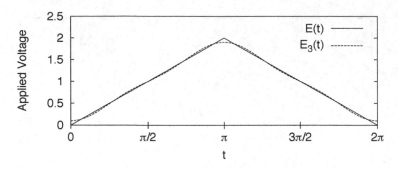

Fig. 2.4   The applied voltage and its approximation shown in one $2\pi$-periodic window.

### 2.1.2   *Periodic forcing of an LCR circuit*

Consider a $2\pi$-periodic applied voltage that represents a ramp up and a ramp down:

$$E(t) = \frac{2}{\pi} \begin{cases} t, & \text{for } 0 \leq t \leq \pi, \\ 2\pi - t, & \text{for } \pi \leq t \leq 2\pi. \end{cases} \tag{2.1.4}$$

This function is displayed in Fig. 2.4 in one periodic window. Just imagine this function extended periodically and you should see a sawtooth pattern that extends indefinitely.

Let us apply the sawtooth voltage to the LCR circuit. Taking advantage of the circuit parameters $\alpha$ and $\omega_0$ introduced in Eq. (1.5.30), the differential equation is

$$\frac{\mathrm{d}^2 Q}{\mathrm{d}t^2}(t) + 2\,\alpha\, \frac{\mathrm{d}Q}{\mathrm{d}t}(t) + \omega_0^2\, Q(t) = \frac{E(t)}{L}. \tag{2.1.5}$$

How can we calculate the particular solution to this equation? We certainly know how to determine a particular solution with linear forcing in each range of $t$, but we would have to patch them together somehow and that would require the need for the homogeneous solution as well. Imagine trying to do this for each piecewise patch as it extends to infinity. Fortunately, there is an easier way.

Introduce a $2\pi$-periodic approximation to $E(t)$! We have at our disposal many choices of sines and cosines and perhaps an appropriate linear combination may approximate $E(t)$ well enough. To be specific, let us consider a combination of $\sin(kt)$ and $\cos(kt)$ but restricted to $k = 1, 2, 3$. Call this approximation $E_3(t)$ and express it in the form

$$E_3(t) = A_0 + A_1 \cos(t) + A_2 \cos(2t) + A_3 \cos(3t)$$
$$+ B_1 \sin(t) + B_2 \sin(2t) + B_3 \sin(3t).$$

There are several important observations that must be made about this approximation.

- $E_3(t)$ is $2\pi$-periodic. Obviously the periodicity of the approximation and $E(t)$ must be the same.

- We pick coefficients with a certain convention: $A$ for cosines and $B$ for sines. They have a subscript $k$ to associate them with the corresponding $\cos(kt)$ and $\sin(kt)$ terms, respectively. Obviously, we could have made another approximation $E_N(t)$ by extending the range of sines and cosines with $k = 1, \ldots, N$.
- We have included a constant – it is obviously $2\pi$-periodic as well – and associated it with the cosines; $A_0 \cos(0 \times t) = A_0$.

The key question is, of course, how do we pick the coefficients $A_k$ and $B_k$ to make $E_3(t)$ a good approximation. The next section will describe how this is done, but for now the result will just be stated: $A_0 = 1$, $A_k = -8/(k^2\pi^2)$ for $k = 1, 3$; the remaining coefficients are all zero. Thus,

$$E_3(t) = 1 - \frac{8}{\pi^2}\cos(t) - \frac{8}{9\pi^2}\cos(3t). \tag{2.1.6}$$

This approximation is also shown in Fig. 2.4 to show that it is not too bad. Instead of $E(t)$, use $E_3(t)$ as the applied voltage in Eq. (2.1.5).

Now the advantage of the approximation Eq. (2.1.6) is clear, because we can easily pick a particular solution. The particular solution takes the form – see Principle 5 and Table 1.4,

$$Q_p(t) = a_0 + a_1\cos(t) + b_1\sin(t) + a_3\cos(3t) + b_3\sin(3t). \tag{2.1.7}$$

Notice that we follow a similar convention for labeling the unknown coefficients; $a_k$ for the cosines and $b_k$ for the sines.

Normally, we would substitute Eq. (2.1.7) into Eq. (2.1.5) and balance terms. But the algebra is going to be complicated. Instead, let us split up the particular solution into several parts, each part corresponding to a specific pair of cosines and sines with the same argument. Thus we replace Eq. (2.1.7) with

$$Q_p(t) = Q_{p,0}(t) + Q_{p,1}(t) + Q_{p,3}(t), \tag{2.1.8}$$

where

$$Q_{p,0}(t) = a_0, \tag{2.1.9}$$
$$Q_{p,1}(t) = a_1\cos(t) + b_1\sin(t), \tag{2.1.10}$$
$$Q_{p,3}(t) = a_3\cos(3t) + b_3\sin(3t). \tag{2.1.11}$$

Notice that the numeral subscript on $Q_p$ coincides with the integer multiple of $t$ that appears in the arguments of the sines and cosines. Of course our intent is to match each $Q_{p,k}$ with the corresponding cosine term in Eq. (2.1.6). Thus we control the algebra![4]

Substitute Eq. (2.1.9) into Eq. (2.1.5) and balance the constant terms;

$$\omega_0^2\, a_0 = \frac{1}{L}.$$

---

[4]Algebra is all about rearranging information to obtain useful results.

Next, substitute Eq. (2.1.10) into Eq. (2.1.5) and balance terms with $\cos(t)$ and $\sin(t)$ separately;

$$-a_1 + 2\alpha\, b_1 + \omega_0^2\, a_1 = -\frac{8}{\pi^2 L}, \tag{2.1.12}$$

$$-b_1 - 2\alpha\, a_1 + \omega_0^2\, b_1 = 0. \tag{2.1.13}$$

Finally, substitute Eq. (2.1.11) into Eq. (2.1.5) and balance the terms with $\cos(3t)$ and $\sin(3t)$ separately;

$$-9a_3 + 6\alpha\, b_3 + \omega_0^2\, a_3 = -\frac{8}{9\pi^2 L}, \tag{2.1.14}$$

$$-9b_3 - 6\alpha\, a_3 + \omega_0^2\, b_3 = 0. \tag{2.1.15}$$

The pattern in these equations shows several interesting features, in particular, they break into pairs of equations for $a_k$, $b_k$, except for $a_0$ which occurs by itself. The pairs of equations show strong similarities; the differences being only the factors $k^2$ and $k$ that appear as a consequence of the second and first derivatives respectively. Indeed, the two systems, Eq. (2.1.12), Eq. (2.1.13) and Eq. (2.1.14), Eq. (2.1.15), can be represented by the single system

$$\begin{bmatrix} (\omega_0^2 - k^2) & 2\,\alpha\,k \\ -2\,\alpha\,k & (\omega_0^2 - k^2) \end{bmatrix} \begin{pmatrix} a_k \\ b_k \end{pmatrix} = -\frac{8}{k^2\pi^2 L} \begin{pmatrix} 1 \\ 0 \end{pmatrix} \tag{2.1.16}$$

where $k = 1, 3$.

The solutions for the coefficients are easily obtained – see Appendix C.

$$a_0 = C\,,$$

$$a_k = -\frac{8\left(\omega_0^2 - k^2\right)}{k^2\pi^2 L\, D}\,,$$

$$b_k = -\frac{16\,\alpha}{k\pi^2 L\, D}\,,$$

where

$$D = \left(\omega_0^2 - k^2\right)^2 + 4\alpha^2 k^2$$

is the determinant of the matrix in Eq. (2.1.16) and $k = 1, 3$. The solutions show a clear pattern, and such patterns give hope that we can systematically make even better approximations by increasing the range in $k$.

As a specific example, pick again $L = 1$ H, $C = 4$ F and $R = 1/5\ \Omega$ as we did in Sec. 1.6.2, which gives $\omega_0 = 0.5\,\mathrm{s}^{-1}$ and $\alpha = 0.1\,\mathrm{s}^{-1}$. All the coefficients $a_k$ and $b_k$ can be evaluated from the equations above. To make a sensible comparison, the response of the circuit is expressed as the voltage on the capacitor $Q_\mathrm{p}(t)/C$ and the result is displayed along with the input voltage in Fig. 2.5. The periodic response $Q_\mathrm{p}(t)/C$ shows a much smoother variation than the input voltage and suggests the approximate approach has value.

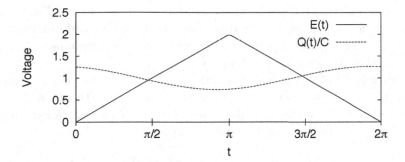

Fig. 2.5 The approximate response $Q_{\mathrm{p}}(t)/C$ to the applied voltage $E(t)$ in one $2\pi$-periodic window.

---

**Reflection**

The success in obtaining an approximation to $Q_{\mathrm{p}}(t)$ depended on finding a periodic approximation $E_3(t)$ to $E(t)$. Indeed, $E_3(t)$ looks rather unusual. There is no $\cos(2t)$ term and why do the amplitudes of the different modes contain a factor $8/\pi^2$? So how is $E_3(t)$ selected? In general, the challenge will be to construct appropriate approximations to periodic functions, and that challenge is faced in the next section.

---

### 2.1.3 *Abstract view*

There are many situations where scientists and engineers exploit periodic behavior. Most machinery has some type of spinning motion: many signals have a repetitive pattern, propagating waves usually have a repetitive pattern. Mathematically, we insist that a periodic function repeats *precisely*.

---

**Principle 10**

*A function $f(t)$ is periodic if there is a positive number $T$ so that*

$$f(t + T) = f(t).$$

*A function that is $T$-periodic is also $nT$-periodic for any integer $n$. To avoid ambiguity, the smallest possible period is usually termed the period.*

---

For the scientist and engineer, we select a range in time or space that contains a repetitive pattern and pretend that it extends infinitely so that the mathematical definition is satisfied.

There is one function $f(t) = C$, a constant, that technically has all $T$ as a period; this is an artificial case since we would not describe a wall painted uniformly with one color as having a periodic pattern. Otherwise, the classic examples of periodic functions are the trigonometric functions.

The combination of two period functions may lead to a new periodic function, but not always. For example, suppose the two functions $f_1(t)$ and $f_2(t)$ are $T_1$ and $T_2$-periodic. Can we find a period $T$ for the sum of the functions? Suppose $T = nT_1 = mT_2$, where $n$ and $m$ are integers. Then

$$f_1(t + T) + f_2(t + T) = f_1(t + nT_1) + f_2(t + mT_2)$$
$$= f_1(t) + f_2(t).$$

This works because of Principle 10 and the result shows the sum is $T$-periodic. But if $T_1/T_2$ is not a rational number ($T_1/T_2 \neq m/n$) then we can never find a common period. This distinction between rational and irrational numbers is somewhat artificial for the scientist and engineer since results are not expected to depend on whether a quantity has a rational or irrational value.[5]

## Principle 11

*If the periods of two periodic functions have a ratio that is a rational number, then their sum is periodic. Obviously, this is also true if the ratio of the frequencies is a rational number.*

This principle allows us to consider the following sequence

$$1, \ \cos(\omega t), \ \sin(\omega t), \ \cos(2\omega t), \ \sin(2\omega t), \ \ldots, \ \cos(n\omega t), \ \sin(n\omega t), \ \ldots$$

The common period for all members is $T = 2\pi/\omega$, and their sum, a linear combination,

$$f(t) = a_0 + \sum_{n=1}^{\infty} \left[ a_n \cos(n\omega t) + b_n \sin(n\omega t) \right] \tag{2.1.17}$$

is called a *Fourier series*. The Fourier series is an infinite sum containing two sets of coefficients, $a_n$ and $b_n$. The subscript $n$ is a dummy counter selecting which of the trigonometric functions are being added to the sum. Since the sum is infinite we are unsure whether it adds up to a function. In other words, the sum may not converge. That will depend on the nature of the coefficients, an issue that we will set aside for now and return to later.

---

[5]Can you imagine an experiment that determines whether a measurement is rational or irrational? Note that if a number is expressed with a finite number of digits, then it is obviously rational.

### 2.1.3.1 *Application to differential equations*

There are many examples in science and engineering where a differential equation is forced with some periodic function and we want to find the periodic response given by the particular solution. Take a simple example to illustrate how we can construct a periodic solution when the forcing function is a Fourier series.

$$\frac{dy}{dt}(t) + \alpha\, y(t) = F(t)\,, \tag{2.1.18}$$

where $F(t)$ is $T$-periodic. The first step is to recast $F(t)$ in the form of a Fourier series,

$$F(t) = a_0 + \sum_{n=1}^{\infty} \left[ a_n \cos(n\omega t) + b_n \sin(n\omega t) \right], \tag{2.1.19}$$

where $\omega = 2\pi/T$. For the moment, put aside the question of how the Fourier coefficients $a_n$ and $b_n$ are determined. Just consider them as known parameters.

An important observation is that the arguments of the cosines and sines are such that they ensure $T$ periodicity to match the periodicity of $f(t)$. It makes no sense to approximate a periodic function by cosines and sines that do not have the same periodicity. The sum Eq. (2.1.19) contains an infinite number of terms. By Principle 5, each term will require a corresponding cosine and sine in the guess for the particular solution. Thus we anticipate the particular solution takes the form,

$$y_{\rm p}(t) = A_0 + \sum_{n=1}^{\infty} \left[ A_n \cos(n\omega t) + B_n \sin(n\omega t) \right]. \tag{2.1.20}$$

Clearly, the particular solution will also have periodicity $T$ as expected. The unknown coefficients $A_n$ and $B_n$ are determined by substituting Eq. (2.1.20) into Eq. (2.1.18) and balancing terms, nothing more than the standard method of undetermined coefficients.

To keep control on the algebra, we can organize the work by following the approach we adopted in Eq. (2.1.8). Pick one value of $n$ in the forcing term and use the corresponding term in the particular solution to achieve the balance. First, the constant term $a_0$ will require a constant term $A_0$ in the particular solution and, after substitution into the differential equation, the balance of the constants will require

$$\alpha\, A_0 = a_0\,, \quad \text{or} \quad A_0 = \frac{a_0}{\alpha}\,. \tag{2.1.21}$$

Next, the combination for a particular choice of $n$,[6]

$$A_n \cos(n\omega t) + B_n \sin(n\omega t)$$

is substituted into Eq. (2.1.18) with the intent to balance

$$a_n \cos(m\omega t) + b_n \sin(n\omega t).$$

---

[6]For the moment, $n$ becomes just a parameter in the calculations. Once the coefficients $A_n$ and $B_n$ are determined and substituted back into Eq. (2.1.20), $n$ reverts back to being a dummy counter in the sum.

The differential equation becomes

$$\left[ n\omega\, B_n + \alpha\, A_n \right] \cos(n\omega t) + \left[ -n\omega\, A_n + \alpha\, B_n \right] \sin(n\omega t)$$
$$= a_n \cos(n\omega t) + b_n \sin(n\omega t)\,.$$

By balancing the cosine terms,

$$n\omega\, B_n + \alpha\, A_n = a_n\,, \qquad (2.1.22)$$

and the sine terms,

$$-\,n\omega\, A_n + \alpha\, B_n = b_n\,. \qquad (2.1.23)$$

The result is two simultaneous equations Eq. (2.1.22) and Eq. (2.1.23) for $A_n$ and $B_n$. They may be expressed in matrix form – see Eq. (C.3.2) – as

$$\begin{bmatrix} \alpha & n\omega \\ -n\omega & \alpha \end{bmatrix} \begin{pmatrix} A_n \\ B_n \end{pmatrix} = \begin{pmatrix} a_n \\ b_n \end{pmatrix}\,.$$

The solution is given by Eq. (C.1.7) and Eq. (C.1.9);

$$A_n = \frac{\alpha\, a_n - n\omega\, b_n}{n^2\omega^2 + \alpha^2}\,, \qquad (2.1.24)$$

$$B_n = \frac{\alpha\, b_n + n\omega\, a_n}{n^2\omega^2 + \alpha^2}\,. \qquad (2.1.25)$$

By substituting the results Eq. (2.1.21), Eq. (2.1.24) and Eq. (2.1.25) into Eq. (2.1.20), we have completed the construction of $y_{\mathrm{p}}(x)$.

The procedure does depend on the order of the differential equation. If the equation is second order, then we still substitute Eq. (2.1.20) into the equation and balance terms as in Eq. (2.1.21), Eq. (2.1.22) and Eq. (2.1.23). Obviously, the details will be different but in the end, we will still obtain pairs of equations for $A_n$ and $B_n$.

---

### Principle 12

*By converting a periodic forcing function into a Fourier series, the solution to a differential equation can be constructed term-by-term.*

---

It is interesting to compare what we have done here with what we did in Sec. 1.3.2 where we used the Taylor series to represent the forcing function – see Eq. (1.3.20) – and then constructed the particular solution as a Taylor series – see Eq. (1.3.21). Here we use the Fourier series to represent the periodic forcing function and obtain the particular solution as a Fourier series. In both cases, we replace the forcing function by expressions containing simple functions for which a particular solution is easily determined. The disadvantage is that it is not easy to understand the nature or properties of the particular solution when it is expressed as a series. Normally, we have to plot the results to assess the consequences of the particular solution.

## 2.1.4  *Exercises*

**Exercises that emphasize periodicity and how to recognize it.**

(1) Check whether the following functions are periodic:

(a) Does $x$ have period 1?

(b) For what values of $k$ does $\sin(kx)$ have period

    i. $2\pi$?

    ii. $4\pi$?

    iii. $2n\pi$?

(c) For what values of $k$ does $\sin(kx)$ have period $\pi/4$?

(d) Does $\cos^2(x)$ have period $\pi$?

(e) Does $\cos^3(x)$ have period $\pi$?

(f) Does $\cos(\sin(x))$ have period $\pi$?

(2) Change independent variables so that the following functions are $2\pi$-periodic:

(a) $\cos(k\pi x)$ which is $2/k$-periodic for any positive integer $k$.

(b) $\cos^2(x/L)$ which has period $\pi L$.

(c) $\cosh\{\sin[(x/(2L)]\}$ which has period $2\pi L$.

**The response of differential equations to periodic forcing.**

(3) The model for the influx of concentration of a pollutant into a dam that was described in Sec. 1.2.1 can be expressed as

$$\frac{d\rho}{dt}(t) = \beta(c(t) - \rho(t)),$$

where the concentration in the dam is $\rho(t)$ (kg/m$^3$) and the concentration entering the dam is $c(t)$ (kg/m$^3$). For this exercise, assume the time is measured in hours and $\beta = 0.158$ h$^{-1}$. Suppose $c(t)$ has a daily variation modulated by a weekly variation:

$$c(t) = 1 + \cos\left(\frac{\pi}{12}t\right)\sin\left(\frac{\pi}{84}t\right).$$

A graph of the variation in $c(t)$ in given in Fig. 2.6. Find the behavior of $\rho(t)$ when it starts with $\rho(0) = 0$. What is the long term behavior of the concentration in the dam?

**Answer:** See Fig. 2.7. Note that the output looks similar to the input but is reduced in the size of its variation.

(4) Suppose the resistance in the standard electric circuit is negligible. The equation for the charge on the capacitor is

$$\frac{d^2Q}{dt^2}(t) + \omega_0^2\, Q(t) = \frac{E(t)}{L},$$

where $\omega_0 = 1/\sqrt{LC}$ (rad/s). Suppose a $T$-periodic voltage $E(t)$ is applied that has the Fourier series

$$E(t) = \sum_{n=1}^{\infty} \frac{1}{n^2+1}\cos(n\omega t),$$

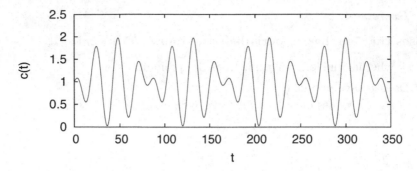

Fig. 2.6    Modulation in the input $c(t)$ as a function of time.

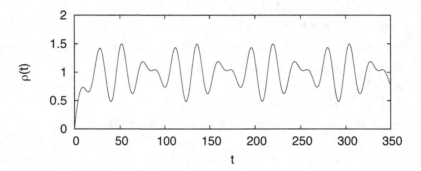

Fig. 2.7    Modulation in the output $\rho(t)$ as a function of time.

where $\omega = 2\pi/T$ (rad/s). Determine $Q(t)$ if the circuit is dead when the forcing is first applied. In other words,

$$Q(0) = 0, \quad \text{and} \quad \frac{dQ}{dt}(0) = 0.$$

What is the long term behavior of $Q(t)$? What happens to the long term behavior of the charge when $\omega_0$ is varied?

**Answer**: An experimentalist provides some data for the case $T = 5$ s, $L = 0.1$ H and $\omega = 3$ rad/s in the Table 2.1.

Table 2.1    Experimental
results.

| Time (s) | Charge (C) |
| --- | --- |
| 2.0 | -1.472 |
| 4.0 | -1.245 |
| 6.0 | -1.031 |
| 8.0 | -0.048 |
| 10.0 | 0.985 |

Resonance occurs whenever there is a choice of $n$ for which $\omega_0 = n\omega$. By selecting a specific frequency with a choice of $n$ and $\omega$, we can tune the circuit by adjusting $\omega_0$ until it matches the selected frequency and enhancing that frequency in the input signal.

### 2.1.5  *Periodic complex functions*

Because of the close connection between trigonometric functions and exponentials with purely imaginary arguments provided by Euler's formula and its companion, the conjugate,

$$e^{i\theta} = \cos(\theta) + i\sin(\theta) \quad \text{and} \quad e^{-i\theta} = \cos(\theta) - i\sin(\theta),$$

it is not surprising that exponentials with purely imaginary arguments are periodic. Specifically, with the choice $\omega = 2\pi/T$,

$$e^{i\omega(t+T)} = e^{i\omega t}\, e^{i\omega T} = e^{i\omega t},$$

since $\omega T = 2\pi$ and $\exp(2\pi i) = 1$; we have confirmed property Eq. (2.1.1).

In Sec. 1.6.3, a method to construct the particular solution for periodic forcing is developed based on the use of exponentials with complex arguments. It can be used here too with great effect. Start by considering a typical term in Eq. (2.1.17) and by rewriting it as

$$a_n \cos(n\omega t) + b_n \sin(n\omega t) = \frac{a_n}{2}\left(e^{in\omega t} + e^{-in\omega t}\right) + \frac{b_n}{2i}\left(e^{in\omega t} - e^{-in\omega t}\right)$$
$$= \frac{a_n - ib_n}{2} e^{in\omega t} + \frac{a_n + ib_n}{2} e^{-in\omega t}$$
$$= c_n e^{in\omega t} + c_{-n} e^{-in\omega t}, \tag{2.1.26}$$

where, obviously, the complex coefficients

$$c_n = \frac{a_n - ib_n}{2} \quad \text{and} \quad c_{-n} = \frac{a_n + ib_n}{2} \tag{2.1.27}$$

have been introduced to accompany the complex exponentials. The subscript $-n$ is used in association with the negative argument in the second exponential. Of course, $c_{-n} = \bar{c}_n$ as required by Eq. (1.5.4) to ensure the result is a purely real function.

From Eq. (2.1.26), we are led to an alternative expression for the Fourier series Eq. (2.1.17).

$$f(t) = a_0 + \sum_{n=1}^{\infty}\left[c_n e^{in\omega t} + c_{-n} e^{-in\omega t}\right],$$

and now the choice of notation for the coefficients helps us see that it can be written as

$$f(t) = \sum_{n=-\infty}^{\infty} c_n e^{in\omega t},$$

which is called – no surprise! – the complex Fourier series. The term for $n = 0$ is just $c_0 = a_0$.

The value in the complex form of the Fourier series is that the particular solution for the differential equation Eq. (2.1.18) can also be expressed as a complex Fourier series

$$y_p(t) = \sum_{n=-\infty}^{\infty} C_n e^{in\omega t}$$

with $\omega = 2\pi/T$.[7] Direct substitution into Eq. (2.1.18) leads to

$$\sum_{n=-\infty}^{\infty} in\omega C_n e^{in\omega t} + \alpha \sum_{n=-\infty}^{\infty} C_n e^{in\omega t} = \sum_{n=-\infty}^{\infty} c_n e^{in\omega t}.$$

The balance of each exponential with a chosen value for $n$ gives

$$C_n = \frac{c_n}{\alpha + in\omega} = \frac{(\alpha\, a_n - n\omega\, b_n) - i(\alpha\, b_n + n\omega\, a_n)}{2(\alpha^2 + n^2\omega^2)}. \tag{2.1.28}$$

From our understanding of the connection between $C_n$ and $A_n$ and $B_n$ as revealed by Eq. (2.1.27), we obtain

$$A_n = \frac{\alpha\, a_n - n\omega\, b_n}{\alpha^2 + n^2\omega^2}, \tag{2.1.29}$$

$$B_n = \frac{\alpha\, b_n + n\omega\, a_n}{\alpha^2 + n^2\omega^2}. \tag{2.1.30}$$

These results agree with Eq. (2.1.24) and Eq. (2.1.25). The work involved to obtain the result Eq. (2.1.28) – hence Eq. (2.1.29) and Eq. (2.1.30) – is much less complicated and has a simple perspective – there are only exponentials.

## 2.2    The Fourier series

In the previous section, we used a linear combination Eq. (2.1.8) of a constant, $\cos(t)$ and $\cos(3t)$ to approximate a $2\pi$-periodic function $E(t)$. But how are the coefficients for each of the terms determined? Were they just inspired guesses? If not, is there a way to add more terms and to make an even better approximation?

Since $\cos(kt)$ and $\sin(kt)$ are all $2\pi$-periodic if $k$ is any integer, adding them all up in a linear combination will produce a $2\pi$-periodic function – Principle 11. Well, that is a good start, but why might it be important? Since each $\cos(kt)$ and $\sin(kt)$ is also $2\pi/k$-periodic, these trigonometric functions exhibit very rapid variations when $k$ is large, and since we plan to add them up for all $k$, we will be capturing all scales in time. Hopefully, we will be able to represent $E(t)$ extremely accurately. Thus an approximation such as

$$E_N(t) = a_0 + a_1\,\cos(t) + b_1\,\sin(t) + a_2\,\cos(2t) + b_2\,\sin(2t) + \cdots$$
$$+ a_N\,\cos(Nt) + b_N\,\sin(Nt)$$

$$= a_0 + \sum_{n=1}^{N}\left[a_n\,\cos(nt) + b_n\,\sin(nt)\right] \tag{2.2.1}$$

---

[7]By now it should be clear that using $\omega$ as a parameter in place of $2\pi/T$ has definite advantages.

might make sense. How big should $N$ be? Just enough to make the approximation accurate to a specified degree. How do we pick the coefficients $a_k$ and $b_k$? Well, we will try to pick them off one at a time. Note again that the choice of symbols for the coefficients helps us identify whether they belong to a cosine ($a_k$) or a sine ($b_k$) term. Further, the subscript is coordinated with the $k$ in the argument of the trigonometric function. The subscript $N$ on $E$ reminds us how many terms in the sum we are keeping.

The task that faces us is a method to determine the coefficients $a_k$ and $b_k$ so that the limiting function[8]

$$\lim_{N \to \infty} E_N(t) = g(t)$$

is the same as $E(t)$. In general, how do we decide whether two functions $f(t)$ and $g(t)$ are the same? What we mean by the same is that they are equivalent; they agree at every point $x$. There are several strategies we can follow:

- There may be analytic tools that establish the equivalence. For example, we have shown that $f(t) = \exp(ix)$ is equivalent to $g(t) = \cos(x) + i\sin(x)$, a very useful identity.
- We could evaluate $f(t)$ and $g(t)$ at the same locations and check whether they agree. This seems the simplest, but how could we check at *every* location?
- We can pick a set of properties and demand that both functions satisfy these properties. Of course the set of properties must define these functions uniquely. For example, we have shown that solutions to a differential equation with appropriate initial conditions are unique, so if we find two solutions in different forms they must agree. That is exactly how we established the equivalence of $f(t) = \exp(it)$ and $g(t) = \cos(t) + i\sin(t)$ – see Sec. 1.5.1.

Another very important example of a set of properties that determines a function uniquely, one more relevant to our present circumstances, is the specification of a function and all its derivatives at a fixed location. This idea is used to determine the Taylor series expansion – see Sec. B.1. Let

$$g(t) = \sum_{n=0}^{\infty} a_n t^n,$$

then matching derivatives at $x = 0$ requires

$$a_n = \frac{1}{n!} \frac{d^n f}{dt^n}(0).$$

We seek a property, or a set of properties, that will determine the coefficients in Eq. (2.2.1) in the same vein as the coefficients in the Taylor series. Clearly the property should reflect the present circumstance, namely, that we are working with periodic functions. The starting point will be the requirement that the average value over a periodic window of $E(t)$ and $E_N(t)$ agree.

---

[8]At some point we will have to confirm that this limit exists!

## 2.2.1   *Construction of a Fourier series*

The first coefficient we can pick off is $a_0$ because it is the average value of $E_N(t)$ over one period and if $E_N(t)$ is to be a good approximation to $E(t)$, then their average values must agree. To drive this point home, suppose $E(t)$ was just a constant function $E_0$. Surely we would want $a_0 = E_0$ in the approximation $E_N(t)$ and all the other coefficients should be set to zero. Another way to state the same result is that the area under the curve $E(t)$ should match the area under the approximation $E_N(t)$.

So let us introduce an operator that performs the action of determining the average over a periodic window:

$$\mathcal{A}_T\{f\} = \frac{1}{T} \int_0^T f(t)\, dt\,. \tag{2.2.2}$$

The symbol $\mathcal{A}$ is called an operator because it performs an action, integration, on a function. Its argument given in {} specifies the function it is acting on, and the subscript tells us the period over which the average is performed.

To determine the average of $E_N(t)$, we will need the average of the following quantities,

$$\mathcal{A}_{2\pi}\{a_0\} = \frac{1}{2\pi} \int_0^{2\pi} a_0\, dt = a_0\,, \tag{2.2.3}$$

$$\mathcal{A}_{2\pi}\{a_k \cos(kt)\} = \frac{1}{2\pi} \int_0^{2\pi} a_k \cos(kt)\, dt = 0\,, \tag{2.2.4}$$

$$\mathcal{A}_{2\pi}\{b_k \sin(kt)\} = \frac{1}{2\pi} \int_0^{2\pi} b_k \sin(kt)\, dt = 0\,, \tag{2.2.5}$$

where $k$ is any non-zero integer. Because the integral of a sum is the sum of the integrals, the average of a sum is the sum of the averages. The average value of $E_N(t)$ will be

$$\mathcal{A}_{2\pi}\{E_N(t)\} = \mathcal{A}_{2\pi}\{a_0 + a_1 \cos(t) + b_1 \sin(t) + \cdots$$
$$+ a_N \cos(Nt) + b_N \sin(Nt)\}$$
$$= \mathcal{A}_{2\pi}\{a_0\} + \mathcal{A}_{2\pi}\{a_1 \cos(t)\} + \mathcal{A}_{2\pi}\{b_1 \sin(t)\} + \cdots$$
$$+ \mathcal{A}_{2\pi}\{a_N \cos(Nt)\} + \mathcal{A}_{2\pi}\{b_N \sin(Nt)\}$$
$$= a_0\,.$$

An amazing thing happens! By taking the average of all the terms in the sum for $E_N(t)$, all the terms drop out except for $a_0$. In other words, the average value of $E_N(t)$ is just $a_0$.

The idea we implement is that the average value of $E_N(t)$ over one period should match the average value of $E(t)$ over one period, or

$$\frac{1}{2\pi} \int_0^{2\pi} E(t)\, dt = \frac{1}{2\pi} \int_0^{2\pi} E_N(t)\, dt = a_0\,. \tag{2.2.6}$$

In other words, $a_0$ must be the average value of $E(t)$ over one period.

To illustrate how to apply this result, consider again the choice Eq. (2.1.6);

$$E(t) = \frac{2}{\pi} \begin{cases} t, & \text{for } 0 \le t \le \pi, \\ 2\pi - t, & \text{for } \pi \le t \le 2\pi. \end{cases} \quad (2.2.7)$$

To apply the matching principle Eq. (2.2.6), we must integrate $E(t)$ to determine $a_0$ and because $E(t)$ is piecewise continuous, we must split up the range of integration into two parts. Thus

$$a_0 = \frac{1}{2\pi} \int_0^{2\pi} E(t)\,dt$$
$$= \frac{1}{2\pi} \int_0^{\pi} \frac{2}{\pi} t\,dt + \frac{1}{2\pi} \int_{\pi}^{2\pi} \frac{2}{\pi}(2\pi - t)\,dt$$
$$= 1.$$

So that is how we find $a_0 = 1$ in the approximation Eq. (2.1.6) to $E(t)$.

How can we apply the property of averaging to the determination of the other coefficients, $a_1$ for example? What other quantities might be worth averaging? Well, what about the difference between $E(t)$ and $E_N(t)$? We would like the average value of this difference to be small, even zero. But unfortunately, a quantity need not be zero in order for its average to be zero. For example, the average values of sines and cosines given in Eq. (2.2.4) and Eq. (2.2.5) are zero. We need to take the average of a positive (or negative) quantity because its average can only be zero if the quantity is zero. So we should consider the average of the square of the difference in $E(t)$ and $E_N(t)$:

$$F = \frac{1}{2\pi} \int_0^{2\pi} [E_N(t) - E(t)]^2\,dt.$$
$$= \frac{1}{2\pi} \int_0^{2\pi} [a_0 + a_1 \cos(t) + b_1 \sin(t) + \cdots - E(t)]^2\,dt. \quad (2.2.8)$$

Now try to find values for $a_k$ and $b_k$ that make $F$ as small as possible.

When we look at Eq. (2.2.8), we recognize that we are seeking a least squares fit of $E_N(t)$ with $E(t)$. That means we consider $F$ to be a function of $a_0, a_1, b_1, \ldots$ and look for values of these coefficients that make $F$ a minimum. At the minimum value of $F$, we must have

$$\frac{\partial F}{\partial a_0} = 0, \quad \frac{\partial F}{\partial a_1} = 0, \quad \frac{\partial F}{\partial b_1} = 0, \quad \ldots \quad (2.2.9)$$

Let us examine the consequence of this sequence of requirements one at a time.

The first requirement in Eq. (2.2.9) leads to

$$\frac{\partial F}{\partial a_0} = \frac{1}{2\pi} \int_0^{2\pi} 2[a_0 + a_1 \cos(t) + b_1 \sin(t) + \cdots - E(t)]\,dt = 0.$$

We may restate the result in terms of averages.

$$\frac{\partial F}{\partial a_0} = 2\,\mathcal{A}_{2\pi}\{a_0 + a_1 \cos(t) + b_2 \sin(t) + \cdots\} - 2\,\mathcal{A}_{2\pi}\{E(t)\} = 0.$$

In other words, we require

$$\mathcal{A}_{2\pi}\{E(t)\} = \mathcal{A}_{2\pi}\{E_N(t)\}$$

which is exactly the same as Eq. (2.2.6) and it determines $a_0$!

Our hopes are high. Maybe all the requirements in Eq. (2.2.9) will determine all the coefficients. Let us see what happens when we consider the second requirement.

$$\frac{\partial F}{\partial a_1} = \frac{1}{2\pi} \int_0^{2\pi} 2\left[a_0 + a_1\cos(t) + b_1\sin(t) + \cdots - E(t)\right]\cos(t)\,\mathrm{d}t = 0\,.$$

The consequence may be expressed in terms of averages.

$$\mathcal{A}_{2\pi}\{\cos(t)E_N(t)\} = \mathcal{A}_{2\pi}\{\cos(t)E(t)\}\,. \tag{2.2.10}$$

Since

$$\mathcal{A}_{2\pi}\{\cos(t)E_N(t)\} = \mathcal{A}_{2\pi}\{a_0\cos(t)\} + \sum_{n=1}^{N}\mathcal{A}_{2\pi}\{a_n\cos(t)\cos(nt)\}$$

$$+ \sum_{n=1}^{N}\mathcal{A}_{2\pi}\{b_n\cos(t)\sin(nt)\}\,, \tag{2.2.11}$$

we are left with many integrals to perform, and except for the one containing $a_0$ they all involve the product of two trigonometric functions. Fortunately, they are easy to do because we can convert the products of sines and cosines into sums of cosines and sines – details are available in Appendix D – and the results are easily integrated.

Consider first the average value of $\cos(t)\cos(nt)$. We have a series of integrals for $n = 1, \ldots, N$, but we can do them all at once by regarding $n$ as a fixed parameter. By using Eq. (D.2.5),

$$2\cos(t)\cos(nt) = \cos\left[(n+1)t\right] + \cos\left[(n-1)t\right],$$

the average of a product of cosines becomes the average of a sum of cosines.

$$2\,\mathcal{A}_{2\pi}\{\cos(t)\cos(nt)\} = \mathcal{A}_{2\pi}\{\cos\left[(n+1)\,t\right]\} + \mathcal{A}_{2\pi}\{\cos\left[(n-1)\,t\right]\}\,,$$

and because of Eq. (2.2.4), all of the integrals are zero, except for $n = 1$. In that case, $\cos\left[(n-1)\,t\right] = 1$, and its average is just one – see Eq. (2.2.3).[9] Thus,

$$\mathcal{A}_{2\pi}\{\cos(t)\cos(nt)\} = \begin{cases} 0 & \text{for } n > 1\,, \\ \dfrac{1}{2} & \text{for } n = 1\,. \end{cases} \tag{2.2.12}$$

Wow, this is an amazing result! It means that there is only one non-zero contribution in the sum of the averages of $\cos(t)\cos(nt)$.

We are left with integrals containing the combination $\cos(t)\sin(kt)$. This time we use Eq. (D.2.7),

$$2\cos(t)\sin(nt) = \sin\left[(n+1)\,t\right] + \sin\left[(n-1)\,t\right],$$

---

[9]The case $n = 1$ corresponds to taking the average of $\cos^2(t)$ and that is obviously not zero.

so that

$$2\,\mathcal{A}_{2\pi}\{\cos(t)\sin(nt)\} = \mathcal{A}_{2\pi}\{\sin[(n+1)\,t]\} + \mathcal{A}_{2\pi}\{\sin[(n-1)\,t]\}$$
$$= 0\,. \tag{2.2.13}$$

This time, Eq. (2.2.5) ensures that all the averages are zero, even for $n = 1$ because $\sin[(n-1)t] = 0$. The consequence is that the sum of averages of $\cos(t)\sin(nt)$ is zero.

The results in Eq. (2.2.12) and Eq. (2.2.13) are truly wonderful because now (2.2.11) becomes

$$\mathcal{A}_{2\pi}\{\cos(t)E_N(t)\} = \frac{a_1}{2}\,.$$

Apply the matching principle Eq. (2.2.10),

$$\frac{a_1}{2} = \frac{1}{2\pi}\int_0^{2\pi}\cos(t)\,E(t)\,\mathrm{d}t\,, \tag{2.2.14}$$

and $a_1$ is determined as twice the average value of $\cos(t)\,E(t)$ over one period.

To illustrate the result Eq. (2.2.14), let us apply it to Eq. (2.2.7).

$$a_1 = \frac{1}{\pi}\int_0^{\pi}\frac{2}{\pi}t\,\cos(t)\,\mathrm{d}t + \frac{1}{\pi}\int_\pi^{2\pi}\frac{2}{\pi}(2\pi - t)\,\cos(t)\,\mathrm{d}t$$
$$= -\frac{8}{\pi^2}\,, \tag{2.2.15}$$

which gives the second term in the approximation $E_3(t)$ – see Eq. (2.1.8). For those that need a reminder, the integrals are performed with integration by parts.

At this point, we could continue the strategy by working through all the partial derivatives in Eq. (2.2.9) one at a time. Instead, proceed in a more general way. Let us consider the general partial derivative,

$$\frac{\partial F}{\partial a_m} = \frac{1}{2\pi}\int_0^{2\pi}2\left(E_N(t) - E(t)\right)\cos(mt)\,\mathrm{d}t = 0\,.$$

Restated in terms of averages,

$$\mathcal{A}_{2\pi}\{\cos(mt)E_N(t)\} = \mathcal{A}_{2\pi}\{\cos(mt)E(t)\}\,. \tag{2.2.16}$$

Note that Eq. (2.2.10) agrees with this result for the choice $m = 1$.

Since

$$\mathcal{A}_{2\pi}\{\cos(mt)E_N(t)\} = \mathcal{A}_{2\pi}\{a_0\cos(mt)\} + \sum_{n=1}^{N}\mathcal{A}_{2\pi}\{a_n\cos(mt)\cos(nt)\}$$
$$+ \sum_{n-1}^{N}\mathcal{A}_{2\pi}\{b_n\cos(mt)\sin(nt)\}\,, \tag{2.2.17}$$

we must consider the averages of $\cos(mt)\cos(nt)$ and $\cos(mt)\sin(nt)$. Incidentally, the first average on the right hand side of Eq. (2.2.17) is zero by Eq. (2.2.3). Once again we will make use of Eq. (D.2.5),

$$2\cos(mt)\cos(nt) = \cos[(n+m)\,t] + \cos[(n-m)\,t],$$

so that

$$2\,\mathcal{A}_{2\pi}\{\cos(mt)\cos(nt)\} = \mathcal{A}_{2\pi}\{\cos\left[(n+m)\,t\right]\} + \mathcal{A}_{2\pi}\{\cos\left[(n-m)\,t\right]\}\,,$$

and because of Eq. (2.2.4), all of the integrals are zero, except for $n = m$. In that case $\cos\left[(n-m)\,t\right] = 1$, and its average is just one – see Eq. (2.2.3). Thus,

$$\mathcal{A}_{2\pi}\{\cos(mt)\cos(nt)\} = \begin{cases} 0 & \text{for } n \neq m\,, \\ \dfrac{1}{2} & \text{for } n = m\,. \end{cases} \tag{2.2.18}$$

Next we use Eq. (D.2.7),

$$2\cos(mt)\sin(nt) = \sin\left[(n+m)\,t\right] + \sin\left[(n-m)\,t\right]\,,$$

so that

$$2\,\mathcal{A}_{2\pi}\{\cos(mt)\sin(nt)\} = \mathcal{A}_{2\pi}\{\sin\left[(n+m)\,t\right]\} + \mathcal{A}_{2\pi}\{\sin\left[(n-m)\,t\right]\}$$
$$= 0\,. \tag{2.2.19}$$

This time, Eq. (2.2.5) ensures that all the averages are zero, even for $n = m$ because $\sin\left[(n-m)t\right] = 0$.

How do these results, Eq. (2.2.18) and Eq. (2.2.19) help? Imagine performing the sum in Eq. (2.2.17) for a particular choice of $m$. As long as the sum counter $n$ does not match $m$, then the averages are zero and do not contribute to the sum. But as $n$ increases, it will match $m$ at some point. Then the average of $\cos^2(mt) = 1/2$ and a single contribution to the sum will be made. The sum of the averages of $\cos(mt)\sin(nt) = 0$. Consequently, Eq. (2.2.17) reduces to

$$\frac{1}{2\pi}\int_0^{2\pi} \cos(mt)\,E_N(t)\,\mathrm{d}t = \frac{a_m}{2}\,.$$

By applying the matching principle Eq. (2.2.16), we end up with the way to determine $a_m$.

$$\frac{a_m}{2} = \frac{1}{2\pi}\int_0^{2\pi} \cos(mt)\,E(t)\,\mathrm{d}t\,. \tag{2.2.20}$$

This result, then, gives us the way to determine all the Fourier coefficients $a_m$ by evaluating the integrals Eq. (2.2.20) for each choice of $m$. By considering $m$ a parameter, we obtain the results for all the averages by performing just one integration. The result in Eq. (2.2.20) can be remembered as follows: The coefficient $a_m$ of the term $a_m\cos(mt)$ in the Fourier series is obtained by taking the average of $\cos(mt)E(t)$ over one periodic window, and multiplying the result by two.

We now anticipate that the other general partial derivatives in Eq. (2.2.9) will lead to the determination of $b_m$. Consider

$$\frac{\partial F}{\partial b_m} = \frac{1}{2\pi}\int_0^{2\pi} 2\left(E_N(t) - E(t)\right)\sin(mt)\,\mathrm{d}t = 0\,.$$

Restated in terms of averages,

$$\mathcal{A}_{2\pi}\{\sin(mt)E_N(t)\} = \mathcal{A}_{2\pi}\{\sin(mt)E(t)\}\,. \tag{2.2.21}$$

Since

$$\mathcal{A}_{2\pi}\{\sin(mt)E_N(t)\} = \mathcal{A}_{2\pi}\{a_0\sin(mt)\} + \sum_{n=1}^{N}\mathcal{A}_{2\pi}\{a_n\sin(mt)\cos(nt)\}$$

$$+ \sum_{n=1}^{N}\mathcal{A}_{2\pi}\{b_n\sin(mt)\sin(nt)\}, \quad (2.2.22)$$

we must consider the averages of $\sin(mt)\cos(nt)$ and $\sin(mt)\sin(nt)$. We have already shown that the first term on the right hand side of Eq. (2.2.22) and the sum of the averages of $\sin(mt)\cos(nt)$ are zero. We need consider only the average of $\sin(mt)\sin(nt)$.

We will make use of Eq. (D.2.6),

$$2\sin(mt)\sin(nt) = \cos[(n-m)t] - \cos[(n+m)t],$$

so that

$$2\mathcal{A}_{2\pi}\{\sin(mt)\sin(nt)\} = \mathcal{A}_{2\pi}\{\cos[(n-m)t]\} - \mathcal{A}_{2\pi}\{\cos[(n+m)t]\},$$

and because of Eq. (2.2.4), all of the averages are zero, except for $n = m$. In that case, $\cos[(n-m)t] = 1$, and its average is just one – see Eq. (2.2.3). Thus,

$$\mathcal{A}_{2\pi}\{\sin(mt)\sin(nt)\} = \begin{cases} 0 & \text{for } n \neq m, \\ \dfrac{1}{2} & \text{for } n = m. \end{cases} \quad (2.2.23)$$

As a consequence of Eq. (2.2.19) and Eq. (2.2.23), Eq. (2.2.2) reduces to

$$\frac{1}{2\pi}\int_0^{2\pi}\sin(mt)\,E_N(t)\,dt = \frac{b_m}{2}.$$

By applying the matching principle Eq. (2.2.21), we end up with the way to determine $b_m$.

$$\frac{b_m}{2} = \frac{1}{2\pi}\int_0^{2\pi}\sin(mt)\,E(t)\,dt. \quad (2.2.24)$$

The calculations of the Fourier coefficients $a_m$ and $b_m$ depend on our ability to perform the integrals in Eq. (2.2.20) and Eq. (2.2.24). As well we know, integrals cannot always be performed in closed form, so it is not surprising that scientists and engineers often choose functions that contain our favorite functions, polynomials, exponentials and sines and cosines. By using piecewise functions with these simple functions, a rich variety of input functions can be created with the certainty that the integrals can be performed exactly.

Even if the input function is too complicated for the integrations to be performed, or perhaps the input function contains only discrete data, arising say from experimental measurements, there is wonderful news. The Fast Fourier Transform allows us to determine the Fourier coefficients numerically in a very speedy way. It has become one of the most used numerical algorithms in all science and engineering.

---

**Reflection**

Let us reflect a little on what we have achieved. We propose an approximation Eq. (2.2.1) to a periodic function, but this approximation requires the determination of many coefficients $a_n$ and $b_n$. The coefficients are subscripted appropriately so that we know which coefficient corresponds to which trigonometric function. We invoke a procedure to minimize the average over one periodic window of the square of the difference between the function and its approximation. The procedure, essentially a least squares fit, leads to matching conditions, Eq. (2.2.6), Eq. (2.2.16) and Eq. (2.2.21), that determine the Fourier coefficients, Eq. (2.2.20) and Eq. (2.2.24). The matching conditions determine each coefficient $a_m$ and $b_m$ separately, but the formulas allow us to calculate these coefficients simultaneously by considering $m$ as a parameter. Once the dependency of the coefficients on $m$ are known, we can replace the coefficients $a_n$ and $b_n$ by their known pattern. The procedure is illustrated in the next subsection.

---

### 2.2.2 *Calculation of the Fourier coefficients*

Let us calculate the Fourier coefficients of $E(t)$ given by Eq. (2.2.7). We have already calculated $a_0 = 1$ from Eq. (2.2.6). Next, from Eq. (2.2.20),

$$a_m = \frac{1}{\pi} \int_0^{2\pi} \cos(mt)\, E(t)\, \mathrm{d}t$$

$$= \frac{2}{\pi^2} \int_0^{\pi} t\, \cos(mt)\, \mathrm{d}t + \frac{2}{\pi^2} \int_\pi^{2\pi} (2\pi - t)\, \cos(mt)\, \mathrm{d}t\,.$$

Integration by parts establishes the results,

$$\int t\, \cos(mt)\, \mathrm{d}t = \frac{t}{m}\, \sin(mt) + \frac{1}{m^2}\, \cos(mt)\,, \qquad (2.2.25)$$

$$\int (2\pi - t)\, \cos(mt)\, \mathrm{d}t = \frac{2\pi - t}{m}\, \sin(mt) - \frac{1}{m^2}\, \cos(mt)\,. \qquad (2.2.26)$$

These expressions must now be evaluated at the endpoints of the integration to give

$$a_m = \frac{2}{\pi^2} \left[ \frac{t}{m} \sin(mt) + \frac{1}{m^2} \cos(mt) \right]_0^\pi$$

$$+ \frac{2}{\pi^2} \left[ \frac{2\pi - t}{m} \sin(mt) - \frac{1}{m^2} \cos(mt) \right]_\pi^{2\pi}$$

$$= \frac{2}{\pi^2 m^2} \left( 2\cos(m\pi) - 1 - \cos(2m\pi) \right). \tag{2.2.27}$$

Note that $m = 1$ gives a result that agrees with Eq. (2.2.15).

The result Eq. (2.2.27) has a feature typical of integration that is split over intervals to calculate the Fourier coefficients, namely, the appearance of trigonometric functions evaluated at integer multiples of $\pi$ or some fraction of $\pi$. In Eq. (2.2.27), the terms are $\cos(m\pi)$ and $\cos(2m\pi)$. The simplest way to see the consequence is to use a table of entries, Table 2.2. First, the table reveals $\cos(2m\pi) = 1$. So $2\cos(m\pi) - 1 - \cos(2m\pi) = 2(\cos(m\pi) - 1)$. Then it is not too difficult to develop a special table, Table 2.3, for the needed expression $\cos(m\pi) - 1$. Incidentally, note that $\cos(m\pi) = (-1)^m$ for any integer $m$.

Table 2.2  Patterns in $\cos(\theta)$ and $\sin(\theta)$.

| $\theta$ | $\cos(\theta)$ | $\sin(\theta)$ |
|---|---|---|
| 0 | 1 | 0 |
| $\pi/2$ | 0 | 1 |
| $\pi$ | -1 | 0 |
| $3\pi/2$ | 0 | -1 |
| $2\pi$ | 1 | 0 |
| $5\pi/2$ | 0 | 1 |
| $3\pi$ | -1 | 0 |
| $7\pi/2$ | 0 | -1 |
| $4\pi$ | 1 | 0 |
| $\vdots$ | $\vdots$ | $\vdots$ |
| $(2m+1)\pi$ | $(-1)^m$ | 0 |
| $(2m+1)\pi/2$ | 0 | $(-1)^m$ |

Table 2.3  Patterns in $\cos(\theta) - 1$.

| $\theta$ | $\cos(\theta) - 1$ |
|---|---|
| 0 | 0 |
| $\pi$ | -2 |
| $2\pi$ | 0 |
| $3\pi$ | -2 |
| $4\pi$ | 0 |
| $\vdots$ | $\vdots$ |
| $2m\pi$ | 0 |
| $(2m+1)\pi$ | -2 |

The pattern is clear, and can be written as

$$a_m = -\frac{8}{m^2\pi^2} \begin{cases} 1 & \text{for } m \text{ odd}, \\ 0 & \text{for } m \text{ even}. \end{cases} \tag{2.2.28}$$

To calculate the Fourier coefficients $b_m$ for the sine terms, we use Eq. (2.2.24).

$$b_m = \frac{1}{\pi} \int_0^{2\pi} \sin(mt)\, E(t)\, dt$$

$$= \frac{2}{\pi^2} \int_0^\pi t \sin(mt)\, dt + \frac{2}{\pi^2} \int_\pi^{2\pi} (2\pi - t) \sin(mt)\, dt.$$

This time, integration by parts establishes the results,

$$\int t \sin(mt)\, dt = -\frac{t}{m}\cos(mt) + \frac{1}{m^2}\sin(mt)\,, \qquad (2.2.29)$$

$$\int (2\pi - t)\sin(mt)\, dt = -\frac{2\pi - t}{m}\cos(mt) - \frac{1}{m^2}\sin(mt)\,. \qquad (2.2.30)$$

These expressions must be evaluated at the endpoints of the integration.

$$b_m = \frac{2}{\pi^2}\left[-\frac{t}{m}\cos(mt) + \frac{1}{m^2}\sin(mt)\right]_0^\pi$$

$$+ \frac{2}{\pi^2}\left[-\frac{2\pi - t}{m}\cos(mt) - \frac{1}{m^2}\sin(mt)\right]_\pi^{2\pi}$$

$$= \frac{2}{m^2\pi^2}\left(2\sin(m\pi) - \sin(2m\pi)\right)\,. \qquad (2.2.31)$$

The entries in Table 2.2 lead to

$$b_m = 0\,. \qquad (2.2.32)$$

Incidentally note that $\sin(m\pi) = 0$ for any integer $m$.

Now that we have calculated the Fourier coefficients we can replace them in Eq. (2.2.20). The first few approximations $E_N(t)$ are easily written down;

$$E_1(t) = 1 - \frac{8}{\pi^2}\cos(t)\,,$$

$$E_3(t) = 1 - \frac{8}{\pi^2}\cos(t) - \frac{8}{9\pi^2}\cos(3t)\,,$$

$$E_5(t) = 1 - \frac{8}{\pi^2}\cos(t) - \frac{8}{9\pi^2}\cos(3t) - \frac{8}{25\pi^2}\cos(5t)\,.$$

There are several observations worth making about the results:

- Note the use of integration by parts in Eq. (2.2.25), Eq. (2.2.26), Eq. (2.2.29) and Eq. (2.2.30). The need for integration by parts is quite typical because there are not many functions $E(t)$ for which the integrals Eq. (2.2.20) and Eq. (2.2.24) can be performed exactly. Scientists and engineers tend to avoid this difficulty by choosing piecewise functions that contain simple functions so that the integrals can be performed, for example, low-order polynomials. If $E(t)$ contains trigonometric functions then Eq. (D.2.5), Eq. (D.2.6) or Eq. (D.2.7) will work.
- Why are $b_m = 0$? We will return to this question in the next section.
- The Fourier coefficients Eq. (2.2.28) and Eq. (2.2.32) are valid for all $m$, including $m > N$. Thus, it is easy to write down the pattern $E_N(t)$ for any $N$. Obviously, the pattern continues with only odd $N$. To ensure that the counter $n$ in Eq. (2.2.1) is only odd, we change the summation counter to $k$ where $n = 2k - 1$. Now as the counter $k$ increase from $k = 1$, the value of $n$ starts at $n = 1$ and increases as odd numbers only. Simply replace $n = 2k - 1$ in the sum leading to

$$E_N = 1 - \frac{8}{\pi^2}\sum_{k=1}^{(N+1)/2}\frac{1}{(2k-1)^2}\cos\big[(2k-1)t\big]$$

with $N$ odd.

We call $E_N(t)$ the partial sum, and we wonder whether the partial sums become better and better approximations to $E(t)$ as we increase $N$. First of all, does the sequence $E_N(t)$ even converge as $N \to \infty$? The only way the sum can converge is if the sum of the remaining terms neglected in $E_N(t)$ makes a smaller and smaller contribution as $N$ is chosen larger and larger. The sum of the remaining terms will be

$$R_N(t) = -\frac{8}{\pi^2} \sum_{n=(N+3)/2}^{\infty} \frac{1}{(2k-1)^2} \cos\big[(2k-1)t\big],$$

which will tend to zero as $N \to \infty$. In other words, $E_N(t)$ converges to some function and the wonderful thing about the Fourier series is that the sum converges to the function $E(t)$ under fairly reasonable conditions, one of the topics in the next subsection. Thus we have established the Fourier series for $E(t)$,

$$E(t) = 1 - \frac{8}{\pi^2} \sum_{k=1}^{\infty} \frac{1}{(2k-1)^2} \cos\big[(2k-1)t\big]. \qquad (2.2.33)$$

---

**Reflection**

We have found an entirely equivalent way to express the $2\pi$-periodic function $E(t)$ as given in Eq. (2.2.7). Obviously the form in Eq. (2.2.7) is easier to graph and understand. But the Fourier series gives us the opportunity to construct a particular solution if $E(t)$ is a forcing term in a differential equation. The disadvantage to the Fourier series representation of a function is that we cannot evaluate the Fourier series because it contains an infinite number of terms. Instead we must evaluate the partial sums and pick $N$ large enough to give the accuracy we require.

---

### 2.2.3 *Abstract view*

We have determined the Fourier series for a $2\pi$-periodic function. What happens if the function $f(t)$ is $T$-periodic? Well, its Fourier series must be $T$-periodic and must be written as

$$g(t) = a_0 + \sum_{k=1}^{\infty} \big[a_k \cos(k\omega t) + b_k \sin(k\omega t)\big] \qquad (2.2.34)$$

with $\omega = 2\pi/T$. The Fourier series is designated as $g(t)$ because we do not know at present whether it is the same as $f(t)$. Under the right conditions it will be, but that is what needs to be clarified.

**Principle 13**

*The periodicity of a Fourier series must be the same as the periodicity of the function.*

The Fourier coefficients are determined by the matching principles Eq. (2.2.6), Eq. (2.2.16) and Eq. (2.2.21), corrected for $T$-periodicity: The following averages of the Fourier series $g(t)$ must match the associated averages of the function $f(t)$.

$$\mathcal{A}_T\{g(t)\} = \mathcal{A}_T\{f(t)\},$$

$$\mathcal{A}_T\{\cos(k\omega t)g(t)\} = \mathcal{A}_T\{\cos(k\omega t)f(t)\},$$

$$\mathcal{A}_T\{\sin(k\omega t)g(t)\} = \mathcal{A}_T\{\sin(k\omega t)f(t)\},$$

for $k = 1, \ldots, \infty$. The first matching requirement can be thought of as a special case $k = 0$ of the second matching requirement.

**Principle 14**

*The averages over one period $T$ of $\cos(k\omega t)\, f(t)$ and $\sin(k\omega t)\, f(t)$ must match the averages over one period of $\cos(k\omega t)\, g(t)$ and $\sin(k\omega t)\, g(t)$, respectively. Here $\omega = 2\pi/T$.*

A direct consequence of these matching principles is the general formulas for the Fourier coefficients:

$$a_0 = \frac{1}{T}\int_0^T f(t)\,\mathrm{d}t, \qquad (2.2.35)$$

$$\frac{a_k}{2} = \frac{1}{T}\int_0^T f(t)\cos(k\omega t)\,\mathrm{d}t, \qquad (2.2.36)$$

$$\frac{b_k}{2} = \frac{1}{T}\int_0^T f(t)\sin(k\omega t). \qquad (2.2.37)$$

It seems that the Fourier coefficients are uniquely determined, but there is no guarantee that they exist! The integrals may not exist if $f(t)$ has certain singular behavior. To proceed, we must place some constraints on the function $f(t)$ so that the integrals for the Fourier coefficients, Eq. (2.2.35), Eq. (2.2.36) and Eq. (2.2.37), can be calculated.

## Principle 15

*The function $f(t)$ must be integrable for the Fourier coefficients to exist.*

Even if we can calculate the Fourier coefficients, there is no guarantee that the sum Eq. (2.2.34) exists (converges). The Fourier coefficients $a_k$ and $b_k$ must decay in magnitude quickly enough as $k \to \infty$ to be sure that the sum Eq. (2.2.34) exists.

### 2.2.3.1 *Convergence of the Fourier series*

The convergence of the sum Eq. (2.2.34) can be studied by asking whether the partial sums converge for any $t$. The partial sums are

$$g_N(t) = a_0 + \sum_{k=1}^{N} \left[ a_k \cos(k\omega t) + b_k \sin(k\omega t) \right]. \tag{2.2.38}$$

The sum of the remaining terms is

$$R_N(t) = \sum_{k=N+1}^{\infty} \left[ a_k \cos(k\omega t) + b_k \sin(k\omega t) \right].$$

The partial sums will converge $g_N(t) \to g(t)$ if we can show $R_N(t) \to 0$ as $N \to \infty$. Consider the magnitude of $R_N(t)$ and by using the fact that the magnitude of a sum is less than the sum of the magnitudes, we can show

$$|R_N(t)| \leq \sum_{k=N+1}^{\infty} \left[ |a_k| \left| \cos(k\omega t) \right| + |b_k| \left| \sin(k\omega t) \right| \right]$$

$$\leq \sum_{k=N+1}^{\infty} \left[ |a_k| + |b_k| \right]. \tag{2.2.39}$$

The last step is possible because $|\cos(\theta)| \leq 1$ and $\sin(\theta)| \leq 1$. Notice that the magnitude of the partial sums is bound by a sum that does not depend on time. That is good news. It means the question of whether the partial sums converge can be addressed without regard to the time.

Obviously, we require $|a_k|$ and $|b_k|$ to become smaller and smaller the larger $k$ becomes otherwise the sum will continue to grow in magnitude and not converge. Imagine an extreme case where $a_k$ and $b_k$ are constants. Then the sum in Eq. (2.2.39) becomes an infinite sum of constants; it clearly does not have a finite value. Also the sum in Eq. (2.2.38) will not converge. Just how rapidly must the coefficients decrease in magnitude so that we can be sure that the sum will converge? First we will find a way to estimate the sums in Eq. (2.2.39) so that their dependency on $N$ becomes clearer.

A very useful approach to find a bound for the sum in Eq. (2.2.39) is to interpret the sum as the sum of rectangular areas as illustrated in Fig. 2.8 for the coefficients

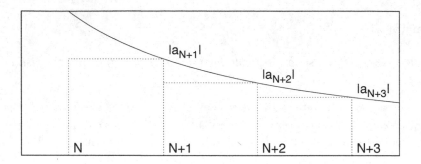

Fig. 2.8  An illustration of the sum as an area bounded by the area under a curve.

$|a_k|$. Consider the rectangle with a dashed boundary lying between $[N, N+1]$. Its area is $|a_{N+1}|$. The rectangle adjacent to it has area $|a_{N+2}|$, and so on. Thus the sum of the areas of the rectangles will be $|R_N(t)|$. Also apparent is that the rectangles lie below a curve $a(t)$ with a specific property, namely, that $a(N) = |a_N|$.[10] There are many curves that could have this property, but the simplest choice is the one where we replace $N$ in $|a_N|$ by $t$. In the end, we have

$$\sum_{N+1}^{\infty} |a_k| \le \int_N^{\infty} a(t)\,\mathrm{d}t\,, \qquad \sum_{N+1}^{\infty} |b_k| \le \int_N^{\infty} b(t)\,\mathrm{d}t\,. \qquad (2.2.40)$$

Obviously, the same approach has been followed for $|b_k|$.

Let us see how this technique works on our example Eq. (2.2.33). Here it is convenient to return to the expression for the Fourier coefficients $a_m$ given in Eq. (2.2.28), because we see easily that they satisfy the bound

$$|a_k| \le \frac{8}{k^2\pi^2} = \frac{A}{k^2}$$

where $A = 8/\pi^2$. As a result, (2.2.40) leads to

$$\sum_{N+1}^{\infty} |a_k| \le \int_N^{\infty} \frac{A}{t^2}\,\mathrm{d}t = \frac{A}{N}\,.$$

Since $b_k = 0$ in the example (2.2.32), we can conclude from (2.2.28) that

$$|R_N(t)| \le \frac{A}{N}$$

and clearly $|R_N(t)| \to 0$ as $N \to \infty$. The Fourier series (2.2.33) does converge.

In general, assume that the coefficients decrease in magnitude as

$$|a_k| \le \frac{A}{k^p} \quad \text{and} \quad |b_k| \le \frac{B}{k^q}\,, \qquad (2.2.41)$$

where $A$ and $B$ are positive constants. Suppose $p, q > 1$, $|R_N| \to 0$ as $N \to \infty$ as a simple consequence of the integral test (2.2.40) and the sum (2.2.34) converges.

---

[10]For this picture to be true, we must assume that $|a_k|$ decrease monotonically; there are no ups and downs. It is often possible to ensure a monotonic decrease by adjusting the estimates for $|a_k|$.

> **Principle 16**
>
> *If the magnitudes of the Fourier coefficients decrease more rapidly than the inverse of the wavenumber $k$, then the Fourier series converges.*

### 2.2.3.2 *Accuracy of the partial sums*

A consequence of our study of the convergence is that we have a bound Eq. (2.2.39) on the error committed by truncating the sum at $N$ terms. Once we find the bounds for $a_k$ and $b_k$ in Eq. (2.2.41), then we can apply the integral approximation for the sums in Eq. (2.2.39) to conclude that the magnitude of the error must be less than

$$|R_N(t)| \le \frac{A}{(p-1)N^{(p-1)}} + \frac{B}{(q-1)N^{(q-1)}} \, .$$

Clearly, the bigger $p$ and $q$ are, the quicker the error becomes small as $N \to \infty$. That means the less terms in the sum we need to keep for accuracy.

### 2.2.3.3 *Functions with finite jumps*

The case $p = 1$ or $q = 1$ is special. The bound on the partial sums Eq. (2.2.39), and the approximation by the integral test Eq. (2.2.40) fail to show convergence, but that does not mean the partial sums do not converge. In other words, the partial sums can converge without the integral approximations going to zero with $N \to \infty$. First, we will show how this special case arises typically in science and engineering.

Suppose that a $T$-periodic function is piecewise continuous. For simplicity, suppose in one window,

$$f(t) = \begin{cases} f_L(t) & \text{for } 0 < t < T_m \, , \\ f_R(t) & \text{for } T_m < t < T \, . \end{cases}$$

There is a jump in $f(t)$ at $t = T_m$, and the functions $f_L(t)$ and $f_R(t)$ are taken to be continuous with continuous derivatives. We have left undecided what value the function takes at $T_m$. In truth it does not usually matter. Then we may split the range of integration to calculate $a_k$.

$$a_k = \frac{2}{T} \int_0^{T_m} f_L(t) \cos(k\omega t) \, \mathrm{d}t + \frac{2}{T} \int_{T_m}^T f_R(t) \cos(k\omega t) \, \mathrm{d}t \, ,$$

and integration-by-parts may be applied to each integral,

$$\frac{2}{T} \int_0^{T_m} f_L(t) \cos(k\omega t) \, dt = \frac{1}{k\pi} \left[ f_L(t) \sin(k\omega t) \right]_0^{T_m}$$
$$- \frac{1}{k\pi} \int_0^{T_m} \frac{df_L}{dt}(t) \sin(k\omega t) \, dt \, ,$$

$$\frac{2}{T} \int_{T_m}^{T} f_R(t) \cos(k\omega t) \, dt = \frac{1}{k\pi} \left[ f_R(t) \sin(k\omega t) \right]_{T_m}^{T}$$
$$- \frac{1}{k\pi} \int_{T_m}^{T} \frac{df_R}{dt}(t) \sin(k\omega t) \, dt \, .$$

Suppose the derivatives of $f_L$ and $f_R$ are bounded in magnitude by some constant $B$, then the integrals with the derivatives can be bounded in magnitude by $C/(k^2)$ where $C$ is some constant. Further,

$$a_k = \frac{1}{k\pi} \left[ f_L(T_m) - f_R(T_m) \right] \sin(k\omega T_m) + \text{correction} \, , \qquad (2.2.42)$$

where the term giving the correction decreases as $1/k^2$. That means the first term on the right hand side is dominant and shows that $a_k$ will vary like $1/k$. We have written $f_L(T_m)$ and $f_R(T_m)$ but what we mean is the limiting values of these functions.

A similar result holds for $b_k$.

$$b_k = \frac{2}{T} \int_0^{T_m} f_L(t) \sin(k\omega t) \, dt + \frac{2}{T} \int_{T_m}^{T} f_R(t) \sin(k\omega t) \, dt \, .$$

By applying integration-by-parts,

$$\frac{2}{T} \int_0^{T_m} f_L(t) \sin(k\omega t) \, dt = \frac{1}{k\pi} \left[ -f_L(t) \cos(k\omega t) \right]_0^{T_m}$$
$$+ \frac{1}{k\pi} \int_0^{T_m} \frac{df_L}{dt}(t) \cos(k\omega t) \, dt \, ,$$

$$\frac{2}{T} \int_{T_m}^{T} f_R(t) \sin(k\omega t) \, dt = \frac{1}{k\pi} \left[ -f_R(t) \cos(k\omega t) \right]_{T_m}^{T}$$
$$+ \frac{1}{k\pi} \int_{T_m}^{T} \frac{df_R}{dt}(t) \cos(k\omega t) \, dt \, .$$

So we obtain

$$b_k = \frac{1}{k\pi} \left[ f_R(T_m) - f_L(T_m) \right] \cos(k\omega T_m) + \frac{f_L(0) - f_R(T)}{k\pi} + \text{correction} \, . \quad (2.2.43)$$

The term giving the correction will decrease as $1/k^2$.

Let us consider the consequence of the results Eq. (2.2.42) and Eq. (2.2.43). Clearly, the jump in values of $f_L(T_m)$ and $f_R(T_m)$ results in the decay in the Fourier coefficients to be $1/k$. If $f_R(T) \neq f_L(0)$, then there is an additional contribution also with a decay of $1/k$. Since $f(0) = f_L(0)$ and $f(T) = f_R(T)$, this mismatch of the values of $f(t)$ at the beginning and end of the periodic window, reveals another jump in the function. Just consider the periodic extension of the function and the jump should be clear. Of course if $f_R(T) = f_L(0)$, there is no jump at the boundaries of the periodic window and no additional contribution to the decay of the Fourier coefficients. In summary, wherever the jumps in the function occur they will cause a $1/k$ decay in the Fourier coefficients.

For this case, the Fourier series only just converges. It matches $f_L(t)$ for $0 < t < T_m$ and $f_R(t)$ for $T_m < t < T$. The difficulty occurs where $f(t)$ jumps in value. At $T_m$, the Fourier series is the average value of the limiting values of $f_L(T_m)$ and $f_R(T_m)$.

$$g(T_m) = \frac{1}{2} \lim_{t \to T_m^-} f_L(t) + \frac{1}{2} \lim_{t \to T_m^+} f_R(t). \qquad (2.2.44)$$

The important point is that $f(t)$ and $g(t)$ match everywhere except possibly at $T_m$; normally we do not care if there is a mismatch at a point.

Time to state the Fourier convergence theorem.

**Fourier Convergence Theorem**: Assume $f(t)$ is $T$-periodic. If $f(t)$ and its derivative are piecewise continuous in $0 < t < T$, then the Fourier series $g(t)$ converges and is $f(t)$ at all places where $f(t)$ is continuous. At any jump in $f(t)$, $g(t)$ equals the midpoint of the jump.

Note the important consequence of this theorem. If the function is continuous and its first derivative is continuous, then the Fourier series and the function are identical.

### 2.2.3.4 *Derivatives of Fourier series*

The last important property that must be established is whether the Fourier series can be differentiated and whether the result is the derivative of $f(t)$. We certainly want this to be true if we plan to substitute Fourier series into differential equations as a guess for a particular solution. We start by simply differentiating the partial sum Eq. (2.2.38); since the sum has a finite number of terms, we apply the fundamental property that the derivative of a sum is the sum of the derivatives. After we differentiate term-by-term, the result is simply

$$\frac{dg_N}{dt}(t) = \frac{2\pi}{T} \sum_{k=1}^{N} \left[ -k \, b_k \cos(k\omega t) + k \, a_k \sin(k\omega t) \right]. \qquad (2.2.45)$$

The sum (2.2.45) is a new Fourier series

$$\frac{dg_N}{dt}(t) = \sum_{k=1}^{N} \left[ A_k \cos(k\omega) + B_k \sin(k\omega t) \right]$$

whose coefficients are $A_k = -k \, b_k$ and $B_k = k \, a_k$. If the Fourier coefficients of $f(t)$ satisfy Eq. (2.2.41), then the magnitude of the Fourier coefficients for its derivative must decrease in magnitude as

$$|A_k| \leq \frac{A}{k^{(p-1)}} \quad \text{and} \quad |B_k| \leq \frac{B}{k^{(q-1)}}.$$

If $p, q \geq 2$, the Fourier series for $\mathrm{d}f(t)/\mathrm{d}t$ will converge and all is well.

As an example, consider the Fourier series Eq. (2.2.33) for $E(t)$ as given in Eq. (2.2.7). The derivative is

$$\frac{\mathrm{d}E}{\mathrm{d}t}(t) = \frac{8}{\pi^2} \sum_{n=0}^{\infty} \frac{1}{2n+1} \sin\left[(2n+1)\,t\right]. \qquad (2.2.46)$$

The magnitude of the coefficients $a_k$ behave like $C/k^2$, and $b_k = 0$. The coefficients of the derivative decrease in magnitude as $C/k$ which means the derivative is a piecewise function, and that is the case! The derivative of $E(t)$ is

$$\frac{\mathrm{d}E}{\mathrm{d}t}(t) = \frac{2}{\pi} \begin{cases} 1, & \text{for } 0 \leq t \leq \pi, \\ -1, & \text{for } \pi \leq t \leq 2\pi, \end{cases}$$

which clearly has a jump discontinuity at $t = \pi$ and at $t = 0, 2\pi$ because of the periodic extension.

---

**Reflection**

Each time we take the derivative of a Fourier series, the Fourier coefficients are multiplied by $k$ and interchanged with some possible sign changes. The important point is that the magnitudes are multiplied by $k$, and there is the possibility that higher enough derivatives may fail to have convergent Fourier series, and that suggests there are singularities in these derivatives.

---

This observation suggests that the behavior of the magnitude of the Fourier coefficients depends on the function and its derivatives. For example, Eq. (2.2.33) and Eq. (2.2.46) demonstrate that if a function is continuous and has a piecewise continuous derivative, then the coefficients $a_k$ and $b_k$ vary as $1/(k^2)$ and there is no difficulty in the convergence of the Fourier series. We can extend the result in the following principle.

**Principle 17**

*If a function and its first $m - 1$ derivatives are all continuous, but the $m$th derivative is only piecewise continuous, then the Fourier coefficients will behave as $1/k^{(m+1)}$ as $k$ becomes large.*

Here are a few specific examples:

$f(t)$ has a jump discontinuity. $\qquad\qquad\qquad\qquad |a_k|, |b_k| \sim 1/k$

$f(t)$ is continuous but its slope has a $\qquad\qquad |a_k|, |b_k| \sim 1/k^2$
jump discontinuity.

$f(t)$ and its derivative are continuous $\qquad |a_k|, |b_k| \sim 1/k^3$
but its second derivative has a jump dis-
continuity.

Why might Principle 17 be important? Suppose the forcing term in the differential equation

$$\frac{dy}{dt} + \alpha\, y = f(t)$$

is $T$-periodic and has the Fourier series

$$f(t) = a_0 + \sum_{n=1}^{\infty} \left[ a_n \cos(n\omega t) + b_n \sin(n\omega t) \right].$$

This is exactly the problem considered in Eq. (2.1.18) and the particular solution

$$y_\mathrm{p}(t) = A_0 + \sum_{n=1}^{\infty} \left[ A_n \cos(n\omega t) + B_n \sin(n\omega t) \right]$$

has Fourier coefficients – see Eq. (2.1.24) and Eq. (2.1.25),

$$A_n = \frac{\alpha\, a_n - n\omega\, b_n}{n^2\omega^2 + \alpha^2},$$

$$B_n = \frac{\alpha\, b_n + n\omega\, a_n}{n^2\omega^2 + \alpha^2}.$$

Note that for large $n$, $A_n \sim b_n/n$ and $B_n \sim a_n/n$. Suppose $f(t)$ is piecewise continuous, then $a_n \sim C/n$ and $b_n \sim C/n$ and $A_n \sim D/n^2$ and $B_n \sim D/n^2$. This means that $y_\mathrm{p}(t)$ has a piecewise continuous derivative. The result is consistent with Principle 2. The jump in the derivative of $y_\mathrm{p}(t)$ balances the jump in $f(t)$.

### Principle 18

*We can predict the nature of the solution to a differential equation by estimating the Fourier coefficients $A_n$ and $B_n$ to be those of the forcing terms $a_n$ and $b_n$ divided by $n^p$ where $p$ is the order of the highest derivative in the differential equation.*

### 2.2.4 *Exercises*

**Matching functions by balancing averages of certain quantities can be used to find approximations based on other forms, such as polynomials, a very popular idea.**

(1) Polynomial approximations to a function in a fixed interval can be constructed in several ways. For example, consider $\exp(x)$ in $0 \le x \le 1$. Information at $x = 0$ can be used to construct the Taylor series for $\exp(x)$. To illustrate, just use the terms up to quadratic,

$$\exp(x) \approx 1 + x + \frac{1}{2}x^2 .$$

Another approach parallels what is done in this section to find the Fourier coefficients. Start with the polynomial with unknown coefficients,

$$p(x) \approx a_0 + a_1 x + a_2 x^2 ,$$

and determine the coefficients by insisting that the following averages match:

$$\int_0^1 p(x)\,\mathrm{d}x = \int_0^1 \exp(x)\,\mathrm{d}x ,$$

$$\int_0^1 x\,p(x)\,\mathrm{d}x = \int_0^1 x\,\exp(x)\,\mathrm{d}x ,$$

$$\int_0^1 x^2\,p(x)\,\mathrm{d}x = \int_0^1 x^2\,\exp(x)\,\mathrm{d}x .$$

These conditions determine $a_0$, $a_1$ and $a_2$. Calculate the coefficients and plot the Taylor series approximation and the polynomial approximation. Which one do you prefer?
**Answer:** The approximation is $e^x \approx 1.013 + 0.851x + 0.839x^2$.

**The following exercises help develop a better understanding of Fourier series.**

(2) Consider three periodic functions of period 4,

$$f_1(t) = \begin{cases} t & \text{for } 0 < t < 1 , \\ (4 - t)/3 & \text{for } 1 < t < 4 , \end{cases}$$

$$f_2(t) = \begin{cases} t/2 & \text{for } 0 < t < 2 , \\ (4 - t)/2 & \text{for } 2 < t < 4 , \end{cases}$$

$$f_3(t) = \begin{cases} t/3 & \text{for } 0 < t < 3 , \\ 4 - t & \text{for } 3 < t < 4 . \end{cases}$$

(a) Plot these functions and notice the nature of the patterns.
(b) Calculate their Fourier series.

(c) How do the Fourier coefficients $a_k$ and $b_k$ depend on $k$ when $k$ is large? Is their behavior consistent with the nature of the functions?

(d) Do you see any interesting connections between the Fourier coefficients of the three functions?

**Answer**: The first few Fourier coefficients are given in Table 2.4.

Table 2.4    Fourier coefficients for the three cases.

| | $f_1(t)$ | | $f_2(t)$ | | $f_3(t)$ | |
|---|---|---|---|---|---|---|
| $n$ | $a_n$ | $b_n$ | $a_n$ | $b_n$ | $a_n$ | $b_n$ |
| 0 | 0.500 | | 0.500 | | 0.500 | |
| 1 | -0.270 | 0.270 | -0.405 | 0.000 | -0.270 | -0.270 |
| 2 | -0.135 | 0.000 | 0.000 | 0.000 | -0.135 | 0.000 |
| 3 | -0.030 | -0.030 | -0.045 | 0.000 | -0.030 | 0.030 |
| 4 | 0.000 | 0.000 | 0.000 | 0.000 | 0.000 | 0.000 |

(3) Consider the $\pi$-periodic functions, given in $0 \le y < \pi$ by

$$f_1(y) = \pi^2 \left| \sin(y) \right|,$$
$$f_2(y) = 4\,y\,(\pi - y),$$

and plot them for comparison. Calculate their Fourier series and compare their coefficients. How alike are they?

**Answer**: The first few Fourier coefficients are given in Table 2.5.

Table 2.5   Fourier coefficients for the two cases.

| | $f_1(y)$ | $f_2(y)$ |
|---|---|---|
| $a_0$ | 6.283 | 6.580 |
| $a_1$ | -4.189 | -4.000 |
| $a_2$ | -0.838 | -1.000 |
| $a_3$ | -0.359 | -0.444 |
| $a_4$ | -0.199 | -0.250 |

(4) The function $f(x) = \cos^2(x)$ is obviously $2\pi$-periodic, but it is also $\pi$-periodic. Construct the Fourier series in two cases, where the period is assumed to be $2\pi$ and $\pi$. Are the series different?

## 2.2.5   *Complex Fourier coefficients*

In Sec. 2.1.5, we show how the real Fourier series for a $T$-periodic function,

$$f(t) = a_0 + \sum_{n=1}^{\infty} \left( a_n \cos(n\omega t) + b_n \sin(n\omega t) \right)$$

can be expressed in complex form,

$$f(t) = \sum_{n=-\infty}^{\infty} c_n \, e^{in\omega t},$$  (2.2.47)

where $\omega = 2\pi/T$, and

$$c_n = \frac{a_n - i\,b_n}{2} \quad \text{and} \quad c_{-n} = \frac{a_n + i\,b_n}{2}.$$  (2.2.48)

We can determine the real coefficients $a_n$ and $b_n$ through the integrals Eq. (2.2.36) and Eq. (2.2.37) and then use the results to determine $c_n$ and $c_{-n}$, but we can also determine $c_n$ directly through the following integral.

$$c_n = \frac{1}{T} \int_0^T f(t)\,\cos(n\omega t)\,dt - \frac{i}{T} \int_0^T f(t)\,\sin(n\omega t)\,dt$$

$$= \frac{1}{T} \int_0^T f(t)\,\big(\cos(n\omega t) - i\,\sin(n\omega t)\big)\,dt$$

$$= \frac{1}{T} \int_0^T f(t)\,e^{-in\omega t}\,dt.$$  (2.2.49)

It is now easy to check that this formula Eq. (2.2.49) is also correct for $c_{-n}$ and $c_0$.

Notice that the complex coefficients $c_n$ are determined by the average over a periodic window without the presence of a factor one half. The factor one half in Eq. (2.2.48) is associated with the connection between the trigonometric functions, cosine and sine, with the exponential with a complex argument and may be viewed as another reason why the half appears in Eq. (2.2.36) and Eq. (2.2.37).

Expressed in complex form, the Fourier series Eq. (2.2.47) is compact and simple. The calculation of the Fourier coefficients is often easier with Eq. (2.2.49). For example, let us use Eq. (2.2.49) to calculate the complex Fourier coefficients of the $2\pi$-periodic function $E(t)$ given in Eq. (2.2.7),

$$E(t) = \frac{2}{\pi} \begin{cases} t, & \text{for } 0 \le t \le \pi, \\ 2\pi - t, & \text{for } \pi \le t \le 2\pi. \end{cases}$$

We must split the range of integration,

$$c_n = \frac{1}{\pi^2} \int_0^\pi t\,e^{int}\,dt + \frac{1}{\pi^2} \int_\pi^{2\pi} (2\pi - t)\,e^{int}\,dt.$$

Integration-by-parts is relatively easy to use.

$$\frac{1}{\pi^2} \int_0^\pi t\,e^{int}\,dt = \left[\frac{t}{in\pi^2}\,e^{int}\right]_0^\pi - \left[\frac{1}{\pi^2(in)^2}\,e^{int}\right]_0^\pi$$

$$= \frac{e^{in\pi}}{in\pi} + \frac{1}{n^2\pi^2}\left(e^{in\pi} - 1\right),$$

$$\frac{1}{\pi^2} \int_\pi^{2\pi} (2\pi - t)\,e^{int}\,dt = \left[\frac{(2\pi - t)}{in\pi^2}\,e^{int}\right]_\pi^{2\pi} + \left[\frac{1}{\pi^2(in)^2}\,e^{int}\right]_\pi^{2\pi}$$

$$= -\frac{e^{in\pi}}{in\pi} - \frac{1}{n^2\pi^2}\left(1 - e^{in\pi}\right).$$

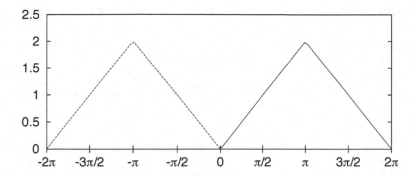

Fig. 2.9 Periodic extension of $E(t)$: the original function (solid line) and the even extension (dashed line).

These results lead to

$$c_n = -\frac{2}{n^2\pi^2}\left(1 - (-1)^n\right)$$

where we use $\exp(in\pi) = (-1)^n$. Note first that $c_n$ is in fact real; so $b_n = 0$ and

$$a_n = \frac{4}{n^2\pi^2}\left(1 - (-1)^n\right)$$

a result that agrees with Eq. (2.2.28) and Eq. (2.2.32).

## 2.3 Symmetry in the Fourier series

If we have a $T$-periodic function and wish to express it in a Fourier series, Principle 13 requires that the Fourier series must have the same periodicity – see Eq. (2.2.34). The Fourier coefficients are determined by the integrals Eq. (2.2.35), Eq. (2.2.36) and Eq. (2.2.37). These integrals require knowledge of $f(t)$ in the periodic window $[0, T]$ only. But sometimes it is helpful to consider $f(t)$ in adjacent regions. Then certain properties of the pattern such as particular symmetries become obvious, and we find in this section that symmetries affect the pattern in the Fourier coefficients. For example, when we compute the Fourier coefficients of $E(t)$ as given in Eq. (2.2.7), repeated here for convenience,

$$E(t) = \frac{2}{\pi}\begin{cases} t, & \text{for } 0 \leq t \leq \pi, \\ 2\pi - t, & \text{for } \pi \leq t \leq 2\pi, \end{cases} \tag{2.3.1}$$

we find that $b_k = 0$ – see Eq. (2.2.39). Why?

### 2.3.1 *Even and odd functions*

Let us look at the function $E(t)$ over two periods, in particular, the periodic window on the left which covers $-T < x < 0$. It is displayed in Fig. 2.9. The pattern exhibits certain symmetries. It is symmetric about its peaks and troughs. In particular, if we fold the graph along the line $t = 0$, the parts will align perfectly. Alternatively, we

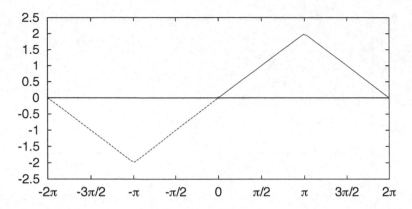

Fig. 2.10   Odd extension of $E(t)$: the original function (solid line) and the odd extension (dashed line).

can rotate the graph around the line $t = 0$, and the result is the exact same function. We call this symmetry *even*, and we can identify this property mathematically by the statement,

$$f(-t) = f(t).$$

Notice what it says: if I go the same distance to the left or right, the value of the function will be the same.

The classic example of an even function is $x^n$ with $n$ an even integer. But the even function we are interested in is a cosine; it satisfies the property $\cos(-\theta) = \cos(\theta)$ as required. Further, a sum of cosines will be even since each one is even and a constant is obviously even. If we consider, then, just the cosine part of the Fourier series,

$$f_{\mathrm{e}}(t) = a_0 + \sum_{k=1}^{\infty} a_k \, \cos(\omega k t),$$

we will have an even function since $f_{\mathrm{e}}(-t) = f_{\mathrm{e}}(t)$.

The other symmetry is motivated by the nature of $x^n$ with $n$ an odd integer. Since $(-x)^n = -(x)^n$ we define a function as *odd* when

$$f(-t) = -f(t).$$

To see what this property means, let us make $E(t)$ odd, and show the result in Fig. 2.10. Visually it appears as though we folded the function over the line $t = 0$ and then folded it over the $t$-axis. Alternatively, we could consider the result as a rotation through $\pi$ about the origin. In either case, an odd function returns to the same function after the two folds or the rotation. Of course, the new function is no longer $2\pi$-periodic, although we can now imagine it to be $4\pi$-periodic.

The odd function we are interested in is the sine function; it has the property $\sin(-\theta) = -\sin(\theta)$ as required. If we consider just the sine part of the Fourier series,

$$f_{\rm o}(t) = \sum_{k=1}^{\infty} b_k \sin(\omega k t)\,,$$

we will have an odd function since $f_{\rm o}(-t) = -f_{\rm o}(t)$.

Curiously then, we can write the Fourier series as the sum of an even part and an odd part. Indeed, we may write any periodic function as the sum of an even part and an odd part,[11]

$$f(t) = \frac{1}{2}\left[f(t) + f(-t)\right] + \frac{1}{2}\left[f(t) - f(-t)\right],$$

and then notice that

$$f_{\rm e}(t) = \frac{1}{2}\left[f(t) + f(-t)\right], \tag{2.3.2}$$

$$f_{\rm o}(t) = \frac{1}{2}\left[f(t) - f(-t)\right], \tag{2.3.3}$$

are even and odd, respectively.

The implications of these symmetries is that if a periodic function is even, then it has just a cosine part to its Fourier series. If the periodic function is odd, then it just has a sine series part to its Fourier series. The consequence then is that $b_k = 0$ for $E(t)$ in Eq. (2.3.1) because it is even. We verify these implications in the next section.

### 2.3.2 *Fourier coefficients for even and odd periodic functions*

When a $T$-periodic function is even or odd, we can take advantage of its properties to simplify the calculation of its Fourier coefficients. First of all, we should shift the range of integration from $[0, T]$ to $[-T/2, T/2]$ because in this range we can exploit the symmetry. Obviously, the integration to calculate the Fourier coefficients must always be over a periodic window, but the result should not depend on where we start the integration. To illustrate this idea, consider performing integration on $E(t)$ as illustrated in Fig. 2.9. The standard range of integration is over one period $[0, 2\pi]$, but note that the integration over $[\pi, 2\pi]$ is exactly the same as $[-\pi, 0]$. So integration over $[-\pi, \pi]$ gives the same result as integration over $[0, 2\pi]$.

Confirmation that we can shift the integration from $[0, T]$ to $[-T/2, T/2]$ is provided by an appropriate change of integration variable. First split the integral,

$$\int_0^T h(t)\,{\rm d}t = \int_0^{T/2} h(t)\,{\rm d}t + \int_{T/2}^T h(t){\rm d}t\,.$$

---

[11]Not every function can be written as the sum of an even and odd part, for example, $f(t) = \sqrt{t}$. The function must be defined for all $x$, which is the case if it is periodic.

In the second integral, introduce the change of variable $p = t - T$.

$$\int_{T/2}^{T} h(t)\, \mathrm{d}t = \int_{-T/2}^{0} h(p+T)\, \mathrm{d}p$$

$$= \int_{-T/2}^{0} h(p)\, \mathrm{d}p.$$

Note the use of the definition of a $T$-periodic function to replace $h(p+T)$ by $h(p)$. By adding the result to the first integral and replacing the dummy integration variable $p$ by $t$, we have

$$\int_{0}^{T} h(t)\, \mathrm{d}t = \int_{-T/2}^{T/2} h(t)\, \mathrm{d}t. \qquad (2.3.4)$$

The new range of integration is very useful when we want the Fourier coefficients of even or odd periodic functions. To exploit the symmetry, we should split the integrals again,

$$\int_{-T/2}^{T/2} h(t)\, \mathrm{d}t = \int_{-T/2}^{0} h(t)\, \mathrm{d}t + \int_{0}^{T/2} h(t)\mathrm{d}t.$$

In the first integral set $p = -t$, then

$$\int_{-T/2}^{0} h(t)\, \mathrm{d}t = \int_{0}^{T/2} h(-p)\mathrm{d}p.$$

Combine this result with the second integral to obtain

$$\int_{-T/2}^{T/2} h(t)\, \mathrm{d}t = \int_{0}^{T/2} \big[h(-t) + h(t)\big]\, \mathrm{d}t. \qquad (2.3.5)$$

Now it is easy to see that if $h(t)$ is even, then the integration over $[-T/2, T/2]$ is twice the integration over $[0, T/2]$. On the other hand, if $h(t)$ is odd then the integral is zero. We are ready to exploit the symmetry in the integral for the Fourier coefficients.

### Even functions

Suppose the $T$-periodic function $f(t)$ is even, $f(-t) = f(t)$. The integrals for the sine coefficients have $h(t) = f(t)\sin(\omega k t)$ as the integrand, which is odd because $h(-t) = f(-t)\sin(-\omega k t) = -f(t)\sin(\omega k t) = -h(t)$. As a consequence, the integrand in Eq. (2.3.5) is $h(-t) + h(t) = 0$ and $b_k = 0$. We have confirmed that the sine coefficients of an even function must be zero!

The integrands of the integrals for the cosine coefficients are $h(t) = f(t)\cos(\omega k t)$ and this time the integrands are even; $h(-t) = f(-t)\cos(-\omega k t) = f(t)\cos(\omega k t) =$

$h(t)$. As a consequence, the integrand in Eq. (2.3.5) is $h(-t) + h(t) = 2h(t)$. In other words,

$$\frac{a_k}{2} = \frac{1}{T} \int_{-T/2}^{T/2} f(t) \cos(\omega kt) \, dt$$

$$= \frac{2}{T} \int_0^{T/2} f(t) \cos(\omega kt) \, dt. \tag{2.3.6}$$

Obviously,

$$a_0 = \frac{2}{T} \int_0^{T/2} f(t) \, dt. \tag{2.3.7}$$

### 2.3.2.1 *Odd functions*

If the $T$-periodic function is odd, then $f(-t) = -f(t)$ and the integrands of the integrals for the cosine coefficients are $h(t) = f(t) \cos(\omega kt)$ and it is easy to verify that they are odd. Thus $a_k = 0$ (including $a_0$). On the other hand, the integrands of the integrals for the sine coefficients are $h(t) = f(t) \sin(\omega kt)$ which are even. As a consequence,

$$\frac{b_k}{2} = \frac{2}{T} \int_0^{T/2} f(t) \sin(\omega kt) \, dt. \tag{2.3.8}$$

**Reflection**

There is a simple way to remember how to calculate the Fourier series. First, make sure that the periodicity matches the periodicity of the function. Then, take the average over the full periodic window if the full Fourier series is needed. For $a_k/2$, take the average of $f(t) \cos(2\pi t/T)$ and for $b_k/2$, take the average of $f(t) \sin(2\pi t/T)$. If the function is odd or even, then the averages may be taken over half the periodic window.

### 2.3.2.2 *An example*

As an example, look how easy it is to calculate the Fourier coefficients of $E(t)$ given in Eq. (2.3.1). From Eq. (2.3.7), using $T = 2\pi$,

$$a_0 = \frac{1}{\pi} \int_0^\pi \frac{2}{\pi} t \, dt = 1,$$

and from Eq. (2.3.6), making use of the result Eq. (2.2.25) from integration-by-parts,

$$a_k = \frac{2}{\pi} \int_0^\pi \frac{2}{\pi} t \cos(kt) \, dt$$

$$= \frac{4}{\pi^2} \left[ \frac{t}{k} \sin(kt) + \frac{1}{k^2} \cos(kt) \right]_0^\pi$$

$$= \frac{4}{\pi^2 k^2} \left( \cos(k\pi) - 1 \right).$$

These results agree with what we obtained before in (2.2.27). The value of exploiting the symmetry of the function is that only half the coefficients need to be determined and the range of integration no longer requires split integrals to be performed as done in Sec. 2.2.2. Much less work!

### 2.3.3 Abstract view

The first step in connecting the symmetries in the graph of $f(t)$ to patterns in the Fourier coefficients is to regard its periodic extensions so that the symmetries can be clearly identified. In particular, it is important to consider the nature of the function over the range $[-T, T]$. Then the question of whether the function is odd or even becomes simpler. If the function is odd or even, then one set of Fourier coefficients must be zero, and the calculation of the other set is simplified.

#### 2.3.3.1 Even functions

An even function $f(x)$ satisfies the property

$$f(-x) = f(x).$$

If the function is also $T = 2L$-periodic, then it has a cosine series,

$$f(x) = a_0 + \sum_{k=1}^{\infty} a_k \cos(k\omega x), \qquad (2.3.9)$$

where $\omega = 2\pi/T = \pi/L$ and the coefficients may be calculated from

$$a_0 = \frac{1}{L} \int_0^L f(x)\,dx, \qquad (2.3.10)$$

$$\frac{a_k}{2} = \frac{1}{L} \int_0^L f(x)\cos(k\omega x)dx. \qquad (2.3.11)$$

#### 2.3.3.2 Odd functions

An odd function $f(x)$ satisfies the property

$$f(-x) = -f(x).$$

If the function is also $T = 2L$-periodic, then it has a sine series,

$$f(x) = \sum_{k=1}^{\infty} b_k \sin(k\omega x), \qquad (2.3.12)$$

where $\omega = \pi/L$ and the coefficients may be calculated by

$$\frac{b_k}{2} = \frac{1}{L} \int_0^L f(x)\sin(k\omega x)dx. \qquad (2.3.13)$$

**Principle 19**

*All periodic functions have an even and/or odd part. Even functions can be represented by a cosine series; odd functions can be represented by a sine series.*

## 2.3.4 Exercises

**Some exercises on basic properties of Fourier series.**

(1) Confirm that the function $f_e(t)$ defined in Eq. (2.3.7) is even, and that the function $f_o(t)$ defined in Eq. (2.3.8) is odd.

(2) Consider the two functions defined only in $0 < x < L$,

    (a) $f_1(x) = x$,
    (b) $f_2(x) = x^2$,

and extend their definition to the range $-L < x < L$ by demanding that they are i) even, ii) odd.

(3) Consider the function

$$g(z) = \frac{4}{\pi^2} z \, (\pi - z), \quad \text{in } 0 < z < \pi,$$

and insist that it is an odd function with period $2\pi$. Determine its sine series.

    (a) How close is this function to $\sin(z)$?
    (b) Explain the decay of the magnitude of the Fourier coefficients.

**Answer:** $b_1 = 1.032$, $b_2 = 0$, $b_3 = 0.038$, $b_4 = 0$, $b_5 = 0.008$.

(4) This exercise draws attention to the impact of the smoothness of a function on the Fourier coefficients. First determine the Fourier coefficients of the 3-periodic function,

$$f_1(t) = \begin{cases} 1, & \text{for } 1 < t < 2, \\ 0, & \text{for } 0 < t < 1 \text{ and } 2 < t < 3. \end{cases}$$

This function represents an abrupt switch on and off.
Now replace the function with

$$f_2(t) = \begin{cases} 1, & \text{for } 1 < t < 2, \\ t, & \text{for } 0 < t < 1, \\ 3 - t, & \text{for } 2 < t < 3. \end{cases}$$

This function ramps up and down so that $f_2(t)$ is continuous (its derivative is not).

The final choice is to ramp up and down so that the derivative is continuous,

$$f_3(t) = \begin{cases} 1\,, & \text{for } 1 < t < 2\,, \\ \frac{1}{2}\left[1 - \cos(\pi t)\right], & \text{for } 0 < t < 1\,, \\ \frac{1}{2}\left[(1 - \cos(\pi(t-3)))\right], & \text{for } 2 < t < 3\,. \end{cases}$$

Compare the behavior of the magnitude of the Fourier coefficients $a_k$ in each case for large values of $k$. Now assess the convergence of the Fourier series by graphing the truncated Fourier series when stopping at $N = 2$ and $N = 4$. Compare the two results with the exact solution.

**Answer**: For $f_1(x)$, the amplitudes decay in magnitude as $1/k$. The convergence is slow as shown in Fig. 2.11:

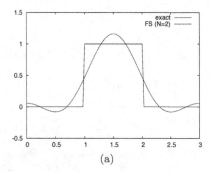 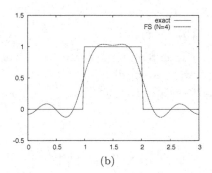

(a)                              (b)

Fig. 2.11   Comparison of the truncated Fourier series (FS) of $f_1(t)$ and its exact definition: (a) $N = 2$. (b) $N = 4$.

For $f_2(x)$, the amplitudes decay in magnitude as $1/k^2$. The convergence is moderate as shown in Fig. 2.12:

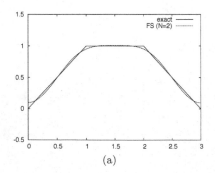 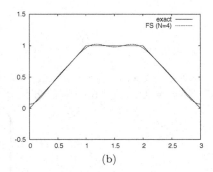

(a)                              (b)

Fig. 2.12   Comparison of the truncated Fourier series (FS) of $f_2(t)$ and its exact definition: (a) $N = 2$. (b) $N = 4$.

For $f_3(x)$, the amplitudes decay in magnitude as $1/k^3$. The convergence is fast as shown in Fig. 2.13:

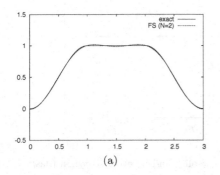

(a)

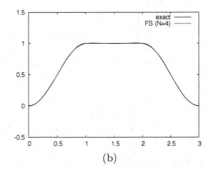

(b)

Fig. 2.13   Comparison of the truncated Fourier series (FS) of $f_3(t)$ and its exact definition: (a) $N = 2$. (b) $N = 4$.

**Other Symmetries**

(5) Consider

$$\cos\left(\frac{\pi}{2}t\right), \quad \cos\left(\frac{3\pi}{2}t\right)$$

and observe their symmetries. Then imagine the symmetries for the function defined by the Fourier series,

$$\sum_{n=1}^{\infty} a_n \cos\left(\frac{(2n+1)\pi}{2}t\right) ?$$

**Hint**: You may find this quite challenging. Read the next subsection and see if you can return to this exercise and complete it.

### 2.3.5   *Other symmetries*

If we reconsider the graphs of $E(t)$ displayed in Fig. 2.9, we note that the function is even and so the Fourier series has only cosines, but why are there only cosines with $n$ odd? See Eq. (2.2.33). Perhaps it has something to do with the symmetries around the peaks, for example, at $t = \pi$? So let us examine the symmetry at $t = \pi$ in more detail. First, let us state the symmetry in words: the function value at a distance $p$ just before $t = \pi$ is the same as the function value at a distance $p$ just beyond $t = \pi$. Mathematically,

$$E(\pi - p) = E(\pi + p). \tag{2.3.14}$$

Now let us check whether the general cosine series – we know the function is even – satisfies this statement. A general cosine term is $\cos(nt)$ – the function is $2\pi$-periodic – so we check whether $\cos(nt)$ satisfies Eq. (2.3.14).

$$\cos\big(n\,(\pi - p)\big) = (-1)^n \cos(p) \quad \text{and} \quad \cos\big(n\,(\pi + p)\big) = (-1)^n \cos(p)$$

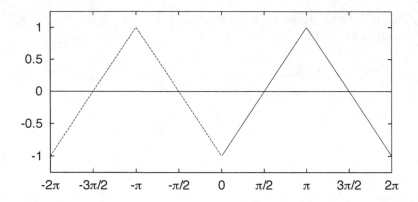

Fig. 2.14   Periodic extension of $F(t)$: original function (solid) and its even extension (dashed).

where we have used the addition angle formulas Eq. (D.2.1) and Eq. (D.2.3). The property Eq. (2.3.14) is satisfied for any integer $n$. Hmmm. In hindsight, this result should have been obvious because the profile in $[-\pi, 0]$ is the reflection of the profile in $[0, \pi]$ and when shifted to the right by $2\pi$ must be the reflection around $t = \pi$ of the profile in $[0, \pi]$. In other words, Eq. (2.3.14) must be true.

Let us look more closely at the Fourier series Eq. (2.2.33) and notice that the constant term $a_0 = 1$ does not fit the pattern. If there are only odd terms in the sum, why should there be a constant term? Perhaps we should look at the function $F(t) = E(t) - 1$; it is displayed in Fig. 2.14. Now another symmetry appears. In words, it appears as though the function is odd about the point $t = \pi/2$. Mathematically,

$$F(\pi/2 - p) = -F(\pi/2 + p). \tag{2.3.15}$$

The function $F(t)$ is even; it has only a cosine series. Consider a general term $\cos(nt)$ and check whether it satisfies Eq. (2.3.15). By using Eq. (D.2.1) and Eq. (D.2.3),

$$\cos\left[n\left(\frac{\pi}{2} - p\right)\right] = \cos\left(\frac{n\pi}{2}\right)\cos(np) + \sin\left(\frac{n\pi}{2}\right)\sin(np),$$

$$\cos\left[n\left(\frac{\pi}{2} + p\right)\right] = \cos\left(\frac{n\pi}{2}\right)\cos(np) - \sin\left(\frac{n\pi}{2}\right)\sin(np).$$

If $n = 2m$ is even,

$$\cos\left[2m\left(\frac{\pi}{2} - p\right)\right] = (-1)^m \cos(2mp),$$

$$\cos\left[2m\left(\frac{\pi}{2} + p\right)\right] = (-1)^m \cos(2mp).$$

The results do not satisfy Eq. (2.3.14). Next, consider $n$ odd and set $n = 2m + 1$.

$$\cos\left[(2m+1)\left(\frac{\pi}{2} - p\right)\right] = (-1)^m \sin\left[(2m+1)p\right],$$

$$\cos\left[(2m+1)\left(\frac{\pi}{2} + p\right)\right] = -(-1)^m \sin\left[(2m+1)p\right],$$

and this time Eq. (2.3.15) is satisfied. As a consequence, $F(t)$ has a Fourier cosine series with odd terms only.

---

**Reflection**

It is often worthwhile shifting a function up or down to seek symmetries. All that is affected is $a_0$. Sometimes the modified function may be even or odd and may exhibit additional symmetries internal to the periodic window. These symmetries affect the pattern of the Fourier coefficients.

# Boundary Value Problems

3

## 3.1 Spatially varying steady states

In the previous two chapters, our attention has been placed on time-varying behavior as described by differential equations. Consequently, the derivatives in the differential equation have been with respect to time, and in order to specify the solutions completely initial conditions must also be supplied. Generally, the initial conditions are assumed to occur at $t = 0$. In other words, the clock is started when the initial conditions take effect. Such problems are called *Initial Value Problems* (IVPs).

There are many situations where quantities vary not only in time but also spatially. By assuming quantities are uniform in space, they change only in time. Recall the example in Sec. 1.1.3 where water was poured into a drum to clear out some contamination. An important assumption was that the water was well stirred so that the concentration of the contaminant could be considered uniform. In the example presented in Sec. 1.2.1, the contaminant in a river flows into a dam. Again the assumption is made that the concentration in the dam is uniform. Obviously, that is not realistic. The contaminant will flow in with the river at one end of the dam and then diffuse slowly throughout the dam resulting in a spatial variation in the concentration.

The mathematical models we now seek must account for this spatial variation, and we can expect that spatial derivatives will arise. In general, then, both temporal and spatial derivatives can arise, but in this chapter we will focus on where the situation has settled into a steady state, and no longer depends on time. The good news is that the construction of particular solutions and homogeneous solutions will proceed as before to obtain the general solution. The difference with IVPs is that the additional information needed to determine a specific solution will be given at the ends of the spatial region, for example, where the river enters the dam and where the water leaves it. Thus these problems are called *Boundary Value Problems* (BVPs) because the additional information is provided at the boundaries of the region of interest.

Boundary value problems introduce new complexities not usually present in IVPs. We will find situations where BVPs do not have solutions or may even have many solutions. There is a strong connection to systems of linear equations where there may be unique solutions, no solutions or many solutions, and the possibilities are discussed in Appendix C. Simple examples will be used to show how and when these difficulties arise and how best to treat them. As usual, we will explore these questions through specific examples.

### 3.1.1 *Steady state transport*

Let us revisit the situation that was considered in Sec. 1.2.1 and illustrated in Fig. 1.6 where a river brings pollution into a dam. This time, imagine that a purification plant is in operation at the dam wall, removing pollution from the water before it leaves the dam. The purification plant is so efficient that we may consider the concentration of the pollution to be zero at the dam wall. Clearly, the concentration of the pollutant can no longer be a constant throughout the dam. Instead it must taper from the river entrance down to zero at the dam wall, and we expect the concentration varies continuously in distance along the dam. Since the water in the dam is essentially stationary, how does the pollution move from the river entrance to the dam wall. It is by the process of diffusion in which the pollution moves from regions of high concentration to regions of low concentration.[1]

How can the movement of pollutant through the water be measured? We record the amount that passes through a unit area in a unit time and call this measurement the flux $q$; its units will be quantity per area per second, for example, pounds per square feet per second. If we pick a spot and measure the flux, we might notice that it changes during the course of time. In other words, the flux can be a function of time. If we then change locations, the flux can also be different. The flux can also depend on the location of the measurement. A simple illustration is traffic flow on a highway. Pick a spot and count the number of cars that pass that spot in a certain period of time. The number of cars depend on the length of time the count is made. So a standard representation is to take the count divided by the time interval, for example, the number of cars passing the spot per second. Suppose the highway has three lanes. The count is likely to be about three times as large. So divide by the number of lanes, in effect the area through which the cars pass. The result is the number of cars per unit time per unit lane (area) which is what we call flux. The flux depends on where the measurement is taken, $x$ say, and the time the measurement is taken, $t$ say. In other words, the flux $q(x, t)$ is a function of the independent variables $x$ and $t$.

For the moment, we are interested in the steady state profile of the concentration in the dam, established presumably after some transient adjustment. An

---

[1] In the absence of the purification plant, diffusion will eventually result in a uniform distribution as assumed before.

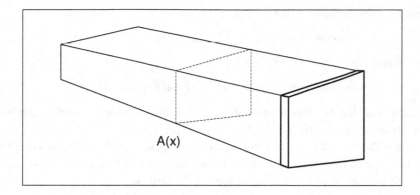

Fig. 3.1 A schematic of the dam showing a cross section.

idealization of the dam is illustrated in Fig. 3.1. Let $x$ measure the distance along the dam, starting at the river entrance and ending at the dam wall. Let us assume that the dam is long and narrow so that concentration is mostly uniform in the cross section of the dam with area $A(x)$, but that it may vary along the dam as $\rho(x)$. We expect the flux too to be a function only of $x$.

While we have drawn the dam as rectangular in shape, this assumption is not necessary. All we require is that the concentration is uniform over the cross section; the cross section can have any shape (within reason). We still need a principle that guides how the spatial variation of concentration $\rho(x)$ is determined. That principle is nothing more than an application of the principle of conservation.

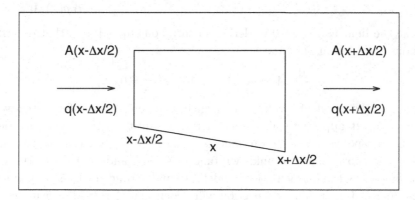

Fig. 3.2 A small segment of the interior of the dam, centered at $x$ and viewed side on.

To set up the application of the principle of conservation, let us consider a small section of the dam located at $x$ with width $\triangle x$ as illustrated in Fig. 3.2. There is a flux of quantity entering from the left and leaving from the right, but for the quantity to be unchanging in time we must have

$$\text{amount entering} = \text{amount leaving}. \tag{3.1.1}$$

Obviously by the definition of the flux,

$$\text{amount entering} = q(x - \triangle x/2)\, A(x - \triangle x/2)\, \triangle t \qquad (3.1.2)$$

during a small time interval $\triangle t$, and

$$\text{amount leaving} = q(x + \triangle x/2)\, A(x + \triangle x/2)\, \triangle t. \qquad (3.1.3)$$

Here we assume that the flux is measured in kilograms per square meters per second and the cross-sectional area of the dam is $A(x)\,(\text{m}^2)$.

To make the model more interesting, let us suppose that pollution can deposit on the bottom of the dam at a rate proportional to the amount in the volume $A(x)\,\triangle x$. Thus the amount deposited during $\triangle t$ will be

$$\text{amount deposited} = \alpha\,\rho(x)\, A(x)\, \triangle x\, \triangle t. \qquad (3.1.4)$$

The parameter $\alpha$ is the constant of proportionality and has units per second.

Of course we must include the amount deposited as part of the amount leaving. Finally, the contributions Eq. (3.1.2), Eq. (3.1.3) and Eq. (3.1.4) to Eq. (3.1.1) lead to

$$q(x - \triangle x/2)\, A(x - \triangle x/2)\, \triangle t =$$
$$q(x + \triangle x/2)\, A(x + \triangle x/2)\, \triangle t + \alpha\,\rho(x)\, A(x)\, \triangle x\, \triangle t.$$

Rearrange the terms, and note that $\triangle t$ is common in all terms and can be dropped;

$$\frac{q(x + \triangle x/2)\, A(x + \triangle x/2) - q(x - \triangle x/2)\, A(x - \triangle x/2)}{\triangle x}$$
$$+ \alpha\,\rho(x)\, A(x) = 0.$$

Now take the limit as $\triangle x \to 0$ to derive a continuous model, one that describes a continuous variation in $\rho(x)$ and $q(x)$.

$$\frac{\mathrm{d}}{\mathrm{d}x}\big(A(x)\, q(x)\big) + \alpha\, A(x)\, \rho(x) = 0. \qquad (3.1.5)$$

This, then, is the result of invoking the principle of conservation for a steady state configuration. It applies to all cases where a quantity with a concentration or density is transported by some flux and so is incomplete in the sense that it is a single equation governing two unknown functions, $\rho(x)$ and $q(x)$. The crucial step then in completing the mathematical model is to understand the possible connection between $q(x)$ and $\rho(x)$. This connection will depend on the real-life situation.

### 3.1.2 *Diffusion*

The next step then is to uncover what mechanism is in effect to cause the flux $q(x)$ of pollutant in the dam. More assumptions must be made. Assume that the water in the dam is not moving (aside from regions close to the river entrance and at the dam outlet). The pollution spreads by diffusion. That means pollution in regions of high concentration will move into regions of low concentration, the bigger the difference, the faster the spread. The simplest mathematical expression for this behavior, called Fick's law, is

$$q(x) = -k \frac{d\rho}{dx}(x) \,. \tag{3.1.6}$$

A positive gradient means that the concentration is larger for larger $x$ and so the flux is negative; the concentration moves to the left where the concentration is lower, thus the minus sign in the law Eq. (3.1.6). The parameter $k$ is a constant of proportionality that express the dependency of the diffusion on the material properties of the process, in this case the mobility of the pollutant through water, and has units square meters per second.

Under the assumption that diffusion governs the transport of the pollutant, Eq. (3.1.6) can be substituted into Eq. (3.1.5) to produce a single equation for the unknown profile $\rho(x)$.

$$\frac{d}{dx}\left( A(x) \frac{d\rho}{dx}(x) \right) - \frac{\alpha}{k} A(x)\, \rho(x) = 0 \,. \tag{3.1.7}$$

This is a second-order differential equation (the first term involves two derivatives of $\rho(x)$) with variable coefficients – the presence of $A(x)$. We expect two independent solutions that can be combined linearly with two unknown coefficients, which means we need two more pieces of information to specify the solution completely.

The information we seek obviously comes from the concentration of the pollution $c\,(\text{kg/m}^3)$ in the river entering the dam and the concentration of the pollution leaving the dam, which we take to be zero. Thus we set,

$$\rho(0) = c \,, \qquad \rho(L) = 0 \,, \tag{3.1.8}$$

where $L$ is the length of the dam.[2]

### 3.1.2.1 *No pollution deposits*

While we expect Eq. (3.1.7) and Eq. (3.1.8) to determine $\rho(x)$ completely, the solution to the differential equation is difficult to construct because of the presence of the variable cross-sectional area $A(x)$. Let us make some simplifications that will help us understand what needs to be done to solve Eq. (3.1.7) with boundary conditions Eq. (3.1.8). First, let us suppose that the deposit of pollution on the dam bottom can be neglected; in other words, $\alpha = 0$. Then, Eq. (3.1.7) becomes

$$\frac{d}{dx}\left( A(x) \frac{d\rho}{dx}(x) \right) = 0 \,. \tag{3.1.9}$$

Obviously the differential equation has been reduced to a derivative that we can simply integrate;

$$A(x) \frac{d\rho}{dx}(x) = c_1 \,, \tag{3.1.10}$$

---

[2] A more careful study would consider how the concentration in the river mixes into the dam at the entrance of the river. We would still expect some value of the concentration of the pollutant to be determined at $x = 0$, and all we would need to do is replace $c$ in the boundary conditions by a more appropriate value.

with $c_1$ some constant, but do not forget that it must have units, in this case, kilograms per square meters. We cannot proceed without specifying how $A(x)$ varies and that depends on the choice of dam. Let us make a simple choice, namely, that the area increases linearly from the river entrance to the dam wall.

$$A(x) = A_0 + (A_L - A_0)\frac{x}{L}, \qquad (3.1.11)$$

where we let the parameters $A_0$ and $A_L$ be the cross-sectional areas at the river entrance and the dam wall, respectively. The linear relation in Eq. (3.1.11) is easy to read and is written to ensure that the units make sense.

We have choices in how we proceed depending on how we treat the algebra and the integration. Probably the most obvious way is to rewrite Eq. (3.1.10) as

$$\frac{d\rho}{dx}(x) = \frac{c_1 L}{A_0 L + (A_L - A_0) x}. \qquad (3.1.12)$$

The intent here is to make the numerator and denominator as simple as possible, but with the consequence that they have different units and that will affect the results of the integration. Typically, engineers and scientists proceed symbolically, that is, units are ignored while mathematical operations are performed; sometimes though performing the operations with units in mind can lead to results expressed in a better way. After integrating Eq. (3.1.12),

$$\rho(x) = \frac{c_1 L}{A_L - A_0} \ln\big[A_0 L + (A_L - A_0) x\big] + c_2. \qquad (3.1.13)$$

Note that the argument of the logarithm has units cubic meters, which does not make sense and so the units of $c_1$ are unclear at present. We proceed in the expectation that once $c_1$ and $c_2$ are determined the results will make sense. These two constants must be determined by the application of the two boundary conditions Eq. (3.1.8). Consequently, we obtain two algebraic equations,

$$\frac{c_1 L}{A_L - A_0} \ln(A_0 L) + c_2 = c,$$

$$\frac{c_1 L}{A_L - A_0} \ln(A_L L) + c_2 = 0.$$

Notice now the units of $c_2$ must be those of $c$. For future reference, these two equations can be written in matrix form – see Appendix C:

$$\begin{bmatrix} L \ln(A_0 L) & A_L - A_0 \\ L \ln(A_L L) & A_L - A_0 \end{bmatrix} \begin{pmatrix} c_1 \\ c_2 \end{pmatrix} = \begin{pmatrix} (A_L - A_0) c \\ 0 \end{pmatrix}.$$

After some algebra – see Appendix C, we obtain the solutions

$$c_1 = \frac{(A_L - A_0)\, c}{L\,[\ln(A_0 L) - \ln(A_L L)]}, \qquad c_2 = \frac{c\, \ln(A_L L)}{\ln(A_L L) - \ln(A_0 L)},$$

leading to the specific solution for the concentration,

$$\rho(x) = \frac{\ln(A_L L) - \ln\left[A_0 L + (A_L - A_0)x\right]}{\ln(A_L L) - \ln(A_0 L)}\, c\,.$$

Notice that the result contains several logarithms with arguments that have units. Use the properties of logarithms, in particular, $\ln(A) - \ln(B) = \ln(A/B)$,[3] to express the result so that all logarithms have arguments without units. There are several possible ways to achieve this, and here is one:

$$\frac{\rho(x)}{c} = -\frac{1}{\ln(A_L/A_0)}\,\ln\left[\frac{A_0}{A_L} + \left(1 - \frac{A_0}{A_L}\right)\frac{x}{L}\right]. \tag{3.1.14}$$

The result has been written so that the quantity $\rho(x)/c$ is dimensionless and expresses the concentration of the pollution in the dam as a fraction of the entering concentration. It is a function of the dimensionless distance $x/L$ and depends on a single dimensionless parameter $A_L/A_0$. The information in the results is compressed into as compact a form as possible and is easy to read. To illustrate the result, the choice $A_L/A_0 = 2$ is made and the profile is shown in Fig. 3.3. The profile is almost linear, and a linear profile is the result we would obtain if the cross-sectional area is just a constant – see one of the exercises.

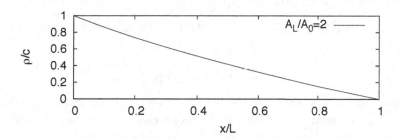

Fig. 3.3   Profile of the dimensionless concentration.

Now that we appreciate the result as written in Eq. (3.1.14) – hindsight is always wonderful – we can ask whether there is a way to obtain the result from Eq. (3.1.12) while preserving expressions with dimensionless quantities. To do so, rewrite Eq. (3.1.12) as

$$\frac{\mathrm{d}\rho}{\mathrm{d}x}(x) = \frac{c_1}{A_L}\left/\left[\frac{A_0}{A_L} + \left(1 - \frac{A_0}{A_L}\right)\frac{x}{L}\right]\right..$$

The denominator on the right is dimensionless. Great! The numerator contains an unknown coefficient $c_1$ which must have the appropriate units to balance the units

---

[3]The logarithm is the only standard function where attention to the units of its argument is necessary. The other standard functions, such as polynomials, exponentials and trigonometric functions should always have arguments that are dimensionless.

on the left. The integration of this form for the derivative leads to

$$\rho(x) = \frac{c_1 L}{A_L - A_0} \ln\left[\frac{A_0}{A_L} + \left(1 - \frac{A_0}{A_L}\right)\frac{x}{L}\right] + c_2 \,. \qquad (3.1.15)$$

The argument of the logarithm has no units as should be. Application of the boundary conditions Eq. (3.1.8) leads to

$$\frac{c_1 L}{A_L - A_0} \ln(A_0/A_L) + c_2 = c \,,$$

$$c_2 = 0 \,.$$

Often when we approach a problem the right way, the algebra proves simpler. Here we see $c_2 = 0$ and $c_1$ is easily determined, leading directly to the solution (3.1.14) as before.

---

**Reflection**

Let us reflect some on the nature of the mathematics in constructing the solution Eq. (3.1.14). The solution to the differential equation has the form given in Eq. (3.1.13) or Eq. (3.1.15). The two unknown coefficients $c_1$ and $c_2$ suggest that the solution may be considered to be the sum of two independent solutions. For example, from Eq. (3.1.13),

$$Y_1(x) = \ln\left[A_0 L + (A_1 - A_0)x\right], \qquad Y_2(x) = 1 \,,$$

so that

$$\rho(x) = c_1 Y_1(x) + c_2 Y_2(x) \,.$$

Of course, this makes sense because we anticipate two solutions to a second-order linear differential equation. In previous chapters, we have considered such equations and the need for two initial conditions (1.4.31) to determine the coefficients $c_1$ and $c_2$. This time it is two boundary conditions (3.1.8) that are used to determine $c_1$ and $c_2$. The important point is that the number of additional conditions must match the order of the differential equation. The same story is played out in the next examples.

---

### 3.1.2.2  *Constant cross-sectional area*

If the cross-sectional area can be taken as constant $A(x) = A_0$, then Eq. (3.1.7) becomes a second-order differential equation with constant coefficients,

$$\frac{d^2\rho}{dx^2}(x) - \beta\rho(x) = 0 \,, \qquad (3.1.16)$$

where $\beta = \alpha/k$ (m$^2$) is a new parameter. This homogeneous equation is solved in the standard way. Assume a solution in the form $\exp(\lambda x)$ and substitute into the differential equation, leading to the characteristic equation,

$$\lambda^2 - \beta = 0.$$

Since $\beta > 0$, there are two solutions,

$$Y_1(x) = e^{\sqrt{\beta}\,x}, \qquad Y_2(x) = e^{-\sqrt{\beta}\,x}, \tag{3.1.17}$$

and the general solution is

$$\rho(x) = c_1\,Y_1(x) + c_2\,Y_2(x)$$
$$= c_1\,e^{\sqrt{\beta}\,x} + c_2\,e^{-\sqrt{\beta}\,x}.$$

The application of the boundary conditions Eq. (3.1.8) gives

$$c_1 + c_2 = c,$$
$$c_1\,e^{\sqrt{\beta}\,L} + c_2\,e^{-\sqrt{\beta}\,L} = 0.$$

In matrix form – see Appendix C,

$$\begin{bmatrix} 1 & 1 \\ e^{\sqrt{\beta}\,L} & e^{-\sqrt{\beta}\,L} \end{bmatrix} \begin{pmatrix} c_1 \\ c_2 \end{pmatrix} = \begin{pmatrix} c \\ 0 \end{pmatrix}.$$

Since the determinant is $e^{-\sqrt{\beta}\,L} - e^{\sqrt{\beta}\,L} \neq 0$, we are sure of a solution and it is

$$c_1 = -\frac{c\,\exp(-\sqrt{\beta}\,L)}{\exp(\sqrt{\beta}\,L) - \exp(-\sqrt{\beta}\,L)},$$
$$c_2 = \frac{c\,\exp(\sqrt{\beta}\,L)}{\exp(\sqrt{\beta}\,L) - \exp(-\sqrt{\beta}\,L)}.$$

Thus the specific solution is

$$\frac{\rho(x)}{c} = -\frac{\exp\left[\sqrt{\beta}\,(x-L)\right]}{\exp(\sqrt{\beta}\,L) - \exp(-\sqrt{\beta}\,L)} + \frac{\exp\left[-\sqrt{\beta}\,(x-L)\right]}{\exp(\sqrt{\beta}\,L) - \exp(-\sqrt{\beta}\,L)}$$
$$= -\frac{\sinh\left[\sqrt{\beta}\,(x-L)\right]}{\sinh(\sqrt{\beta}\,L)}$$
$$= \frac{\sinh\left[\sqrt{\beta}\,L\,(1 - x/L)\right]}{\sinh(\sqrt{\beta}\,L)}. \tag{3.1.18}$$

The final result is written in a similar way to the result in Eq. (3.1.14). The concentration is expressed as a fraction of the entering concentration that varies with the dimensionless distance $x/L$. The parameter $\sqrt{\beta}\,L$ is a single dimensionless parameter that determines the profile completely. Two choices $\sqrt{\beta}\,L = 10$ and $\sqrt{\beta}\,L = 0.1$ produce the two profiles shown in Fig. 3.4. They highlight two extreme cases: for $\sqrt{\beta}\,L = 10$, the pollution deposits on the bottom more readily than diffuses through the water and the concentration falls off exponentially; for $\sqrt{\beta}\,L = 0.1$, the opposite is true and the pollution diffuses more rapidly through the water than deposits on the bottom of the dam with the consequence that the profile falls off much more slowly (looks linear).

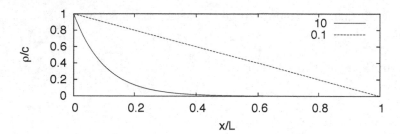

Fig. 3.4   The profile of the concentration in the dam for $\sqrt{\beta}\, L = 10,\ 0.1$.

### 3.1.3   *Abstract view*

The examples considered in this section show no difficulties in constructing specific solutions. In particular, the differential equation Eq. (3.1.16) is second order thus possessing two independent solutions Eq. (3.1.17). Two boundary conditions Eq. (3.1.8) are then used to determine the unknown coefficients in the general solution and the solution is complete Eq. (3.1.18).

Let us consider a more general case. Suppose the differential equation is

$$\frac{\mathrm{d}^2 y}{\mathrm{d}x^2}(x) + a\,\frac{\mathrm{d}y}{\mathrm{d}x}(x) + b\,y(x) = 0\,. \qquad (3.1.19)$$

Since the equation is linear and homogeneous, the solution is exponential in form; $y(x) = \exp(\lambda x)$, leading to the characteristic equation,

$$\lambda^2 + a\,\lambda + b = 0\,.$$

For the moment, we restrict our attention to the case that the roots are real, $\lambda = \lambda_1, \lambda_2$. Then the solutions are

$$Y_1(x) = \mathrm{e}^{\lambda_1 x}\,, \qquad Y_2(x) = \mathrm{e}^{\lambda_2 x}\,,$$

and the general solution is

$$y(x) = c_1\,Y_1(x) + c_2\,Y_2(x)\,. \qquad (3.1.20)$$

Now it is the boundary conditions, expressed in the general form,

$$y(0) = \alpha\,, \qquad y(L) = \beta\,, \qquad (3.1.21)$$

that will determine $c_1$ and $c_2$. Applying Eq. (3.1.21) to Eq. (3.1.20) produces two equations,

$$c_1 + c_2 = \alpha\,,$$
$$c_1\,\mathrm{e}^{\lambda_1 L} + c_2\,\mathrm{e}^{\lambda_2 L} = \beta\,.$$

They may be written in matrix form,

$$\begin{bmatrix} 1 & 1 \\ \mathrm{e}^{\lambda_1 L} & \mathrm{e}^{\lambda_2 L} \end{bmatrix} \begin{pmatrix} c_1 \\ c_2 \end{pmatrix} = \begin{pmatrix} \alpha \\ \beta \end{pmatrix}\,.$$

Since the determinant is $e^{\lambda_2 L} - e^{\lambda_1 L} \neq 0$, there are unique solutions

$$c_1 = \frac{\beta - \alpha\, e^{\lambda_2 L}}{e^{\lambda_1 L} - e^{\lambda_2 L}},$$

$$c_2 = -\frac{\beta - \alpha\, e^{\lambda_1 L}}{e^{\lambda_1 L} - e^{\lambda_2 L}}.$$

With these values, a specific solution to Eq. (3.1.19) subject to the boundary conditions Eq. (3.1.21) has been constructed.

Let us adopt a more abstract view by imagining the general solution to a linear, homogeneous second-order differential equation is given by Eq. (3.1.20). When the differential equation has constant coefficients, then the solutions are exponentials, but Eq. (3.1.20) is still the form of the solution when the differential equation has variable coefficients, such as Eq. (3.1.7), although simple expressions are not likely available for the solutions $Y_1(x)$ and $Y_2(x)$. Chances are we will have to compute these solutions numerically.

What is important is the consequence of applying the boundary conditions Eq. (3.1.21).

$$c_1\, Y_1(0) + c_2\, Y_2(0) = \alpha,$$
$$c_1\, Y_1(L) + c_2\, Y_2(L) = \beta.$$

In matrix form,

$$\begin{bmatrix} Y_1(0) & Y_2(0) \\ Y_1(L) & Y_2(L) \end{bmatrix} \begin{pmatrix} c_1 \\ c_2 \end{pmatrix} = \begin{pmatrix} \alpha \\ \beta \end{pmatrix}.$$

There is a solution for the coefficients – see Appendix C,

$$c_1 = \frac{\alpha\, Y_2(L) - \beta\, Y_2(0)}{Y_1(0)\, Y_2(L) - Y_1(L)\, Y_2(0)}, \tag{3.1.22}$$

$$c_2 = \frac{\beta\, Y_1(0) - \alpha\, Y_1(L)}{Y_1(0)\, Y_2(L) - Y_1(L)\, Y_2(0)}, \tag{3.1.23}$$

provided that the determinant that appears as the denominator in both Eq. (3.1.22) and Eq. (3.1.23) is not zero. In other words, we require

$$Y_1(0)\, Y_2(L) - Y_1(L)\, Y_2(0) \neq 0, \tag{3.1.24}$$

then there are always values for $c_1$ and $c_2$ for any choices of $L$ and $\alpha$ and $\beta$.

It is interesting to compare the determinant in Eq. (3.1.24) with the Wronskian Eq. (1.4.36). Whereas the Wronskian is not zero if the solutions to the differential equation are independent, we can find situations where the determinant in Eq. (3.1.24) is zero even though the solutions $Y_1(x)$ and $Y_2(x)$ are independent. In the next section, we will look at such cases and develop an understanding of how and why it occurs.

## Principle 20: Extension of Principle 9

*Additional information is required to select a specific solution to a differential equation. If the equation is second order, we need two more pieces of information, typically initial conditions for the solution and its derivative or boundary conditions specified at the ends of the interval of interest.*

### 3.1.4   *Exercises*

(1) Repeat the calculation of the concentration profile determined by Eq. (3.1.9) but where the cross-sectional area is constant, $A(x) = A_0$. Does the answer agree with the result in Eq. (3.1.14) when $A_L = A_0$?

**The following exercises provide some practice in solving boundary value problems.**

(2) Construct the solution to the following differential equations and the given boundary conditions:

(a)

$$\frac{d^2 f}{dz^2}(z) + 5\frac{df}{dz}(z) + 6\,f(z) = 0\,,$$

subject to $f(-L) = 1$ and $f(L) = 1$.
**Answer:** For the choice $L = 1$, $f(0.5) = 2.447$.

(b)

$$\frac{d^2 w}{dy^2}(y) + 2\pi\frac{dw}{dy}(y) + 5\pi^2\,w(y) = 0\,,$$

subject to $w(0) = 0$ and $w(1/4) = 2$.
**Answer:** $w(1/8) = 2.094$.

(c)

$$\frac{d^2 y}{dx^2}(x) - k^2\,y(x) = 0\,,$$

subject to $y(0) = 0$ and $y(L) = 0$.
**Answer:** $y(L/2) = 0$.

(3) Consider the differential equation,

$$\frac{d^2 y}{dx^2}(x) - k^2 y(x) = 0\,,$$

subject to the boundary conditions

$$y(0) = a\,, \qquad y(L) = b\,.$$

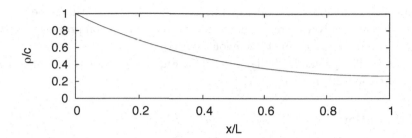

Fig. 3.5 Profile of the relative concentration $\rho/c$ as a function of the dimensionless distance $x/L$ with $\sqrt{\beta}\,L = 2$.

Express your answer in the form,

$$y(x) = a\,Y_1(x) + b\,Y_2(x)\,.$$

The functions $Y_1(x)$ and $Y_2(x)$ do not depend on $a$ or $b$. In other words $Y_1(x)$ is describing the influence of the boundary value $a$, and $Y_2(x)$ is describing the influence of boundary value $b$. You may also find it interesting to compare your results with those in part (c) of Exercise 2.

Provide an interpretation of the result by describing the behavior of the functions $Y_1(x)$ and $Y_2(x)$.

**Answer:** The two functions have special properties: $Y_1(0) = 1$, $Y_1(L) = 0$; $Y_2(0) = 0$, $Y_2(L) = 1$.

**The next exercises consider different types of boundary conditions.**

(4) Suppose the concentration of pollutant is not allowed to leave the dam. The diffusion equation Eq. (3.1.18) for the concentration in the dam still holds, and the concentration at the river entrance is still $\rho(0) = c$, but the boundary condition at the dam wall must be changed to insist that there is no flux of pollutant leaving the dam. Construct the solution with this new boundary condition and plot the profile of the concentration. Where does the pollutant go?

**Answer:** The typical profile is shown in Fig. 3.5: At the wall, $\rho(L)/c = 0.266$.

(5) Suppose we install a filtering system at the dam wall that removes pollution at a rate proportional to the concentration at the wall. Assume that the flux at the dam wall is the flux into the filtering system and that the remaining pollution leaves with the discharge from the dam. We may model the situation by assuming that the flux of pollutant at the dam wall is proportional to the amount of concentration there. In other words, the boundary condition takes the form:

$$\frac{\mathrm{d}\rho}{\mathrm{d}x}(L) = -\gamma\rho(L)\,.$$

Using the diffusion equation Eq. (3.1.18) as before and the same boundary condition at the river entrance, determine the concentration profile and compare to the profile obtained in the previous exercise.

**Answer:** The profile is similar to that in Fig. 3.5 but lower in value near the wall. At the wall, $\rho(L)/c = 0.179$.

## 3.2 Bifurcation

We may be tempted to conclude from the examples in the previous section that boundary value problems are much like initial value problems. Simply construct two solutions to a second-order differential equation, add them together with some unknown coefficients and apply the boundary conditions at two different locations. Solve the resulting system of equations for the two unknown coefficients and we are done. For initial value problems, the solution is guaranteed to exist if the Wronskian is non-zero, but for boundary value problems there is no such guarantee!

As usual, it is an important example from engineering that highlights the possibilities.

### 3.2.1 *Column buckling*

This example will highlight a new possibility in BVP's, namely, that the construction of the solution may fail! Imagine there is a column supporting a load as shown in Fig. 3.6. Our interest lies in understanding the conditions when the column may buckle; obviously, we want to avoid such an occurrence. The question we raise, then, is there a possible steady state configuration where the column is buckled as shown on the right in Fig. 3.6?

For the column to be in a steady state, the moments along the column must be in balance. Let $x(y)$ (m) be the displacement of the center line of the column when it is buckled. Obviously, $x$ (m) refers to a horizontal distance measured from the

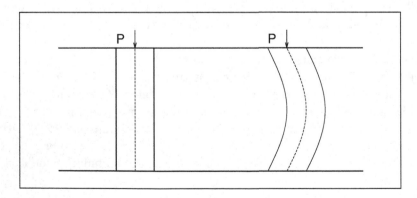

Fig. 3.6   A schematic of a column supporting a load: on the left, straight column, on the right a buckled column. The dashed curve signifies the center line.

center line of the straight column, and $y$ (m) measures the vertical distance from the bottom of the column. The bending moment $M(y)$ (N·m) that is produced by the deformation of the column is $M(y) = EI\kappa(y)$ where $E$ (N/m$^2$) is Young's modulus of elasticity and $I$ (m$^4$) is the moment of inertia of a cross section (assumed uniform) about the center line. The curvature $\kappa(y)$ may be approximated by $\kappa(y) = $ d$^2x(y)/$d$y^2$ when the deformation is not too large. The bending moment must balance the applied moment $-Px(y)$ where $P$ (N) is the load exerted on the top of the column. We are led to the equation,

$$M(y) = EI \frac{\mathrm{d}^2 x}{\mathrm{d}y^2}(y) = -Px(y), \quad \text{or} \quad \frac{\mathrm{d}^2 x}{\mathrm{d}y^2}(y) + \frac{P}{EI} x(y) = 0. \tag{3.2.1}$$

We must now address the choice for the boundary conditions and the natural one is to assume that the column is hinged (able to rotate, but not slide) at the bottom and top. In other words,

$$x(0) = 0, \quad x(L) = 0. \tag{3.2.2}$$

We are now in position to solve Eq. (3.2.1) and Eq. (3.2.2) to seek the shape of a buckled column.

Since Eq. (3.2.1) is a second-order differential equation with constant coefficients, it may be solved in the standard way by assuming solutions in the form $\exp(\lambda y)$, leading to the characteristic equation,

$$\lambda^2 + \frac{P}{EI} = 0.$$

The roots are purely imaginary, so the general solution is

$$x(y) = c_1 \cos\left(\sqrt{\frac{P}{EI}}\, y\right) + c_2 \sin\left(\sqrt{\frac{P}{EI}}\, y\right). \tag{3.2.3}$$

Time to apply the boundary conditions (3.2.2). The boundary condition $x(0) = 0$ requires $c_1 = 0$. The other boundary condition leads to

$$c_2 \sin\left(\sqrt{\frac{P}{EI}}\, L\right) = 0. \tag{3.2.4}$$

Obviously, one choice is to set $c_2 = 0$, with the consequence that $x(y) = 0$; in other words, no buckling. But there are other possibilities that satisfy Eq. (3.2.4) and they arise because of the nature of the sine function. The sine function has many zeros, at any integer multiple of $\pi$. Suppose magically,

$$\sqrt{\frac{P}{EI}}\, L = \pi, \tag{3.2.5}$$

then Eq. (3.2.5) becomes $0 \times c_2 = 0$ so the boundary condition is automatically satisfied for any value of $c_2$: $c_2$ is not determined! For a specific column, $E$, $I$ and $L$ are all known, so (3.2.5) reflects a special choice for $P$,

$$P_1 = EI\left(\frac{\pi}{L}\right)^2. \tag{3.2.6}$$

How are we to interpret the result Eq. (3.2.6)? Let us imagine several scenarios. First suppose $P$ is quite small, much smaller than $P_1$. Then the column remains straight since the only possible solution is $c_2 = 0$ and $x(y) = 0$. Even if the load $P$ is larger but still smaller than $P_1$, the column remains straight. But as we increase the load $P$ some more until $P = P_1$, suddenly there are many possible solutions in that $c_2$ can be any value;

$$x(y) = c_2 \sin\left(\frac{\pi}{L} y\right).$$

Since all values of $c_2$ are possible, we say the solution is not unique. Before trying to assess the implications of this solution, let us note in passing that non-unique solutions arise whenever

$$\sqrt{\frac{P_n}{EI}} L = n\pi, \quad \text{or} \quad P_n = EI\left(\frac{n\pi}{L}\right)^2 \quad \text{and} \quad x(y) = c_2 \sin\left(\frac{n\pi}{L} y\right).$$

Non-unique solutions are not satisfying! Surely, there must be some additional information or some inadequacy of the mathematical model that has produced this result. Let us consider some possible changes in the model that might help determine a unique solution. First, what about allowing the top of the column to shift during buckling? Replace the boundary conditions with

$$x(0) = 0, \quad x(L) = \delta.$$

We leave aside for the moment how $\delta$ might the determined; we simply want to see if it cures the difficulty of non-unique solutions. The solution to the differential equation remains unchanged and the boundary condition at $x = 0$ still leads to $c_1 = 0$. It is the last boundary condition that has changed and it now reads

$$c_2 \sin\left(\sqrt{\frac{P}{EI}} L\right) = \delta. \tag{3.2.7}$$

Clearly, if $P \neq P_n$, then

$$c_2 = \frac{\delta}{\sin(\sqrt{P/EI} L)}$$

and all is well; however, if $P = P_n$, then the denominator is zero and $c_2 = \infty$. Even worse! No solution exists. We have not avoided the difficulty in obtaining a specific solution.

The deficiency lies in the mathematical model. It comes about because we approximated the true curvature

$$\kappa(y) = \frac{\mathrm{d}^2 x}{\mathrm{d}y^2}(y) \Bigg/ \left[1 + \left(\frac{\mathrm{d}x}{\mathrm{d}y}(y)\right)^2\right]^{3/2} \approx \frac{\mathrm{d}^2 x}{\mathrm{d}y^2}(y),$$

on the grounds that the slope of $x(y)$ will be much smaller than one for small displacements. Using the full expression for the curvature leads to

$$\frac{\mathrm{d}^2 x}{\mathrm{d}y^2}(y) + \frac{P}{EI}\left[1 + \left(\frac{\mathrm{d}x}{\mathrm{d}y}(y)\right)^2\right]^{3/2} x(y) = 0,$$

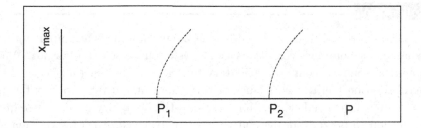

Fig. 3.7   An illustration of the nature of the results to a buckled column.

which is a nonlinear equation. Unlike linear equations, nonlinear equations may have several distinct solutions and that is the case here.

The specific solutions for the full nonlinear differential equation form an interesting and important pattern as shown in Fig. 3.7. There are several distinct solutions and they are characterized by the measurement $x_{max} = \max |x(y)|$. Note first that $x(y) = 0$ is always a solution (no buckling). For $P < P_1$, there is only one solution $x_{max} = 0$, but for $P_1 < P < P_2$, there is an additional solution where the column is buckled. That solution is displayed as a dashed curve in Fig. 3.7.[4] The difference between the solutions to the linear and nonlinear differential equations is now clear. For the liner solutions, $x_{max} = c_2$ has any value and the dashed curve would be a vertical line. The effect of the nonlinear term is to bend the vertical line into a curve with the important consequence that there is only one solution along the curve for each $P$.

Let us appreciate the consequences of the nonlinear solutions through a simple thought experiment. Imagine that we apply a load $P$ that is small, $P < P_1$, then there is only the unbuckled column. This is good news if we are designing a structure supported by the column; we must simply make sure that the values for $E$ and $I$ are large enough to ensure that the largest expected value for $P$ is less than $P_1$. If we now consider cases where $P$ is larger, indeed $P_1 < P < P_2$, then there is another solution where the column is buckled.

The value $P_1$ where the multiple solutions first start is called a *bifurcation* point. Which solution occurs in practice for $P_1 < P < P_2$ now depends on the initial conditions and a transient adjustment that requires a mathematical model which includes temporal behavior as well as spatial variation. Such models have both time and space partial derivatives and lead to *partial differential equations*, the topic of the next chapter. For the buckled column, the straight column is unstable, meaning the slightest initial disturbance causes the column to buckle and take the shape corresponding to the solution on the dashed curve. For even larger values of $P$, there are even more shapes the buckled column can take.

---

[4]The dashed curve is called a branch of solutions because it seems to sprout from the unbuckled solutions at $P = P_1$.

---

**Reflection**

The nature of the solution to the differential equation plays an important role in whether the boundary conditions lead to well-defined solutions or not. When the general solution contains oscillatory functions as in Eq. (3.2.3), then the general solution has the possibility of being zero at both end points for any value of one of the coefficients. In this case, if the boundary conditions require the solution to be zero at the end points, for example, Eq. (3.2.2), then the general solution automatically satisfies the boundary conditions and there are many solutions. On the other hand, if the boundary condition requires a non-zero value, for example, Eq. (3.2.7), then the boundary condition can never be satisfied and solution will not exist. Clearly, BVPs can be more challenging then IVPs.

---

### 3.2.2 *Abstract viewpoint*

Solutions to homogeneous, linear differential equations with constant coefficients are exponentials which may contain real or complex arguments. The possibilities are determined by the roots to the characteristic equation. If the roots are real, then the exponentials have real arguments and the exponentials can never be zero. The expectation, then, is that the BVP will have well-defined solutions. On the other hand, if the roots are complex, the solution contains oscillatory functions (sines and cosines) that have many zeros and can affect the possibility of solutions.

To place the matter on a clear footing, consider a second-order differential equation

$$\frac{\mathrm{d}^2 y}{\mathrm{d}x^2}(x) + a\,\frac{\mathrm{d}y}{\mathrm{d}x}(x) + b\,y(x) = 0\,.$$

Because the equation is homogeneous, its solutions have the form $\exp(\lambda x)$ where

$$\lambda^2 + a\,\lambda + b = 0$$

is the characteristic equation. The roots of the characteristic equation determine two solutions, referred to here as just $Y_1(x)$ and $Y_2(x)$. The general solution is

$$y(x) = c_1\,Y_1(x) + c_2\,Y_2(x)\,.$$

If the roots are real, $Y_1(x)$ and $Y_2(x)$ are exponentials, while if the roots are complex, $Y_1(x)$ and $Y_2(x)$ contain cosines and sines.

Time to consider the application of the general boundary conditions,

$$y(0) = \alpha\,, \quad y(L) = \beta\,. \tag{3.2.8}$$

Consequently, we obtain two equations for $c_1$ and $c_2$;

$$c_1\,Y_1(0) + c_2\,Y_2(0) = \alpha\,,$$
$$c_1\,Y_1(L) + c_2\,Y_2(L) = \beta\,.$$

These two equations form a system that may also be expressed in matrix form,

$$\begin{bmatrix} Y_1(0) & Y_2(0) \\ Y_1(L) & Y_2(L) \end{bmatrix} \begin{pmatrix} c_1 \\ c_2 \end{pmatrix} = \begin{pmatrix} \alpha \\ \beta \end{pmatrix}.$$

Either way, the solution is – see Appendix C,

$$c_1 = \frac{\alpha Y_2(L) - \beta Y_2(0)}{Y_1(0) Y_2(L) - Y_1(L) Y_2(0)}, \tag{3.2.9}$$

$$c_2 = \frac{\beta Y_1(0) - \alpha Y_1(L)}{Y_1(0) Y_2(L) - Y_1(L) Y_2(0)}, \tag{3.2.10}$$

where the denominator is the determinant of the matrix.

There are several possible outcomes to the results in (3.2.9) and (3.2.10). The first consideration is whether the denominator

$$D = Y_1(0) Y_2(L) - Y_1(L) Y_2(0) \neq 0. \tag{3.2.11}$$

If this is true, then clearly there are values for $c_1$ and $c_2$ for any choice of $\alpha$ and $\beta$. In particular, if $\alpha = 0$ and $\beta = 0$, then the result is $c_1 = 0$ and $c_2 = 0$ giving the solution as $y(x) = 0$. We call this the *trivial* solution, and it is the solution when the boundary conditions are homogeneous, $y(0) = 0$ and $y(L) = 0$. This observation encourages us to include the boundary conditions with the differential equation when we characterize the problem as homogeneous or not.

## Principle 21

*A homogeneous boundary value problem is composed of a homogeneous differential equation and homogeneous boundary conditions. The prime property is that the trivial solution is always a solution. If there is a non-trivial solution, then any multiple of the solution is a solution to the differential equation and the boundary conditions; thus the occurrence of many solutions.*

Now suppose that $D = 0$. Then there are two possibilities: At least one of the numerators in Eq. (3.2.9) or Eq. (3.2.10) is non-zero and division by zero means at least one of the coefficients $c_1$, $c_2$ is infinite. This means the solution does not exist! Alternatively, if both the numerators are zero, the coefficients are undefined – the result appears as $0/0$, and we have many solutions.

All these cases can and do occur. Here are the consequences for the different types of solutions $Y_1(x)$ and $Y_2(x)$:

**Real roots.** The solutions are

$$Y_1(x) = e^{\lambda_1 x}, \qquad Y_2(x) = e^{\lambda_2 x},$$

and

$$D = e^{\lambda_2 L} - e^{\lambda_1 L}.$$

Since $D \neq 0$, we have the case of unique solutions.

**Double roots.** The solutions are

$$Y_1(x) = e^{\lambda x}, \qquad Y_2(x) = x\, e^{\lambda x},$$

and

$$D = L\, e^{\lambda L},$$

which is also non-zero, and the solution is unique.

**Complex roots.** The solutions are

$$Y_1(x) = e^{\lambda_r x} \cos(\lambda_i x), \qquad Y_2 = e^{\lambda_r x} \sin(\lambda_i x),$$

and

$$D = e^{\lambda_r L} \sin(\lambda_i L).$$

Provided $\lambda_i L \neq n\pi$ for some integer $n$, $D \neq 0$ and the solution is unique. If $\lambda_i L = n\pi$, then $D = 0$ and either

(1) the numerator in Eq. (3.2.10)

$$\beta - \alpha\, e^{\lambda_r L} \cos(n\pi) \neq 0,$$

and the solution does not exist. Note the other numerator in Eq. (3.2.9)

$$\alpha\, Y_2(L) - \beta\, Y_2(0) = 0,$$

no matter what the values of $\alpha$ and $\beta$ are.

(2) the numerator in Eq. (3.2.10)

$$\beta - \alpha\, e^{\lambda_r L} \cos(n\pi) = 0,$$

and the solution is non-unique.

---

**Principle 22**

*If the differential equation has solutions that are oscillatory, then there is the possibility that the boundary value problem may not have a solution, or may have many solutions.*

---

At this point, the situation may seem confusing, but there is a simple pattern.

### 3.2.3   *Summary*

For the boundary conditions Eq. (3.2.8), the determinant of the system of equations for the coefficients $c_1$ and $c_2$ is given by $D$ in Eq. (3.2.11).

**If $D \neq 0$**, then the following statements are true:

(a) The homogeneous boundary value problem ($\alpha = \beta = 0$) has only the trivial solution.

(b) The inhomogeneous boundary value problem ($\alpha \neq 0$ and/or $\beta \neq 0$) has a unique solution.

**If $\mathbf{D} = \mathbf{0}$**, then the following cases can occur:

(a) The homogeneous boundary value problem has many solutions.
(b) The inhomogeneous boundary value problem has either no solutions or many solutions depending on the boundary values.

These results may be restated in a simple way.

### Principle 23

*If a homogeneous boundary value problem has only the trivial solution, then the inhomogeneous problem has a unique solution. If the homogeneous boundary value problem has many solutions, then the inhomogeneous problem may have no solutions or many solutions.*

The great value to Principle 23 is that it is also true when other boundary conditions are considered. For example, suppose the boundary conditions are

$$\frac{dy}{dx}(0) = \gamma, \qquad \frac{dy}{dx}(L) = \delta. \tag{3.2.12}$$

Then the system of equations for the coefficients in the general solution becomes

$$c_1 \frac{dY_1}{dx}(0) + c_2 \frac{dY_2}{dx}(0) = \gamma,$$

$$c_1 \frac{dY_1}{dx}(L) + c_2 \frac{dY_2}{dx}(L) = \delta,$$

with the determinant given by

$$D = \frac{dY_1}{dx}(0) \frac{dY_2}{dx}(L) - \frac{dY_1}{dx}(L) \frac{dY_2}{dx}(0). \tag{3.2.13}$$

As before, there are three cases to consider depending on the nature of the roots to the characteristic equation:

**Real roots.** The solutions are

$$Y_1(x) = e^{\lambda_1 x}, \qquad Y_2(x) = e^{\lambda_2 x},$$

and

$$D = \lambda_1 \lambda_2 \, e^{\lambda_2 L} - \lambda_1 \lambda_2 \, e^{\lambda_1 L}.$$

Since $D \neq 0$, we have the case of unique solutions. In other words, the homogeneous problem ($\gamma = \delta = 0$) has only the trivial solution and the inhomogeneous problem has a unique solution, consistent with Principle 23.

**Double roots.** The solutions are

$$Y_1(x) = e^{\lambda x}, \qquad Y_2(x) = x\,e^{\lambda x},$$

and

$$D = \lambda^2 L\,e^{\lambda L},$$

which is also non-zero, and the solution is unique. The situation is the same as the case of real roots.

**Complex roots.** The solutions are

$$Y_1(x) = e^{\lambda_r x}\cos(\lambda_i x), \qquad Y_2 = e^{\lambda_r x}\sin(\lambda_i x),$$

and

$$D = \left(\lambda_r^2 + \lambda_i^2\right)e^{\lambda_r L}\sin(\lambda_i L).$$

Provided $\lambda_i L \neq n\pi$ for some integer $n$, $D \neq 0$ and the solution is unique. If $\lambda_i L = n\pi$, then $D = 0$. The system of equations for the coefficients becomes

$$\lambda_r c_1 + \lambda_i c_2 = \gamma,$$
$$\left(\lambda_r c_1 + \lambda_i c_2\right)e^{\lambda_r L}\cos(n\pi) = \delta,$$

and unless $\gamma\,e^{\lambda_r L}\cos(n\pi) = \delta$ there are no solutions. Otherwise, there are many solutions. In other words, if $D = 0$, the homogeneous boundary value problem ($\gamma = \delta = 0$) has many solutions and the inhomogeneous problem either has no solution or many solutions, consistent with Principle 23.

### 3.2.4   *Exercises*

**The following exercises provide some more practice in solving boundary value problems.**

(1) Construct the solution to the following differential equations and the given boundary conditions:

(a)

$$\frac{\mathrm{d}^2 f}{\mathrm{d}z^2}(z) + 5\,\frac{\mathrm{d}f}{\mathrm{d}z}(z) + 6\,f(z) = 0,$$

subject to

$$\frac{\mathrm{d}f}{\mathrm{d}z}(-L) = 0, \qquad \frac{\mathrm{d}f}{\mathrm{d}z}(L) = 0.$$

**Answer:** $f(0) = 0$.

(b)

$$\frac{\mathrm{d}^2 w}{\mathrm{d}y^2}(y) + 2\pi\,\frac{\mathrm{d}w}{\mathrm{d}y}(y) + 5\pi^2\,w(y) = 0,$$

subject to

$$w(0) = 0, \qquad \frac{\mathrm{d}w}{\mathrm{d}y}(1/4) = 0.$$

**Answer:** $w(\pi/16) = 0$.

(c)

$$\frac{d^2y}{dx^2}(x) + k^2\,y(x) = 0\,,$$

subject to $y(0) = 0$ and $y(L) = 0$. The parameter $k > 0$.

**Answer:** Trivial solution only if $k \neq n\pi/L$, $n$ any positive integer.

(2) Consider the differential equation,

$$\frac{d^2y}{dx^2}(x) + k^2 y(x) = 0\,,$$

with the parameter $k > 0$, subject to the boundary conditions,

$$y(0) = a\,, \qquad y(L) = b\,.$$

Express your answer in the form,

$$y(x) = a\,Y_1(x) + b\,Y_2(x)\,,$$

where $Y_1(x)$ and $Y_2(x)$ do not depend on either $a$ or $b$. Provide an interpretation of the result by describing the behavior of the functions $Y_1(x)$ and $Y_2(x)$.
For what values of $k$ are there no solutions?
It is worthwhile to compare the results of this exercise with those of Exercise 3 in Sec. 3.1.4.
**Answer:** The functions are oscillatory with the special properties: $Y_1(0) = 1$, $Y_1(L) = 0$; $Y_2(0) = 0$, $Y_2(L) = 1$. The solutions do not exist if $kL = n\pi$.

**The next exercise examines the choice of interval.**

(3) The vertical coordinate is measured from the bottom in the model for column buckling. Another reasonable choice is to set the origin in the middle of the column. The differential equation does not change,

$$\frac{d^2x}{dx^2}(y) + \beta^2 x(y) = 0\,,$$

but the boundary conditions are shifted,

$$x(-L/2) = x(L/2) = 0\,.$$

The parameter $\beta^2 = P/(EI)$ is introduced to simplify algebra.

(a) Determine the conditions under which there are many solutions.
(b) Establish whether these solutions are the same as in Sec. 3.2.1. To do so, consider a change of independent variable $s = y + L/2$. The results in this exercise in terms of $s$ should agree with the results in Sec. 3.2.1.

**The next exercises consider different types of boundary conditions.**

(4) Consider a vertical column as we did in Eq. (3.2.1), but now assume that the column cannot bend at its ends. In other words, the column must have zero slope at the ends,

$$\frac{\mathrm{d}x}{\mathrm{d}y}(0) = \frac{\mathrm{d}x}{\mathrm{d}y}(L) = 0.$$

Do the new boundary conditions affect where the column will buckle?

**Answer:** Same bifurcation points as before but the profile is different. For $P = \pi^2 EI/(4L^2)$, $x(L/2) = 0$.

(5) Repeat the previous exercise but assume that the column is hinged at the top, but cannot bend at the bottom.

**Answer:** Bifurcation points occur at $P = (2n-1)^2\pi^2 EI/(4L^2)$ where $n$ is a positive integer.

## 3.3  Forcing effects

The first two sections of this chapter have introduced the perspective that both the differential equation and the boundary conditions determine whether the problem is homogeneous or not – see Principle 21. This perspective is not only of value mathematically, but also makes sense physically in that the system can be forced in different ways. There may be forcing in the process governing the system, leading to forcing terms in the differential equation. At the boundaries of the system, there may be forcing effects that influence the interior, for example, the flow of pollutants in a river that enters a dam or the removal of pollutants at the dam wall. Incorporating the boundary conditions with the differential equation becomes a natural way to study the consequences of forcing.

From the mathematical perspective, Principle 23 informs us that if the homogeneous problem has only trivial solutions we can expect a unique solution when forcing is present. So far in our considerations, the forcing has occurred only at the boundaries through the specification of inhomogeneous boundary conditions. We now consider the possibility that the differential equation also has a forcing term, and explore the various ways we incorporate multiple contributions to forcing effects in the construction of solutions.

From a practical point of view, we simply seek a particular solution to the differential equation, add the homogeneous solution to the differential equation to obtain the general solution, and then apply the boundary conditions to take care of the forcing effects at the boundaries. The overall strategy is the same as always: particular solutions, homogeneous solutions, general solutions and specific solutions. The difference is only in the application of the boundary conditions to the general solution to determine the specific solution.

Construction of a particular solution follows from what we have done before. If the forcing term in the differential equation is of the appropriate form, we can rely on the method of undetermined coefficients. What can we do if it is not? Fortunately, there are some wonderful ways we can adapt the Fourier series to handle any forcing term, at least formally. As usual, if we cannot perform the integration in close form to calculate the Fourier coefficients, then we must resort to numerical integration. Nevertheless, there are many cases where we can calculate the Fourier coefficients and complete the construction of the solution. All along the way, we will note how Principle 23 remains valid.

### 3.3.1 *Cooling fins*

A simple, but important example helps us get started.

Cooling fins are used to remove heat from sources of heat such as engines and motors. They are typically thin rectangular plates attached to the heating source to allow heat to conduct along the plate while heat is transfer to the ambient air through Newton's law of cooling. A small segment of the plate is shown in Fig. 3.8. The width of the plate is $W$ (cm) and its thickness is $H$ (cm) and we consider a small segment of length $\triangle x$ (cm). We anticipate a steady state temperature profile $T(x)$ (K) when the running of the motor has reached normal operating conditions. Our task is to determine that temperature profile, and subsequently design the cooling fin for optimal operating conditions.

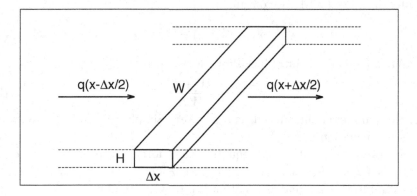

Fig. 3.8    A small segment of the cooling fin located at $x$.

There is a flux of heat $q(x)$ $(\text{J}/(\text{cm}^2 \cdot \text{K} \cdot \text{s}))$ that measures the energy transferred by conduction. The conservation of energy in the plate under steady state conditions requires that the flux of energy leaving the small segment of the plate from the right must balance the amount entering from the left minus the amount leaving through the top, bottom and sides through Newton's law of cooling. Since during a small time interval $\triangle t$ s,

$$\text{amount leaving in } \triangle t = q(x + \triangle x/2)\, WH\, \triangle t\,,$$
$$\text{amount entering in } \triangle t = q(x - \triangle x/2)\, WH\, \triangle t\,,$$

and the amount lost to cooling is proportional to the surface area, the temperature difference and the time interval,

$$\text{amount lost to cooling} = 2k \left( W + H \right) \triangle x \left( T(x) - T_0 \right) \triangle t \,,$$

where $k$ $(J/(cm^2 \cdot s))$ is the constant of proportionality and the ambient temperature of the air is $T_0$ (K) – assumed a constant for the moment. The energy balance becomes

$$W H \, q(x + \triangle x/2) = W H \, q(x - \triangle x/2) - 2k \left( W + H \right) \triangle x \left( T(x) - T_0 \right) .$$

Taking the limit as usual, we obtain

$$\frac{dq}{dx}(x) = -2k \, \frac{W + H}{WH} \left( T(x) - T_0 \right) . \tag{3.3.1}$$

The conduction of heat in a solid is akin to the diffusive process. Energy is transported from regions of higher temperature (where molecules are more agitated) to regions of lower temperature (where molecules are less agitated). Fourier expressed this idea as

$$q(x) = -\alpha \, \frac{dT}{dx}(x) \,, \tag{3.3.2}$$

where $\alpha$ $(J/(cm \cdot K \cdot s))$ is a constant. Both $\alpha$ and $k$ are constants that depend on the material properties of the cooling fin. When this relationship Eq. (3.3.2) is substituted into Eq. (3.3.1), we obtain

$$\frac{d^2T}{dx^2}(x) - \beta^2 \, T(x) = -\beta^2 \, T_0 \,. \tag{3.3.3}$$

The parameter $\beta$ $cm^{-1}$ is introduced for convenience;[5]

$$\beta^2 = 2 \frac{k}{\alpha} \, \frac{W + H}{WH} \,. \tag{3.3.4}$$

Why is the parameter introduced as a square? Simply to make the roots of the characteristic equation just $\pm \beta$.

In this example, the differential equation has a forcing term. It is just a constant so the guess for a particular solution is just $T_p(x) = A$, a constant, and after substitution into the differential equation Eq. (3.3.3), we obtain

$$T_p(x) = T_0 \,.$$

The homogeneous solution to the differential equation is

$$y_h(x) = C_1 \, e^{\beta x} + C_2 \, e^{-\beta x} \,,$$

leading to the general solution,

$$y(x) = T_0 + C_1 \, e^{\beta x} + C_2 \, e^{-\beta x} \,. \tag{3.3.5}$$

The approach, so far, is the standard one of constructing a particular solution to the differential equation and adding the homogeneous solution to the differential

---

[5]All of the geometric parameters, $W$ and $H$, and the transport parameters, $k$ and $\alpha$, have been combined into a single parameter $\beta$, so much more convenient!

equation to generate a general solution to the differential equation. The difference with initial value problems is only that we now apply boundary conditions. What are the boundary conditions? Where the fin attaches to the motor at $x = 0$, we assume the temperature is that of the motor, $T_a$, while at the end of the fin $x = L$, there can be no flux. Thus,

$$T(0) = T_a, \qquad \frac{dT}{dx}(L) = 0. \qquad (3.3.6)$$

The fact that one of the boundary conditions is homogeneous should not distract us from the fact that the set of boundary conditions Eq. (3.3.6) is inhomogeneous.

Application of the boundary conditions to the general solution Eq. (3.3.5) produces two equations for the two unknown coefficients, $C_1$ and $C_2$:

$$C_1 + C_2 = T_a - T_0,$$
$$\beta C_1 e^{\beta L} - \beta C_2 e^{-\beta L} = 0.$$

Written as a system,

$$\begin{bmatrix} 1 & 1 \\ \beta \exp(\beta L) & -\beta \exp(-\beta L) \end{bmatrix} \begin{pmatrix} C_1 \\ C_2 \end{pmatrix} = \begin{pmatrix} T_a - T_0 \\ 0 \end{pmatrix}.$$

Notice how both information from the particular solution ($T_0$) and the inhomogeneous boundary condition ($T_a$) appear as known information on the right hand side of these equations. The influence of the particular solution on the right hand side of the equations does not change the considerations in Sec. 3.2.3 on whether there are solutions for $C_1$ and $C_2$. Since the determinant of the matrix

$$D = -\beta \left( e^{\beta L} + e^{-\beta L} \right),$$

is always non-zero, there are always unique solutions for $C_1$ and $C_2$.

$$C_1 = \frac{(T_a - T_0) \exp(-\beta L)}{\exp(\beta L) + \exp(-\beta L)}, \qquad C_2 = \frac{(T_a - T_0) \exp(\beta L)}{\exp(\beta L) + \exp(-\beta L)}. \qquad (3.3.7)$$

After some simple algebra, the solution can be written as

$$T(x) = T_0 + (T_a - T_0) \frac{\cosh\left[\beta(L - x)\right]}{\cosh(\beta L)}. \qquad (3.3.8)$$

Clearly, if the differential equation were homogeneous $T_0 = 0$ and the boundary conditions were homogeneous $T_a = 0$, then the only solution is the trivial solution $T(x) = 0$. In other words, with no heating from the motor nor cooling from the air, there is no temperature in the cooling fin.

A form of the solution that contains only dimensionless variables is

$$\frac{T(x)}{T_0} = 1 - \left(1 - \frac{T_a}{T_0}\right) \frac{\cosh\left[\beta L(1 - x/L)\right]}{\cosh(\beta L)},$$

where $T(x)/T_0$ is the temperature profile as a fraction of the temperature at the motor, $x/L$ is the dimensionless distance along the plate, $T_a/T_0$ is the ratio of the ambient temperature to the temperature of the motor and $\beta L$ is a dimensionless parameter representing all of the geometric and transport parameters. Normally, we expect $T_a/T_0 < 1$. This form of the result allows an engineer to focus on the parameter $\beta L$ as an important part in designing a fin to meet requirements, for example, the temperature at the end of the fin is "close" to ambient temperature.

### 3.3.2  *Heated plate; homogeneous boundary conditions*

Let us consider the rectangular plate as lying horizontally between two vertical support structures made of concrete. The ambient temperature is now allowed to be spatially variable, reflecting, for example, hot air produced by a furnace below. The differential equation is the same, but the boundary conditions now reflect an assumed property of the concrete, namely, that it acts as an insulator. That means no heat can flow through either end of the plate and the boundary conditions are homogeneous derivative conditions,

$$\frac{\mathrm{d}T}{\mathrm{d}x}(0) = \frac{\mathrm{d}T}{\mathrm{d}x}(L) = 0. \qquad (3.3.9)$$

How do we now construct the solution for the differential equation with a forcing term that varies spatially?

$$\frac{\mathrm{d}^2 T}{\mathrm{d}x^2}(x) - \beta^2\, T(x) = -\beta^2\, T_0(x). \qquad (3.3.10)$$

Of course, if $T_0(x)$ is of the form that allows the particular solution to be constructed by the method of undetermined coefficients, then the way forward is the approach we used in the previous example. What if it is not? Maybe it is a piecewise function, or involves a rational expression like $1/(1 + x^2)$. It would be nice if we could somehow use the ideas behind periodic forcing; replace $T_0(x)$ by a Fourier series. Then we can proceed by writing the particular solution as a Fourier series in the spirit of the method of undetermined coefficients. But obviously, $T_0(x)$ is not necessarily periodic. But wait! We do not know anything about $T_0(x)$ outside of the range $0 \leq x \leq L$. That certainly suggest some flexibility in defining $T_0(x)$ elsewhere but how to exploit this possibility?

There are three ways we can extend the function $T_0(x)$ outside of the range $0 < x < L$. We can insist that the function is $L$-periodic and construct its Fourier series. We can pretend that it is even, $T_0(-x) = T_0(x)$, and extend it as a $2L$-periodic function; it will have a cosine series. The third choice is to insist the extended function is even, $T_0(-x) = -T_0(x)$, and $2L$-periodic; it can be expressed as a sine series.

Which of these choices should we make? Recall that the decay in $n$ of the magnitude of the Fourier coefficients $a_n$ or $b_n$ depends on the continuity of the function, in this case the extended version of $T_0(x)$: you may need to review Principle 17 and the discussion surrounding it in Sec. 2.2.3. If $T_0(0) \neq T_0(L)$, then a $L$-periodic extension will have a jump discontinuity and the Fourier coefficients will behave as $1/n$ which is a very slow convergence. Now consider the even extension: there will be no jump in the function at $x = 0$ and since $T_0(-L) = T_0(L)$ also, there is no jump discontinuity in $T_0(x)$ and the Fourier cosine coefficients will decay at least as fast as $1/n^2$, which is a better rate of convergence. The third choice makes $T_0(x)$ odd, and unless $T_0(0) = 0$ and $T_0(L) = 0$, the extended function will have jump discontinuities and the Fourier sine coefficients will decay as $1/n$ which is the same as the $L$-periodic extension. The best choice then is a cosine series representation.

Let the cosine representation for $T_0(x)$ be

$$T_0(x) = A_0 + \sum_{n=1}^{\infty} A_n \cos\left(\frac{n\pi}{L}x\right), \tag{3.3.11}$$

where the coefficients are determined by

$$A_0 = \frac{1}{L} \int_0^L T_0(x)\,rdx, \qquad A_n = \frac{2}{L} \int_0^L T_0(x) \cos\left(\frac{n\pi}{L}x\right)rdx.$$

Note that the determination of these coefficients depends on integrals that cover the range $0 < x < L$, just where the function $T_0(x)$ is known.

Now we proceed as before in constructing a particular solution to the differential equation. We guess

$$T_{\mathrm{p}}(x) = a_0 + \sum_{n=1}^{\infty}\left[a_n \cos\left(\frac{n\pi}{L}x\right) + b_n \sin\left(\frac{n\pi}{L}x\right)\right]. \tag{3.3.12}$$

The next steps then are to substitute Eq. (3.3.12) and Eq. (3.3.11) into Eq. (3.3.10) and balance terms containing cosines and sines.

$$-\sum_{n=1}^{\infty} \frac{n^2\pi^2}{L^2}\left[a_n \cos\left(\frac{n\pi}{L}x\right) + b_n \sin\left(\frac{n\pi}{L}x\right)\right]$$

$$-\beta^2 a_0 - \beta^2 \sum_{n=1}^{\infty}\left[a_n \cos\left(\frac{n\pi}{L}x\right) + b_n \sin\left(\frac{n\pi}{L}x\right)\right]$$

$$= -\beta^2 A_0 - \beta^2 \sum_{n=1}^{\infty} A_n \cos\left(\frac{n\pi}{L}x\right).$$

Balancing constants produces $a_0 = A_0$, and balancing the cosine terms with arguments $n\pi x/L$ produces

$$a_n = \frac{\beta^2 L^2 A_n}{\beta^2 L^2 + n^2\pi^2}. \tag{3.3.13}$$

Since there are no sine terms on the right hand side, $b_n = 0$ and the particular solution is completely determined: It turns out to be simply a cosine series.

**Reflection**

Since there is no first derivative in the differential equation, there is no need to add the sine terms in the particular solution. The second derivative of a cosine series returns only a cosine series and that is all we need to balance a forcing term that is a cosine series.

Since the homogeneous solution is

$$T_{\text{h}}(x) = C_1 e^{\beta x} + C_2 e^{-\beta x} \, ,$$

the general solution is

$$T(x) = A_0 + \beta^2 L^2 \sum_{n=1}^{\infty} \frac{A_n}{\beta^2 L^2 + n^2 \pi^2} \cos\left(\frac{n\pi}{L} x\right) + C_1 e^{\beta x} + C_2 e^{-\beta x} \, . \qquad (3.3.14)$$

The final steps are the substitution of Eq. (3.3.14) into the boundary conditions Eq. (3.3.9). Note first that the derivative of Eq. (3.3.14) is

$$\frac{dT}{dx}(x) = -\beta^2 L^2 \sum_{n=1}^{\infty} \frac{n\pi}{L} \frac{A_n}{\beta^2 L^2 + n^2 \pi^2} \sin\left(\frac{n\pi}{L} x\right) + \beta C_1 e^{\beta x} - \beta C_2 e^{-\beta x} \, .$$

Since the sine functions are all zero at $x = 0$ and $x = L$, the only contribution to the boundary conditions are those from the homogeneous solution.

$$\beta C_1 - \beta C_2 = 0 \, ,$$
$$\beta C_1 e^{\beta L} - \beta C_2 e^{\beta L} = 0 \, ,$$

and clearly, $C_1 = C_2 = 0$. The homogeneous solution to the differential equation must also satisfy homogeneous boundary conditions Eq. (3.3.9); in other words, we have a homogeneous boundary value problem which has only the trivial solution. There is a unique specific solution given by the particular solution.

---

**Reflection**

It is great that we could assume an extension of the forcing function outside of the range $0 < x < L$ to ensure a Fourier series. There are several ways we can do this, but our choice of a cosine series means that the homogeneous boundary conditions are automatically satisfied: no need to add a homogeneous solution.

---

### 3.3.3  *Heated plate; inhomogeneous boundary conditions*

This time imagine that we can control the temperatures at the junctions of the plate with the walls. That means we can specify the temperature,

$$T(0) = T_a \, , \qquad T(L) = T_b \, . \qquad (3.3.15)$$

How does this change in boundary conditions affect what we did in Sec. 3.3.2? Now we have both forcing effects in the differential equations and in the boundary conditions – we are forcing the temperature to take on certain values. The standard approach will be to construct a particular solution to the differential equation as we did in the previous section and then add the homogeneous solution to the differential equation to create a general solution to which we can apply the inhomogeneous boundary conditions. Once again there are different choices for the Fourier series representation of the forcing term in the differential equation.

First, let us proceed with the choice made in the previous section; the forcing term is expressed as a Fourier cosine series. The particular solution is also a cosine Fourier series and to that we add the homogeneous solution to obtain Eq. (3.3.14), rewritten here as

$$T(x) = a_0 + \sum_{n=1}^{\infty} a_n \cos\left(\frac{n\pi}{L} x\right) + C_1 e^{\beta x} + C_2 e^{-\beta x}. \tag{3.3.16}$$

The Fourier coefficients $a_n$ are known Eq. (3.3.13) but it is easier to proceed without substituting their values.

Now apply the boundary conditions Eq. (3.3.15):

$$a_0 + \sum_{n=1}^{\infty} a_n + C_1 + C_2 = T_a,$$

$$a_0 + \sum_{n=1}^{\infty} (-1)^n a_n + C_1 e^{\beta L} + C_2 e^{-\beta L} = T_b.$$

For convenience, define

$$S_{\mathrm{p}} = \sum_{n=1}^{\infty} a_n, \qquad S_{\mathrm{m}} = \sum_{n=1}^{\infty} (-1)^n a_n.$$

These sums are just numbers, although they depend on the parameters $\beta$ and $L$. The equations for $C_1$ and $C_2$ may now be written as

$$C_1 + C_2 = T_a - a_0 - S_{\mathrm{p}}, \tag{3.3.17}$$

$$C_1 e^{\beta L} + C_2 e^{-\beta L} = T_b - a_0 - S_{\mathrm{m}}, \tag{3.3.18}$$

with the solution,

$$C_1 = \frac{T_b - T_a e^{-\beta L} - \left(1 - e^{-\beta L}\right) a_0 - S_{\mathrm{m}} + S_{\mathrm{p}} e^{-\beta L}}{e^{\beta L} - e^{-\beta L}}, \tag{3.3.19}$$

$$C_2 = \frac{T_a e^{\beta L} - T_b + \left(1 - e^{\beta L}\right) a_0 - S_{\mathrm{p}} e^{\beta L} + S_{\mathrm{m}}}{e^{\beta L} - e^{-\beta L}}. \tag{3.3.20}$$

With the known values for $a_n$ Eq. (3.3.15), $C_1$ Eq. (3.3.19) and $C_2$ Eq. (3.3.20), the general solution Eq. (3.3.16) becomes the specific solution. Its drawback is that the coefficients $C_1$ and $C_2$ are messy algebraic expressions, difficult to interpret. In particular, they contain sums $S_{\mathrm{p}}$ and $S_{\mathrm{m}}$ whose dependence on the parameters $\beta$ and $L$ is difficult to interpret. The reason for the appearance of the sums in the boundary conditions Eq. (3.3.17) and Eq. (3.3.18) is that the particular solution does not satisfy the homogeneous boundary conditions,

$$T(0) = 0, \qquad T(L) = 0. \tag{3.3.21}$$

To drive home the importance of the homogeneous boundary conditions, suppose we find a particular solution that satisfies Eq. (3.3.21), that is, $T_p(0) = 0$, $T_p(L) = 0$. Now when the inhomogeneous boundary conditions are applied to the general solution $T(x) = T_p(x) + T_h(x)$ we obtain

$$T(0) = T_p(0) + C_1 + C_2 = C_1 + C_2 = T_a \,, \tag{3.3.22}$$

$$T(L) = T_p(L) + C_1 e^{\beta L} + C_2 e^{-\beta L} = C_1 e^{\beta L} + C_2 e^{-\beta L} = T_b \,. \tag{3.3.23}$$

The consequence is that the equations for $C_1$ and $C_2$ are straightforward; there are no sums arising from the particular solution.

Can we find a Fourier series that satisfies Eq. (3.3.21)? Yes, a sine series

$$T_p(x) = \sum_{n=1}^{\infty} b_n \sin\left(\frac{n\pi}{L}x\right) \tag{3.3.24}$$

works because all the sine terms are zero at $x = 0$ and $x = L$! The role of the particular solution is to satisfy the forcing term in the differential equation and we must now express $T_0(x)$ as a sine series so that we can match the solution with the forcing term.

$$T_0(x) = \sum_{n=1}^{\infty} B_n \sin\left(\frac{n\pi}{L}x\right). \tag{3.3.25}$$

By substituting Eq. (3.3.24) into Eq. (3.3.10) and by balancing the sine terms with Eq. (3.3.25), we obtain

$$b_n = \frac{\beta^2 L^2 B_n}{\beta^2 L^2 + n^2 \pi^2} \,.$$

The calculation follows very closely the calculation of Eq. (3.3.13) except we are balancing sine series rather than cosine series. Notice again the importance of the lack of a first derivative in the differential equation. The derivative of a sine series would create cosine terms and the balance would fail.

With the successful calculation of a particular solution as a sine series Eq. (3.3.24) that satisfies Eq. (3.3.21), we may calculate $C_1$ and $C_2$ from Eq. (3.3.22) and Eq. (3.3.23).

$$C_1 = \frac{T_b - T_a e^{-\beta L}}{e^{\beta L} - e^{-\beta L}} \,, \qquad C_2 = \frac{T_a e^{\beta L} - T_b}{e^{\beta L} - e^{-\beta L}} \,.$$

By gathering all the results, the specific solution becomes

$$T(x) = \beta^2 L^2 \sum_{n=1}^{\infty} \frac{B_n}{\beta^2 L^2 + n^2 \pi^2} \sin\left(\frac{n\pi}{L}\right) + \frac{T_a \sinh\left[\beta(1-x)\right] + T_b \sinh(\beta x)}{\sinh(\beta L)} \,.$$

The interpretation of the results is much easier than Eq. (3.3.16); the particular solution to the differential equation is a sine series that balances the forcing term expressed as a sine series, and the homogeneous solution to the differential equation captures the influence of the inhomogeneous boundary conditions. The slight drawback is the Fourier coefficients $B_n$ are likely to decay in magnitude as $1/n$, and then $b_n$ will decay in magnitude as $1/n^3$, a slower pace than the coefficients $a_n$ in Eq. (3.3.13).

> **Reflection**
>
> There are two considerations when we plan to replace a forcing term by a Fourier series. One is the decay in magnitude of the Fourier coefficients. The other is the form of the solution. If it is possible to choose a Fourier series that satisfies the associated homogeneous boundary conditions, then the form of the solution is easier to understand.

### 3.3.4 *Abstract viewpoint*

It is time to put together all the results of this chapter into a summary. As a start, we classify the problem as being homogeneous if both the differential equation and the boundary conditions are homogeneous. If either the differential equation or the boundary conditions are inhomogeneous then the problem is considered inhomogeneous; see Principle 21.

The problem in general is inhomogeneous with a forcing term in the differential equation,

$$\frac{d^2 y}{dx^2}(x) + a\,\frac{dy}{dx}(x) + b\,y(x) = r(x)\,, \tag{3.3.26}$$

and inhomogeneous boundary conditions,

$$y(0) = \alpha\,, \qquad y(L) = \beta\,. \tag{3.3.27}$$

There are other types for boundary conditions, containing derivatives, but the abstract view does not change.

#### 3.3.4.1 *Homogeneous problem*

Our ability to solve this problem, Eq. (3.3.26) together with Eq. (3.3.27), depends on the nature of the solutions to the associated homogeneous problem, the homogeneous differential equation

$$\frac{d^2 y}{dx^2}(x) + a\,\frac{dy}{dx}(x) + b\,y(x) = 0\,, \tag{3.3.28}$$

together with homogeneous boundary conditions,

$$y(0) = 0\,, \qquad y(L) = 0\,. \tag{3.3.29}$$

As stated in Principle 23, the inhomogeneous problem has a unique solution if the homogeneous problem has only the trivial solution.

The solution to Eq. (3.3.28) depends on the roots of the characteristic equation and there are two important cases to consider, two real roots or two complex conjugate roots.[6]

---

[6] The case of a double root gives results similar to the case of two real roots.

**Real roots.** The solution to Eq. (3.3.28) is

$$y(x) = C_1 e^{\lambda_1 x} + C_2 e^{\lambda_2 x}, \tag{3.3.30}$$

and the application of the homogeneous boundary conditions Eq. (3.3.29) produces the two equations,

$$C_1 + C_2 = 0, \tag{3.3.31}$$
$$C_1 e^{\lambda_1 L} + C_2 e^{\lambda_2 L} = 0. \tag{3.3.32}$$

The substitution of $C_2 = -C_1$ into the second boundary condition gives

$$C_1 \left( e^{\lambda_1 L} - e^{\lambda_2 L} \right) = 0.$$

We are faced with two choices, $C_1 = 0$ or $\exp(\lambda_1 L) = \exp(\lambda_2 L)$. Since $\lambda_1 \neq \lambda_2$, the second choice is not possible and we are forced to conclude that $C_1 = C_2 = 0$. In other words, the only possible solution to the homogeneous problem, Eq. (3.3.28) with Eq. (3.3.29) is $y(x) = 0$, the trivial solution.

**Complex roots.** The solution now has the form,

$$y(x) = e^{\lambda_r x} \left( C_1 \cos(\lambda_i x) + C_2 \sin(\lambda_i x) \right). \tag{3.3.33}$$

The application of the homogeneous boundary conditions Eq. (3.3.29) produces the two equations,

$$C_1 = 0, \tag{3.3.34}$$
$$C_2 e^{\lambda_r L} \sin(\lambda_i L) = 0. \tag{3.3.35}$$

We are faced with two choices, $C_2 = 0$ or $\sin(\lambda_i L) = 0$. It might so happen that $\lambda_i L = n\pi$ for some integer $n$, in which case $C_2$ is undetermined. In other words, we expect $y(x) = 0$, except for special circumstances $\lambda_i L = n\pi$ when the solution is not unique (letting $C_2 = C$)

$$y(x) = C \sin\left( \frac{n\pi}{L} x \right).$$

These special circumstances cause serious difficulties in constructing a solution for the inhomogeneous problem.

### 3.3.4.2  *Inhomogeneous problem*

We spilt the solution for the inhomogeneous problem into two parts, $y(x) = y_p(x) + y_h(x)$ and require $y_p(x)$ to be a particular solution for the forcing term in the differential equation and $y_h(x)$ is the homogeneous solution to the differential equation.

The boundary conditions become

$$y_p(0) + y_h(0) = \alpha, \qquad y(L) = y_p(L) + y_h(L) = \beta \tag{3.3.36}$$

but now we must decide how to choose $y_p(x)$.

*Undetermined coefficients*

How we proceed depends on our ability to construct the particular solution to Eq. (3.3.28). If $r(x)$ is of the form that allows us to use the method of undetermined coefficients, then we construct $y_p(x)$ in the standard way. The construction of $y_h(x)$, the solution to the homogeneous differential equation, depends on the roots of the characteristic equation, and we list the two possibilities.

**Real roots.** The solution has the form

$$y(x) = y_p(x) + C_1 e^{\lambda_1 x} + C_2 e^{\lambda_2 x}. \tag{3.3.37}$$

Then application of the inhomogeneous boundary conditions (3.3.36) gives

$$C_1 + C_2 = \alpha - y_p(0), \tag{3.3.38}$$

$$C_1 e^{\lambda_1 L} + C_2 e^{\lambda_2 L} = \beta - y_p(L). \tag{3.3.39}$$

This system of equations for $C_1$ and $C_2$ is an inhomogeneous version of the system Eq. (3.3.31) and Eq. (3.3.32). Since Eq. (3.3.31) and Eq. (3.3.32) have only trivial solutions, unique solutions for $C_1$ and $C_2$ can always be found.

**Complex roots.** Now the solution looks like

$$y(x) = y_p(x) + e^{\lambda_r x} \left[ C_1 \cos(\lambda_i x) + C_2 \sin(\lambda_i x) \right]. \tag{3.3.40}$$

The application of the inhomogeneous boundary conditions Eq. (3.3.36) yields

$$C_1 = \alpha - y_p(0), \tag{3.3.41}$$

$$e^{\lambda_r L} \left[ C_1 \cos(\lambda_i L) + C_2 \sin(\lambda_i L) \right] = \beta - y_p(L). \tag{3.3.42}$$

This system is an inhomogeneous version of Eq. (3.3.34) and Eq. (3.3.35). When $\lambda_i L \neq n\pi$, there are only trivial solutions to Eq. (3.3.34) and Eq. (3.3.35) and so unique solutions to Eq. (3.3.41) and Eq. (3.3.42). To see the results more clearly, substitute Eq. (3.3.41) into Eq. (3.3.42) to obtain

$$C_2 e^{\lambda_r L} \sin(\lambda_i L) = \beta - y_p(L) - \left( \alpha - y_p(0) \right) e^{\lambda_r L} \cos(\lambda_i L).$$

Provided $\lambda_i L \neq n\pi$, one of the special values that causes multiple solutions for the homogeneous problem, then $C_2$ can be determined and all is well. If $\lambda_i L = n\pi$ then $C_2 = \infty$ and there is no solution, unless $y_p(L) = \beta$ and $y_p(0) = \alpha$, in which case, there are multiple solutions – $C_2$ can be any value.

*Fourier series – First derivative in the differential equation*

Finally, let us consider what to do if $r(x)$ is some general function without an obvious guess for the particular solution; we considered such a situation in Sec. 3.3.3. The idea is that we would love to convert $r(x)$ into a Fourier series somehow and then construct the particular solution as a companion Fourier series. The challenge here is the choice of Fourier series, and the appropriate extension of $r(x)$. Of the three choices, a $L$-periodic function with a full Fourier series, an even $2L$-periodic extension with a cosine Fourier series, or an odd $2L$-periodic extension with a Fourier sine series, the choice that gives the best decay in the Fourier coefficients is the cosine series, because the extension will produce a continuous function.

Suppose we make this choice and express

$$r(x) = A_0 + \sum_{n=1}^{\infty} A_n \cos\left(\frac{n\pi}{L} x\right),$$

where

$$A_0 = \frac{1}{L} \int_0^L r(x)\,\mathrm{d}x,$$

$$A_n = \frac{2}{L} \int_0^L r(x) \cos\left(\frac{n\pi}{L} x\right) \mathrm{d}x.$$

We follow the normal strategy and set the particular solution to be

$$y_{\mathrm{p}}(x) = a_0 + \sum_{n=1}^{\infty}\left[a_n \cos\left(\frac{n\pi}{L} x\right) + b_n \sin\left(\frac{n\pi}{L} x\right)\right]. \tag{3.3.43}$$

Note the presence of sine terms in the Fourier series; we are obliged to include them because of the presence of a first derivative in the differential equation, $a \neq 0$. Now substitute Eq. (3.3.43) into the differential equation Eq. (3.3.28) and insist on the balance of constants, cosines and sines to determine $a_n$ and $b_n$.

The final step is to satisfy the boundary conditions Eq. (3.3.36). If the roots to the characteristic equation are real, then we must satisfy Eq. (3.3.38) and Eq. (3.3.39). If the roots are complex, then we must satisfy and Eq. (3.3.41) and Eq. (3.3.42). Unfortunately, when Eq. (3.3.43) is evaluated at $x = 0$, $x = L$, the results are still infinite sums,

$$y_{\mathrm{p}}(0) = a_0 + \sum_{n=1}^{\infty} a_n, \qquad y_{\mathrm{p}}(L) = a_0 + \sum_{n=1}^{\infty}(-1)^n a_n.$$

Consequently, the solutions for $C_1$ and $C_2$ from either Eq. (3.3.38) and Eq. (3.3.39) or Eq. (3.3.41) and Eq. (3.3.42) will contain these infinite sums making the specific solution quite complicated to interpret. There is a special case that allows a simpler approach.

*Fourier series – No first derivative in the differential equation*

The differential equation is now

$$\frac{\mathrm{d}^2 y}{\mathrm{d}x^2}(x) + b\,y(x) = r(x). \tag{3.3.44}$$

If $r(x)$ is expressed as cosine series, then the particular solution requires only a cosine series because two derivatives of a cosine series produces a cosine series. Similarly, a sine series representation for $r(x)$ requires only a sine series for the particular solution. How does this help? It opens up choices for a Fourier series representation of $r(x)$ that leads to a simpler solution.

The choice of Fourier series now rests on the form of the boundary conditions. For Eq. (3.3.36), we would prefer a particular solution in a Fourier series that satisfies the homogeneous boundary conditions,

$$y_p(0) = 0, \qquad y_p(L) = 0 \tag{3.3.45}$$

because the boundary conditions Eq. (3.3.38) and Eq. (3.3.39) or Eq. (3.3.41) and Eq. (3.3.42) no longer contain contributions from the particular solution – no sums!

Given the property Eq. (3.3.45), the appropriate choice of Fourier series is a sine series. If the boundary conditions are of the type in Eq. (3.3.9), then the appropriate choice of the particular solution is a cosine series Eq. (3.3.11).

### 3.3.5   *Exercises*

**The following exercises provide some more practice in solving boundary value problems.**

(1) Construct the solution to the following differential equations and the given boundary conditions. Note that homogeneous versions of these exercises appear in Exercise 1 of Sec. 3.2.4.

(a)

$$\frac{d^2 f}{dz^2}(z) + 5\frac{df}{dz}(z) + 6\,f(z) = z\,,$$

subject to

$$\frac{df}{dz}(-L) = 1\,, \qquad \frac{df}{dz}(L) = 1\,.$$

**Answer:** With $L = 1$, $f(1/2) = -1.171$.

(b)

$$\frac{d^2 w}{dy^2}(y) + 2\pi\frac{dw}{dy}(y) + 5\pi^2\,w(y) = \alpha\,e^{-2y}\,,$$

subject to

$$w(0) = 0\,, \qquad \frac{dw}{dy}(1/4) = 0\,.$$

**Answer:** For $\alpha = 20$, $w(1/8) = -0.26$.

(c)

$$\frac{d^2 y}{dx^2}(x) + k^2\,y(x) = \alpha\,\sin(3x)\,,$$

subject to $y(0) = 0$ and $y(L) = 0$. The parameter $k > 0$.
**Answer:** For $k = 3$, $L = 1$ and $\alpha = 6$, $y(0.5) = -7.03$.

(2) Consider what happens when the characteristic equation has just one root. Solve

$$\frac{d^2y}{dx^2}(x) + 4\frac{dy}{dx}(x) + 4\,y(x) = \alpha,$$

subject to $y(0) = y(L) = 0$.
**Answer:** For $L = 1$, $y(1/2) = 0.452\,\alpha$.

**The next exercises consider various applications of boundary value problems.**

(3) Let us consider some specific examples of the influence of the ambient temperature on the heated plate. The differential equation is Eq. (3.3.10) and assume that the ends are insulated Eq. (3.3.9). Consider the following profiles for the ambient temperature.
(a)

$$T_0(x) = \alpha + \alpha \, \sin\left(\frac{\pi}{L}\,x\right).$$

(b)

$$T_0(x) = \begin{cases} \alpha, & \text{for } L/2 < |2x - L| < L, \\ 2\alpha - 4\alpha\,(2x/L - 1)^2, & \text{for } |2x - L| < L/2. \end{cases}$$

Construct the solutions and compare the results by graphing them.
**Answer:** The temperature profiles for the two cases are shown below for the choice $\beta L = 5$.

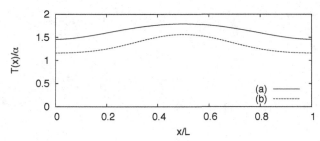

(4) Repeat the previous exercise but now assume the boundary conditions are given by Eq. (3.3.21).
**Answer:** The temperature profiles for the two cases are shown below for the choice $\beta L = 5$, $T_a/\alpha = 1$ and $T_b/\alpha = 2$.

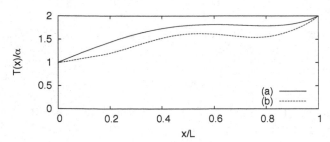

### 3.3.6 *Flowchart: Linear, second-order boundary-value problems*

The flowchart given here assumes a specific choice of boundary conditions. The strategy remains the same for other boundary conditions, but the details will be different. Also, there are no checks listed for the cases where the homogeneous solution is non-unique.

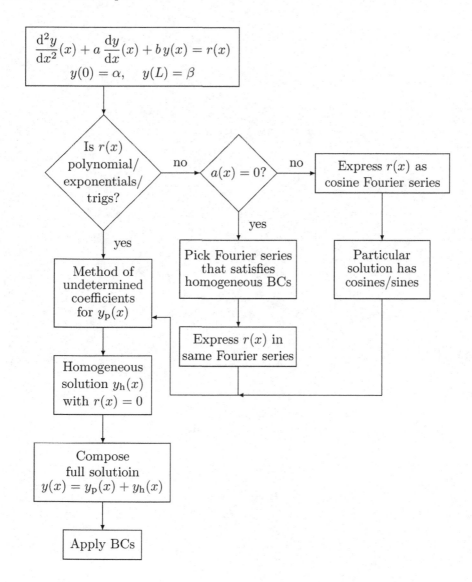

# Linear Partial Differential Equations

**4**

## 4.1 Diffusion: Part I

Ordinary differential equations arise in two natural circumstances: quantities that change only in time without spatial variation and quantities fixed in time but vary spatially. Chapter 1 considered cases where spatial variation is ignored and quantities only change in time, for example, in situations where concentrations may be considered uniform in space because they are well stirred. Chapter 3 considered cases where quantities do not change in time, but have spatial variation, for example, in steady state distributions of concentrations. Now we are ready to face the differential equations that describe how spatially varying profiles can change in time.

If we are going to address how quantities move through space and change in time, then we will need mathematical models that contain functions of both space and time. It is not surprising then, that these models will contain derivatives both in space and in time and that these derivatives will be partial derivatives. The differential equations that arise will contain partial derivatives and hence are called partial differential equations.

By now it should be clear that mathematical models require more than just differential equations. They require initial conditions and boundary conditions, and both are needed for partial differential equations, but there is an important difference in the nature of these conditions that must be acknowledged. Suppose we have a partial differential equation that contains a single time derivative and two spatial derivatives. Because we have a single time derivative, we anticipate the need for an initial condition. So how do we describe the initial state? Since the unknown quantity has a spatial profile that changes in time, we must know the initial spatial profile! In other words, the initial condition requires the initial profile, a function of space.

Similar considerations are in play when we contemplate the boundary conditions that corresponds to the second-order spatial derivatives. We anticipate two boundary conditions, one at each end. Ah, but these conditions may change in time! So the boundary conditions can be functions of time! It is useful to retain the idea of a homogeneous problem introduced by Principle 21 for boundary value problems; a partial differential equation without forcing terms and with homogeneous boundary conditions – no dependency on time – will be considered as constituting a homogeneous problem.

Over the course of time, many techniques have been developed to solve partial differential equations and today they are often solved by numerical techniques. But the "bread and butter" approach is known as separation of variables, and solutions constructed with this technique still serve to introduce the basic concepts underlying the solution to partial differential equations. Separation of variables applies to homogeneous partial differential equations with homogeneous boundary conditions, and is the topic of this section.

### 4.1.1   *Transport equations*

The diffusion equation, like several other partial differential equations, arises as the consequence of the transport of quantities in some region of space. To illustrate the principle of conservation that underlies the derivation of transport equations, let us consider a simple situation. Imagine a cylindrical pipe filled with water that is stationary, and some quantity, a contaminant say, is dumped along a certain section of the pipe. The expectation is that the contaminant will diffuse in the water and the contaminant will spread. Intuitively, the contaminant moves from regions of high concentrations to regions of low concentration, the idea behind diffusion.

Clearly the flux $q(x, t)$ of the contaminant plays an important part in how the concentration changes in time. As the concentration passes from high to low regions the concentration in the high region will decrease and that in the low will increase. But all of this must happen without the loss of any contaminant, and it is this statement of the conservation of contaminant that guides our approach to deriving an equation that governs the change in concentration.

Consider a small segment of the pipe as illustrated in Fig. 4.1. Distance along the pipe is denoted by $x$ (cm), and the concentration of the contamination is denoted by $\rho(x, t)$ (g/cm$^3$). Note that the concentration depends on the location $x$ along the pipe and the time $t$ (s) it is recorded. In addition, the contaminant moves through the pipe and its motion is measured by the flux $q(x, t)$ (g/(cm$^2 \cdot$s)). It is the amount that passes through a unit area in a unit of time. The flux also depends on the location $x$ along the pipe and the time $t$ (s) it is recorded.

We appeal to the conservation of the amount of contaminant as the underlying principle in deriving an equation for the change in concentration. Consider the small volume contained in the segment $[x - \triangle x/2, x + \triangle x/2]$. Let $Q(x, t)$ (g) be the quantity of contaminant in the volume $V = A\triangle x$ centered at $x$, where $A$ (cm$^2$) is the cross-sectional area of the pipe. The amount of contaminant is the concentration $\rho(x, t)$ multiplied by the volume;

$$Q(x, t) = A \triangle x \, \rho(x, t).$$

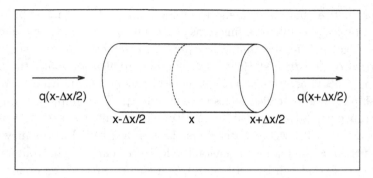

Fig. 4.1   A small segment of the cylindrical pipe, centered at $x$.

The change in the quantity $Q(x,t)$ in a small time interval $\triangle t$ (s) occurs from the difference in the amount entering at $x - \triangle x/2$ and the amount leaving at $x + \triangle x/2$ during $\triangle t$.[1] The increase in the amount is

$$\text{increase} = Q(x, t + \triangle t) - Q(x, t) = A \triangle x \left[ \rho(x, t + \triangle t) - \rho(x, t) \right].$$

The amount entering at $x - \triangle x/2$ during $\triangle t$ is

$$\text{in} = q(x - \triangle x/2, t)\, A \triangle t,$$

and the amount leaving at $x + \triangle x/2$ during $\triangle t$ is

$$\text{out} = q(x + \triangle x/2, t)\, A \triangle t.$$

By invoking the conservation principle "increase = in − out," we are led to

$$A \triangle x \left[ \rho(x, t + \triangle t) - \rho(x, t) \right] = q(x - \triangle x/2, t)\, A \triangle t - q(x + \triangle x/2, t)\, A \triangle t,$$

or

$$\frac{\rho(x, t + \triangle t) - \rho(x, t)}{\triangle t} = -\frac{q(x + \triangle x/2, t) - q(x - \triangle x/2, t)}{\triangle x}. \tag{4.1.1}$$

In the limit as $\triangle x \to 0$ and $\triangle t \to 0$, the expressions in Eq. (4.1.1) become partial derivatives that satisfy

$$\frac{\partial \rho}{\partial t}(x, t) = -\frac{\partial q}{\partial x}(x, t). \tag{4.1.2}$$

This then is the *fundamental statement of the conservation of any quantity,* chosen here to be some contaminant. It is intuitive: if there is a positive gradient in the flux, more leaves than enters and the density will decrease.[2]

---

[1]The direction of the flux is assigned in the same way the direction of velocity is usually assigned. A positive flux indicates movement in the positive direction of $x$. Indeed, flux is a vector much like velocity.

[2]The differential equation describes the changes in density, but it is not the density that is conserved but the quantity.

You may notice that the transport equation Eq. (4.1.6) is incomplete. It is a single equation for two unknown functions $\rho(x,t)$ and $q(x,t)$. There is a good reason why this is the case. *The equation is true for the conservation of any quantity under rather general circumstances!* How does a specific application affect the equation? It does so by specifying a connection between the flux $q(x,t)$ and the density $\rho(x,t)$, a relationship sometimes called a closure relationship. The simplest example arises from the transport (advection) of a quantity with a uniform flow, an example we considered in Sec. 1.2.1 where water flows through a drum. Imagine now that the water flows though the pipe at a constant velocity $v$ (cm/s). The volume of water crossing $x$ in a second will be $vA$ (cm$^3$/s), and the amount of the quantity in this volume will be $\rho vA$ (g/s). Thus the flux $q$, which is the amount crossing $x$ per unit area per second must be

$$q(x,t) = \rho(x,t)v\,, \qquad (4.1.3)$$

leading to the standard *advection equation*,

$$\frac{\partial \rho}{\partial t}(x,t) + v\frac{\partial \rho}{\partial x}(x,t) = 0\,, \qquad (4.1.4)$$

as a result of substituting Eq. (4.1.3) into Eq. (4.1.2). We will explore this equation further in a later section. Note that it is an equation for a single unknown function $\rho(x,t)$ which depends on two independent variables $x$ and $t$ and it contains a single parameter $v$. It is a partial differential equation with single derivatives in $x$ and $t$. Hence it is called a first-order partial differential equation.

Another way flux may arise is the transport of a quantity from high to low concentrations, the larger the difference, the faster the transport, and the concentration quickly diffuses. The standard assumption is that the flux $q(x,t)$ is proportional to the gradient of the concentration, except that the flux is in the opposite direction:

$$q(x,t) = -D\frac{\mathrm{d}\rho}{\mathrm{d}x}(x,t)\,, \qquad (4.1.5)$$

where $D$ (cm$^2$/s) is a constant that depends on the medium through which the quantity is diffusing. When Eq. (4.1.5) is substituted into Eq. (4.1.2) we obtain the standard *diffusion equation*,

$$\frac{\partial \rho}{\partial t}(x,t) = D\frac{\partial^2 \rho}{\partial x^2}(x,t)\,. \qquad (4.1.6)$$

This is a partial differential equation with just one time derivative but two spatial derivatives. Mathematicians call this a second-order equation because there are two derivatives in $x$, even though there is only one derivative in $t$. In general, the highest partial derivative determines the order of the partial differential equation.

### 4.1.2 *Initial and boundary conditions*

A single time derivative in Eq. (4.1.6) suggests that we will need just a single initial condition. In other words, we will need to know the concentration at $t = 0$.

Obviously, this concentration can vary along the pipe. Thus we will need to know $\rho(x, 0)$. For example, if the pipe is of length $L$ (cm), an initial condition might be

$$\rho(x, 0) = \frac{\rho_m}{2}\left[1 - \cos\left(\frac{2\pi x}{L}\right)\right]. \tag{4.1.7}$$

The parameter $\rho_m$ (g/cm$^3$) measures the maximum concentration which occurs at the point $x = L/2$ in the middle of the pipe.

Is one initial condition enough? Let us think back to the time we argued that a single initial condition will determine a specific solution to a first-order differential equation. What we did was to construct the Taylor series. First, we used the differential equation and differentiated it recursively to determine all of the time derivatives of the solution – see Eq. (4.1.26). Then we evaluated these time derivatives at $t = 0$ – see Eq. (1.1.28).

Okay then, let us follow the same approach. Start by recursively differentiating the differential equation Eq. (4.1.6) with respect to time. The first derivative gives

$$\frac{\partial^2 \rho}{\partial t^2}(x, t) = D\frac{\partial}{\partial t}\frac{\partial^2 \rho}{\partial x^2}(x, t)$$

$$= D\frac{\partial^2}{\partial x^2}\frac{\partial \rho}{\partial t}(x, t).$$

All we have done is to interchange the order of the partial derivatives. Repeating the process, we obtain the recursive formula,

$$\frac{\partial^n \rho}{\partial t^n}(x, t) = D\frac{\partial^2}{\partial x^2}\frac{\partial^{n-1} \rho}{\partial t^{n-1}}(x, t). \tag{4.1.8}$$

Now we are ready to evaluate these derivatives at $t = 0$. Since the partial derivatives in $x$ are conducted with $t$ held fixed – in this case at $t = 0$ – we may differentiate the initial condition to obtain

$$\frac{\partial \rho}{\partial t}(x, 0) = D\frac{\partial^2 \rho}{\partial x^2}(x, 0)$$

$$= \frac{\rho_m}{2} D \frac{4\pi^2}{L^2} \cos\left(\frac{2\pi x}{L}\right).$$

Thus we have determined the initial time derivative. By evaluating Eq. (4.1.8) at $t = 0$, we may proceed recursively. For example,

$$\frac{\partial^2 \rho}{\partial t^2}(x, 0) = D\frac{\partial^2}{\partial x^2}\frac{\partial \rho}{\partial t}(x, 0)$$

$$= -\frac{\rho_m}{2} D^2 \left(\frac{4\pi^2}{L^2}\right)^2 \cos\left(\frac{2\pi x}{L}\right).$$

We are able to complete this step because we know the first time derivative as a function of $x$. Continuing the process, we obtain

$$\frac{\partial^n \rho}{\partial t^n}(x, 0) = (-1)^{(n-1)}\frac{\rho_m}{2} D^n \left(\frac{4\pi^2}{L^2}\right)^n \cos\left(\frac{2\pi x}{L}\right). \tag{4.1.9}$$

So now we are in position to evaluate the Taylor series.

$$\rho(x,t) = \sum_{n=0}^{\infty} \frac{\partial^n \rho}{\partial t^n}(x,0) \frac{t^n}{n!}$$

$$= \frac{\rho_m}{2} \left[ 1 - \cos\left(\frac{2\pi x}{L}\right) \right.$$

$$\left. + \sum_{n=1}^{\infty} (-1)^{(n-1)} D^n \left(\frac{4\pi^2}{L^2}\right)^n \cos\left(\frac{2\pi x}{L}\right) \frac{t^n}{n!} \right]. \qquad (4.1.10)$$

This Taylor series looks different. That is because the coefficients depend on $x$. That is okay because we just hold $x$ fixed while we sum up the Taylor series in $t$. The whole procedure to construct the Taylor series depends on judicious choices of when to hold $t$ or $x$ fixed. Just to make this perfectly clear, note that we cannot differentiate the initial condition Eq. (4.1.7) in time; it does not depend on time! Instead we use the differential equation and differentiate the initial condition in $x$ twice. We then proceed recursively by differentiating the differential equation, changing the order of differentiation and then holding $t = 0$ fixed while we complete the calculation of the time derivatives by differentiating in $x$ twice. Subsequently, we sum up the Taylor series while holding $x$ fixed. Very clever!

What happens at the ends of the pipe? For simplicity, let us consider the situation where the pipe is sealed. This may arise because the spill of contaminant in the pipe caused an immediate shutdown at the ends of the segment of the pipe. Another view is that it is a drum lying on its side. Obviously, the concentrate cannot pass through the ends. Since no contaminant can pass through the sealed ends, there can be no flux at the ends; hence from Eq. (4.1.5)

$$\frac{\partial \rho}{\partial x}(0,t) = \frac{\partial \rho}{\partial x}(L,t) = 0. \qquad (4.1.11)$$

These two conditions, written as one statement for convenience, are called *boundary conditions* because they impose conditions on the solution at the ends of the region $0 < x < L$. Why do we need these extra boundary conditions? Well, imagine a different circumstance. One end of the pipe, the one at $x = 0$, is not sealed but is open to a constant source of contamination. Now the concentration is specified, $\rho(0,t) = \rho_m$ say, a different boundary condition from Eq. (4.1.8), and the subsequent concentration profiles will be very different. So, boundary conditions play an important role in specifying the physical situation and the mathematical model is incomplete until they have been given.

How do boundary conditions influence the construction of a specific solution? When we constructed the solution Eq. (4.1.10) by a Taylor series, we paid no attention to the boundary conditions. If the boundary conditions are given by Eq. (4.1.11), then we are in luck because the Taylor series satisfies the boundary conditions. To show this, differentiate the Taylor series Eq. (4.1.10) with respect to $x$,

$$\frac{\partial \rho}{\partial x}(x,t) = \frac{\rho_m}{2} \sum_{n=0}^{\infty} (-1)^n \, D^n \left(\frac{2\pi}{L}\right)^{2n+1} \sin\left(\frac{2\pi x}{L}\right) \frac{t^n}{n!}.$$

Notice that each term contains a sine function that is zero when $x = 0, L$ – the Taylor series will automatically satisfy the boundary conditions!

Unfortunately, the construction of a Taylor series in time will not always work. Indeed, most of the time it does not. The example used here served only to suggest that a single initial condition is all that we need to construct a specific solution and it worked provided we have the boundary conditions Eq. (4.1.11). If the boundary conditions are different, there is no reason why the Taylor series Eq. (4.1.10) will satisfy them; the Taylor series will not be the solution. Clearly then, there is an interplay between the initial conditions and the boundary conditions that raises a fresh challenge in how to proceed.

Just one final note. We see in Chapter 3 how different types of boundary conditions influence the solution to boundary value problems. A new feature here is that the boundary conditions can vary in time! The boundary conditions in Eq. (4.1.11) are homogeneous and do not vary in time, but it is easy to imagine a situation where the contamination at one end does change in time, for example $\rho(0,t) = \exp(-\beta t)$.

The challenge before us, then, is to construct a solution to the partial differential equation Eq. (4.1.6) subject to an initial condition that is some function of $x$ and two boundary conditions that may be functions of $t$. We have already seen that there is an interplay between the boundary conditions and initial condition and our task is to understand how that influences the solution. A good start is to assume a homogeneous partial differential equation and homogeneous boundary conditions so that we can better understand that interplay.

### 4.1.3  *Exponential solutions*

The partial differential equation Eq. (4.1.6) and the boundary conditions Eq. (4.1.11) are homogeneous; there are no forcing effects; all terms contain only the unknown concentration $\rho(x,t)$. Whenever we are faced with homogeneous problems, we know exponential solutions will arise. So it is natural to seek exponential solutions to Eq. (4.1.6). Of course we must take into account that there are two independent variables. So let us try the guess

$$\rho(x,t) = e^{\lambda t + \alpha x} \tag{4.1.12}$$

where $\lambda$ and $\alpha$ are two parameters we hope to determine by substitution into Eq. (4.1.6).

$$\frac{\partial \rho}{\partial t}(x,t) - D \frac{\partial^2 \rho}{\partial x^2}(x,t) = \lambda \, e^{\lambda t + \alpha x} - D \alpha^2 \, e^{\lambda t + \alpha x}$$

$$= (\lambda - D\alpha^2) \, e^{\lambda t + \alpha x}.$$

For the partial differential equation to be satisfied, $\lambda = D\alpha^2$.

Whereas before when an exponential solution is substituted into an ordinary differential equation, the characteristic equation determines the choices for $\lambda$ completely, and we can proceed to the construction of the general solution and the application of the initial conditions. This time, there is only a single equation that connects the choice for $\lambda$ and $\alpha$. Apparently, any choice for $\alpha$ is possible, $\alpha = 2$ say, then $\lambda$ is determined, in this case $\lambda = 4D$, and

$$\rho(x,t) = \mathrm{e}^{4Dt+2x}$$

will be a solution. Unfortunately, this solution does not satisfy the boundary conditions Eq. (4.1.11).

As the situation presently stands, we have many possible choices for $\alpha$ and therefore many possible contributions to the general solution. However, there is no reason to expect this general solution to satisfy the boundary conditions. Somehow we hope that the homogeneous boundary conditions will select appropriate values for $\alpha$ and then the "characteristic equation" $\lambda = D\alpha^2$ will determine $\lambda$. This is precisely what happens in the method of separation of variables, described next.

### 4.1.4   *Separation of variables*

The way forward is based on the observation that the exponential solution can be written as

$$\rho(x,t) = \mathrm{e}^{\lambda t}\,\mathrm{e}^{\alpha x} = T(t)\,X(x)\,,$$

a product of two separate functions. This then, is the starting point for the method of separation of variables. Of course, we still expect the functions $T(t)$ and $X(x)$ to contain exponentials and we want them to be determined by the usual process – remember Principle 1 – of direct substitution into the partial differential equation. What is new is that we also want them to satisfy the boundary conditions and it is this step we hope selects the relevant exponentials.

Now it is time to construct the solution. The process is lengthy, involving several steps.

### 1. Separate the variables.

The starting point is to assume the solution can be decomposed into a product of functions,

$$\rho(x,t) = X(x)\,T(t)\,. \tag{4.1.13}$$

There is no reason yet to believe this is a general solution; after all, many choices of $\alpha$ in Eq. (4.1.9) are possible. The hope is that somewhere in the procedure, specific $\alpha$ will be determined so that the boundary conditions Eq. (4.1.11) are satisfied and then a combination of all the possible choices for $\alpha$ can be used to satisfy the initial conditions.

The choice of capital symbols to denote the functions in Eq. (4.1.13) is purely a matter of convenience. It helps remind us which function is associated with which independent variable. The main value in this representation of the solution is its impact on the partial derivatives.

$$\frac{\partial \rho}{\partial t}(x,t) = X(x)\frac{\mathrm{d}T}{\mathrm{d}t}(t)\,, \tag{4.1.14}$$

$$\frac{\partial \rho}{\partial x}(x,t) = T(t)\frac{\mathrm{d}X}{\mathrm{d}x}(x)\,, \tag{4.1.15}$$

$$\frac{\partial^2 \rho}{\partial x^2}(x,t) = T(t)\frac{\mathrm{d}^2 X}{\mathrm{d}x^2}(x)\,. \tag{4.1.16}$$

In each case, one of the functions is a constant when the partial derivative is performed. The result is the conversion of partial derivatives to ordinary derivatives.

According to Principle 1, we must substitute Eq. (4.1.14) and Eq. (4.1.16) into the partial differential equation Eq. (4.1.6) and demand that the equation is satisfied. Thus

$$X(x)\frac{\mathrm{d}T}{\mathrm{d}t}(t) = D\,T(t)\frac{\mathrm{d}^2 X}{\mathrm{d}x^2}(x)\,. \tag{4.1.17}$$

How can we find $X(x)$ and $T(t)$ so that this equation is always satisfied? We seem to be stuck.

## 2. Separate the equation.

The way forward is to rewrite Eq. (4.1.17) so that the unknown functions $T(t)$ and $X(x)$ appear on separate sides; this is easily accomplished by dividing Eq. (4.1.17) by $X(x)\,T(t)$.

$$\frac{1}{T(t)}\frac{\mathrm{d}T}{\mathrm{d}t}(t) = D\frac{1}{X(x)}\frac{\mathrm{d}^2 X}{\mathrm{d}x^2}(x)\,. \tag{4.1.18}$$

Notice the emphasis on the arguments of all the functions. Why? Because we see clearly that the left hand side of Eq. (4.1.18) depends only on $t$, while the right hand side depends only on $x$. Well, how can that be? This appears to be nonsense. There is only one way this can be true. Both sides can only be some constant $c$!

It is worth exploring this idea carefully. Suppose we have two functions that may represent the left and right hand sides of Eq. (4.1.18); call them $f(t)$ for the left hand side and $g(x)$ for the right hand side. Then we may regard the separated equation as

$$f(t) = g(x)\,. \tag{4.1.19}$$

This appears to be an equation! The variable $t$ is determined by $x$. For example, if $f(t) = t$ and $g(x) = x^2$, then we find $t = x^2$. What is wrong with that? What is wrong is that $t$ and $x$ are independent variables! In other words, Eq. (4.1.19) is *not* an equation linking $t$ and $x$, but is a statement that must be true for all *choices* of

$x$ and $t$. Aaah! This seems hopeless until we see that only the choice $f(t) = c$ and $g(x) = c$ will satisfy Eq. (4.1.19) for all values of $x$ and $t$.[3]

After arranging Eq. (4.1.14) into Eq. (4.1.15), we may conclude that either side must be some constant $c$, called the separation constant.

$$\frac{1}{T(t)}\frac{dT}{dt}(t) = D\frac{1}{X(x)}\frac{d^2X}{dx^2}(x) = c.$$

The result is two separate ordinary differential equations for $T(t)$ and $X(x)$.

$$\frac{dT}{dT}(t) - cT(t) = 0,\qquad(4.1.20)$$

and

$$\frac{d^2X}{dx^2}(x) - \frac{c}{D}X(x) = 0.\qquad(4.1.21)$$

This is great news because we know how to solve these equations! That is exactly what we learnt to do in the previous chapters. But we also know the solution would not be complete unless we add some conditions – Principles 3 and 8. For Eq. (4.1.20), we will need initial conditions, but what about Eq. (4.1.21)? We must apply Eq. (4.1.11) somehow.

### 3. Separate the boundary conditions.
Substitute the separated form for the solution Eq. (4.1.13) into the boundary conditions Eq. (4.1.11). Of course, Eq. (4.1.15) is needed to evaluate the boundary conditions,

$$T(t)\frac{dX}{dx}(0) = 0,\qquad(4.1.22)$$

$$T(t)\frac{dX}{dx}(L) = 0.\qquad(4.1.23)$$

There is no real choice here.[4] We certainly do not want to set $T(t) = 0$, so the other factors must be zero. Thus,

$$\frac{dX}{dx}(0) = \frac{dX}{dx}(L) = 0.\qquad(4.1.24)$$

These boundary conditions, together with the differential equation Eq. (4.1.21), give us hope that we can determine $X(x)$ completely.

---

[3]One way to argue this conclusion is to pick some point $x_1$ and note that $f(t) = a = g(x_1)$ indicates that $f(t)$ is a constant. Now pick another point $x_2$ and this time $f(t) = b = g(x_2)$ is a different constant, which is not possible unless $a = b$. The result does not depend on the choices for $x_1$ or $x_2$. We must conclude that $f(t) = g(x) = c$.

[4]Whenever the product of two quantities must be zero, then either one or both must be zero.

> **Reflection**
>
> This is one of the places where the fact that the boundary conditions are homogeneous is very important. If, for example, the right hand sides of the boundary conditions Eq. (4.1.11) contain functions of time, then the right hand sides of Eq. (4.1.22), Eq. (4.1.23) would contain functions of $t$. We would not be able to satisfy these boundary conditions with a single choice for $T(t)$ nor would we have any way to choose the derivatives of $X(x)$ at the endpoints; the procedure would fail.

## 4. Solve the homogeneous boundary value problem.

The differential equation Eq. (4.1.21) and the boundary conditions Eq. (4.1.24) constitute a homogeneous boundary value problem – see Sec. 3.3.4. As stated in Principle 21, we expect only the trivial solution $X(x) = 0$ in general. However that is of no use to us. Our only hope is that we can find values for the separation constant $c$ for which the homogeneous problem has multiple solutions.

Since Eq. (4.1.21) is a second-order ordinary differential that is linear with constant coefficients, it may be solved with an exponential function $\exp(\alpha x)$ – Principle 8. Upon substitution into Eq. (4.1.21), the characteristic equation becomes

$$\alpha^2 = \frac{c}{D}. \tag{4.1.25}$$

The nature of the general solution depends on whether $\alpha$ is real or complex, and that depends on the sign of $c$. In our search for multiple solutions, we are forced to consider all three cases, $c > 0, = 0, < 0$.

## 5. Consider all possible choices for the separation constant.

(a) Assume $c > 0$. Then the general solution to Eq. (4.1.21) is

$$X(x) = a_1 \, e^{\sqrt{c/D}\,x} + a_2 \, e^{-\sqrt{c/D}\,x}.$$

Now apply the boundary conditions Eq. (4.1.24),

$$a_1 - a_2 = 0,$$
$$a_1 \, e^{\sqrt{c/D}\,L} - a_2 \, e^{-\sqrt{c/D}\,L} = 0.$$

The first condition sets $a_2 = a_1$, and substitution into the second condition leads to

$$a_1 \left( e^{\sqrt{c/D}\,L} - e^{-\sqrt{c/D}\,L} \right) = 0.$$

The only possible solution is $a_1 = 0$ and $a_2 = 0$. This is no good! It means $X(x) = 0$, the trivial solution; our guess for the solution becomes $\rho(x,t) = T(t)\,X(x) = 0$ and the procedure has failed to produce a non-trivial solution.

(b) Assume $c = 0$. The solution to Eq. (4.1.21)[5] is

$$X(x) = a_1 + a_2\, x\,.$$

Now apply the boundary conditions Eq. (4.1.24).

$$\frac{\mathrm{d}X}{\mathrm{d}x}(0) = \frac{\mathrm{d}X}{\mathrm{d}x}(L) = a_2\,.$$

In both cases, $a_2 = 0$, but $a_1$ remains undetermined. This is good; we have found a non-trivial solution. This is not a particularly exciting solution, but at least we have one; $X(x) = a_1$.

(c) Finally, assume $c < 0$. Given the need for the square root of $c$, it is convenient to set

$$c = -D\,r^2\,. \tag{4.1.26}$$

The factor $D$ just simplifies the expression under the square root for the roots of the characteristic equation Eq. (4.1.25), $\alpha = \sqrt{c/D} = \pm i\,r$. All we have done is to replace the unknown $c$ with a new unknown $r$. Once we figure out $r$, we will know $c$.

The general solution to Eq. (4.1.21) may be written in the form,

$$X(x) = a_1\, \cos(rx) + a_2\, \sin(rx)\,. \tag{4.1.27}$$

Apply the boundary conditions Eq. (4.1.24),

$$r\, a_2 = 0\,, \tag{4.1.28}$$

$$-r\, a_1\, \sin(rL) + r\, a_2\, \cos(rL) = 0\,. \tag{4.1.29}$$

Since $r \neq 0$, Eq. (4.1.28) sets $a_2 = 0$, which simplifies Eq. (4.1.29). Either $a_1 = 0$ or $\sin(rL) = 0$, and we do not want $a_1 = 0$. Unlike the situation with exponential solutions, we can find many choices for $r$ that succeed; $rL = n\pi$, for any integer $n = 1, 2, \ldots$ – remember Table 2.2. We have now found many possible solutions! For each choice of $r = n\pi/L$, we have from Eq. (4.1.26) and Eq. (4.1.27),

$$c = -D\,r^2 = -\frac{D\,n^2\pi^2}{L^2} \quad \text{and} \quad X(x) = a_1 \cos\left(\frac{n\pi x}{L}\right)\,.$$

In summary, we have found appropriate choices for $c$ that allow multiple solutions to the homogeneous boundary value problem, (4.1.21) and (4.1.24). They are:

$$X(x) = a_1 \qquad\qquad \text{with } c = 0\,, \tag{4.1.30}$$

$$X(x) = a_1 \cos\left(\frac{n\pi x}{L}\right) \qquad \text{with } c = -\frac{D\,n^2\pi^2}{L^2}\,, \tag{4.1.31}$$

for any positive integer $n$.

---

[5]Strictly speaking, Eq. (4.1.21) is no longer a differential equation. We simply have to integrate twice to obtain the result.

Given our experience with homogeneous boundary value problems in Chapter 3, it is not surprising that we find trigonometric functions that give non-trivial solutions. In this example, they are cosines – the constant may be considered a cosine with $n = 0$.

## 6. Solve the remaining differential equation.

We are tasked with finding solutions to Eq. (4.1.20) and Eq. (4.1.21) for the same choice of $c$. We have succeeded in finding several solutions to Eq. (4.1.21) subject to the boundary conditions Eq. (4.1.24) with different choices for $c$. Now we must determine the corresponding function $T(t)$ by solving Eq. (4.1.20) for each successful choice of $c$.

With $c = 0$, the solution to Eq. (4.1.20) is obviously

$$T(t) = c_1 \,,$$

some constant.

For each choice of $n$ in Eq. (4.1.31), Eq. (4.1.20) becomes

$$\frac{\mathrm{d}T}{\mathrm{d}t}(t) + \frac{Dn^2\pi^2}{L^2} T(t) = 0 \,.$$

The solution is an exponential $\exp(\lambda t)$ with $\lambda = -Dn^2\pi^2/L^2$,

$$T(t) = c_1 \, e^{-Dn^2\pi^2 t/L^2} \,.$$

In summary, we have found the following solutions for $T(t)$.

$$T(t) = c_1 \qquad\qquad \text{with } c = 0 \,, \qquad\qquad (4.1.32)$$

$$T(t) = c_1 \, e^{-Dn^2\pi^2 t/L^2} \qquad \text{with } c = -\frac{Dn^2\pi^2}{L^2} \,, \qquad (4.1.33)$$

for any positive integer $n$.

Looking back to the exponential solution Eq. (4.1.12), we now see that $\alpha = \pm i\, n\pi/L$ is the appropriate choice to satisfy the partial differential equation and the boundary conditions, and consequently $\lambda = -Dn^2\pi^2/L^2$. The method of separation of variables has led us to these choices.

## 7. Compose a homogeneous solution.

Now that $X(x)$ and $T(t)$ are known we can multiply them together and obtain the following solutions $\rho(x,t) = T(t)\,X(x)$ for each successful choice of $c$.

First, corresponding to the choice $c = 0$, we use Eq. (4.1.30) and Eq. (4.1.32) to obtain

$$\rho(x,t) = a_1 \, c_1 = a_0 \,, \tag{4.1.34}$$

where we invoke Principle 6 by replacing a product of unknown coefficients with a single unknown coefficient. The choice $a_0$ reflects the choice $c = 0$.

Next, for each positive $n$ we use Eq. (4.1.31) and Eq. (4.1.33) to obtain

$$\rho(x,t) = a_n \, \mathrm{e}^{-Dn^2\pi^2 \, t/L^2} \cos\left(\frac{n\pi}{L} x\right) \tag{4.1.35}$$

for each positive integer $n$. The product of the unknown coefficients $a_1$ and $c_1$ has been replaced – Principle 6 – by the unknown coefficient $a_n$, the subscript reminding us that the coefficient is associated with the choice of $n$ in $c$.

What do we do with all these solutions? We follow the standard idea and compose a homogeneous solution by invoking Principle 4. We simply add them all together.

$$\rho(x,t) = a_0 + \sum_{n=1}^{\infty} a_n \mathrm{e}^{-D\,n^2\pi^2 t/L^2} \cos\left(\frac{n\pi x}{L}\right). \tag{4.1.36}$$

What remains undetermined are all the coefficients $a_n$, and that is where the last condition, the initial condition Eq. (4.1.7), comes into play.

---

**Reflection**

A glance at the solution Eq. (4.1.36) suggests that it has the form of a cosine Fourier series. That perspective is brought more clearly into focus if we write the solution as

$$\rho(x,t) = a_0 + \sum_{n=1}^{\infty} a_n(t) \cos\left(\frac{n\pi x}{L}\right),$$

where

$$a_n(t) = a_n \mathrm{e}^{-D\,n^2\pi^2 t/L^2}.$$

The solution is a Fourier series where the Fourier coefficients change in time.

---

**8. Satisfy the initial condition.**
By setting $t = 0$ in Eq. (4.1.36), we require Eq. (4.1.7),

$$\rho(x,0) = a_0 + \sum_{n=1}^{\infty} a_n \cos\left(\frac{n\pi x}{L}\right) = \frac{\rho_m}{2}\left[1 - \cos\left(\frac{2\pi x}{L}\right)\right]. \tag{4.1.37}$$

We can make a perfect match if we set $a_0 = \rho_m/2$ and $a_2 = -\rho_m/2$. All other values for $a_n$ should be set to $a_n = 0$. With the coefficients $a_n$ thus determined we have completed the solution!

$$\frac{\rho(x,t)}{\rho_m} = \frac{1}{2}\left[1 - e^{-4D\pi^2 t/L^2}\cos\left(\frac{2\pi x}{L}\right)\right]. \qquad (4.1.38)$$

**Reflection**

Since the initial condition Eq. (4.1.7) is already expressed as a simple Fourier series, the match of the general solution Eq. (4.1.36) to the initial condition is easy.

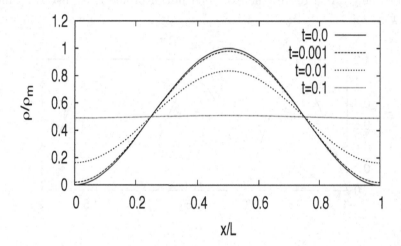

Fig. 4.2   Profiles of the concentration $\rho/\rho_m$ shown at the times indicated in the legend in seconds; $D/L^2 = 1\ \mathrm{s}^{-1}$.

Now we are in a position to graph the results. The profiles of $\rho(x,t)/\rho_m$ at different times are shown in Fig. 4.2 with the choice $D/L^2 = 1\ \mathrm{s}^{-1}$. Noticeable is the smoothing of the profile in time until it becomes essentially flat. Diffusion tends to smooth out any variations until the state is uniform. This behavior is reflected in the exponential decay in time of the amplitude of the cosine term. The time scale of the decay is $L^2/(4\pi^2 D)$. Eventually only $a_0 = 0.5\,\rho_m$ survives. It is, of course, the average value of the initial concentration, and this constant value must not change in time because of the conservation of the concentrate.

Our ability to match the constant and cosine term in Eq. (4.1.37) seems rather fortuitous. What can be done if the initial condition is some other function, for example,

$$\rho(x,0) = \rho_m \sin\left(\frac{\pi x}{L}\right)? \qquad (4.1.39)$$

Now the initial condition becomes

$$a_0 + \sum_{n=1}^{\infty} a_n \cos\left(\frac{n\pi x}{L}\right) = \rho_m \sin\left(\frac{\pi x}{L}\right). \tag{4.1.40}$$

Let us look carefully at what Eq. (4.1.40) suggests. On the left appears a cosine series for a function that must be even and $2L$-periodic. On the right appears a sine function that is odd and $2L$-periodic. But wait! The function on the left is the initial profile for $\rho(x,0)$ and that is only specified in the pipe $0 < x < L$. We have no idea what the initial condition is anywhere else, because there is no pipe elsewhere. The way out of the difficulty is to specify that the initial profile for the concentration be extended as an even function that is $2L$-periodic so that it has a cosine series only. As a result, the extended initial profile looks like the function shown in Fig. 4.3.

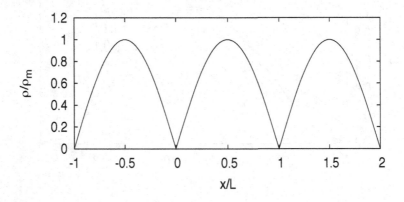

Fig. 4.3   Profile of the initial concentration $\rho/\rho_m$ over a range that includes adjacent extensions.

For Eq. (4.1.39) the even extension to a $2L$-periodic function is easily done;

$$\rho(x,0) = \rho_m \left| \sin\left(\frac{\pi}{L} x\right) \right| \tag{4.1.41}$$

which has the Fourier cosine series,

$$\rho_m \left| \sin\left(\frac{\pi}{L} x\right) \right| = A_0 + \sum_{n=1}^{\infty} A_n \cos\left(\frac{n\pi x}{L}\right)$$

where the Fourier coefficients are determined by Eq. (2.3.10) and Eq. (2.3.11):

$$A_0 = \frac{\rho_m}{L} \int_0^L \sin\left(\frac{\pi x}{L}\right) dx$$

$$= \frac{2\rho_m}{\pi},$$

and by making use of the trigonometric addition formula Eq. (D.2.7) and the entries in Table 2.2,

$$
\begin{aligned}
A_n &= \frac{2\rho_m}{L} \int_0^L \sin\left(\frac{\pi x}{L}\right) \cos\left(\frac{n\pi x}{L}\right) dx \\
&= \frac{\rho_m}{L} \int_0^L \sin\left[\frac{(n+1)\pi x}{L}\right] dx - \frac{\rho_m}{L} \int_0^L \sin\left[\frac{(n-1)\pi x}{L}\right] dx \\
&= \frac{\rho_m}{\pi(n+1)} \left[1 + (-1)^n\right] - \frac{\rho_m}{\pi(n-1)} \left[1 + (-1)^n\right] \quad (\text{if } n \neq 1) \\
&= -\frac{2\rho_m}{\pi(n^2 - 1)} \left[1 + (-1)^n\right].
\end{aligned}
$$

Only terms with $n$ even are non-zero. Notice also we should treat $n = 1$ differently since we cannot divide by $n - 1$.

$$
\begin{aligned}
A_1 &= \frac{2\rho_m}{L} \int_0^L \sin\left(\frac{\pi x}{L}\right) \cos\left(\frac{\pi x}{L}\right) dx \\
&= \frac{\rho_m}{L} \int_0^L \sin\left(\frac{2\pi x}{L}\right) dx \\
&= 0.
\end{aligned}
$$

But the conclusion is the same; only $A_n$ with $n$ even are non-zero.

How does the Fourier cosine representation for the extended initial concentration profile Eq. (4.1.41) help us? Wonderfully, because the initial condition becomes

$$
\begin{aligned}
a_0 + \sum_{n=1}^{\infty} a_n \cos\left(\frac{n\pi x}{L}\right) &= \rho_m \sin\left(\frac{\pi}{L} x\right) \\
&= \frac{2\rho_m}{\pi} - \frac{2\rho_m}{\pi} \sum_{n=1}^{\infty} \frac{1 + (-1)^n}{n^2 - 1} \cos\left(\frac{n\pi x}{L}\right)
\end{aligned}
$$

and now a match of the two series gives

$$
a_0 = A_0 = \frac{2\rho_m}{\pi},
$$

$$
a_n = A_n = -\frac{2\rho_m}{\pi} \frac{1 + (-1)^n}{n^2 - 1}.
$$

Finally, by replacing $a_n$ in Eq. (4.1.36), the final solution is completely determined.

$$
\frac{\rho(x,t)}{\rho_m} = \frac{2}{\pi} - \frac{2}{\pi} \sum_{n=1}^{\infty} \frac{1 + (-1)^n}{n^2 - 1} e^{-Dn^2\pi^2 t/L^2} \cos\left(\frac{n\pi x}{L}\right). \tag{4.1.42}
$$

Now we are in a position to graph the results. The profiles of $\rho(x,t)/\rho_m$ at different times are shown in Fig. 4.4 with the choice $D/L^2 = 1 \text{ s}^{-1}$. Once again the profile smoothes in time until it becomes essentially flat. Diffusion tends to

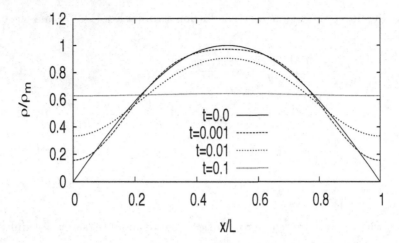

Fig. 4.4   Profiles of the concentration $\rho/\rho_m$ shown at the times indicated in the legend in seconds; $D/L^2 = 1$ s$^{-1}$.

smooth out any variations until the state is uniform. This behavior is reflected in the exponential decay in time of all the terms containing a cosine. The slowest decay occurs with the choice $n = 2$, and the time scale of the decay is $L^2/(4\pi^2 D)$. Eventually only $a_0$ survives. It is, of course, the average value of the concentration initially, and this constant value must not change in time because of the conservation of the pollutant.

### 4.1.5   *Abstract view*

A completely abstract view is not yet possible because there has been only one example, but it is possible to note what assumptions are necessary to make the procedure work, the starting point to a more abstract view. To that end, take the partial differential equation to be the diffusion equation,

$$\frac{\partial u}{\partial t}(x,t) - D\frac{\partial^2 u}{\partial x^2}(x,t) = 0. \tag{4.1.43}$$

Often the diffusion equation is written as in Eq. (4.1.6).[6] Here it is written with all terms containing the unknown function $u(x,t)$ on the left hand side. This form shows clearly that the equation is linear and homogeneous.

     The consequence is that if we can find two different solutions then their sum is a solution. Let $u_1(x,t)$ and $u_2(x,t)$ be two solutions. In other words,

$$\frac{\partial u_1}{\partial t}(x,t) - D\frac{\partial^2 u_1}{\partial x^2}(x,t) = 0, \tag{4.1.44}$$

$$\frac{\partial u_2}{\partial t}(x,t) - D\frac{\partial^2 u_2}{\partial x^2}(x,t) = 0. \tag{4.1.45}$$

---

[6]One reason that scientists and engineers like to write the diffusion equation in the form Eq. (4.1.6) is that they view the diffusive effects as "forcing" the evolution of the concentration.

Now substitute their sum, $u(x,t) = u_1(x,t) + u_2(x,t)$ into Eq. (4.1.43).

$$\frac{\partial u}{\partial t}(x,t) - D\frac{\partial^2 u}{\partial x^2}(x,t) = \frac{\partial(u_1 + u_2)}{\partial t}(x,t) - D\frac{\partial^2(u_1 + u_2)}{\partial x^2}(x,t)$$

$$= \frac{\partial u_1}{\partial t}(x,t) + \frac{\partial u_2}{\partial t}(x,t) - D\frac{\partial^2 u_1}{\partial x^2}(x,t) - D\frac{\partial^2 u_2}{\partial x^2}(x,t)$$

$$= \left[\frac{\partial u_1}{\partial t}(x,t) - D\frac{\partial^2 u_1}{\partial x^2}(x,t)\right]$$

$$+ \left[\frac{\partial u_2}{\partial t}(x,t) - D\frac{\partial^2 u_2}{\partial x^2}(x,t)\right]$$

$$= 0.$$

All that has been used is the fact that a partial derivative of a sum is a sum of the partial derivatives, and that $u_1(x,t)$ and $u_2(x,t)$ are solutions to the equation as expressed in Eq. (4.1.44) and Eq. (4.1.45).

Of course, if there are more solutions $u_n(x,t)$ then we can keep adding them to create new solutions. For example, the solutions Eq. (4.1.34) and Eq. (4.1.35) are added together to create Eq. (4.1.36). A word of warning though. If there are an infinite number of solutions, as in Eq. (4.1.35), their sum will contain infinitely many terms, as in Eq. (4.1.36), and we must check at some point that the sum converges!

What about the influence of the boundary conditions? The ones used in our study of the diffusion of the concentrate in a pipe, Eq. (4.1.11), are examples of homogeneous boundary conditions because there are no forcing terms. There are many ways for boundary conditions to arise, but for the moment, we concentrate on two choices: boundary conditions specifying the solution,

$$u(0,t) = 0, \qquad u(L,t) = 0, \tag{4.1.46}$$

or boundary conditions specifying the derivative,

$$\frac{\partial u}{\partial x}(0,t) = 0, \qquad \frac{\partial u}{\partial x}(L,t) = 0. \tag{4.1.47}$$

Both sets of boundary conditions are considered homogeneous because there are no forcing effects: The right hand sides in Eq. (4.1.46) and Eq. (4.1.47) are zero.

The importance of boundary conditions being homogeneous is the same as the importance of a partial differential equation being homogeneous. The sum of two solutions is also a solution. Suppose we have two solutions $u_1(x,t)$ and $u_2(x,t)$ that satisfy the partial differential equation Eq. (4.1.43). We have already verified that the sum $u(x,t) = u_2(x,t) + u_2(x,t)$ satisfies Eq. (4.1.43). In addition, suppose they satisfy the boundary conditions Eq. (4.1.47). In other words,

$$\frac{\partial u_1}{\partial x}(0,t) = 0, \qquad \frac{\partial u_1}{\partial x}(L,t) = 0,$$

$$\frac{\partial u_2}{\partial x}(0,t) = 0, \qquad \frac{\partial u_2}{\partial x}(L,t) = 0.$$

As a consequence, the sum $u(x,t) = u_1(x,t) + u_2(x,t)$ will satisfy the same boundary conditions,

$$\frac{\partial u}{\partial x}(0,t) = \frac{\partial u_1}{\partial x}(0,t) + \frac{\partial u_2}{\partial x}(0,t) = 0\,,$$

$$\frac{\partial u}{\partial x}(0,t) = \frac{\partial u_1}{\partial x}(0,t) + \frac{\partial u_2}{\partial x}(0,t) = 0\,.$$

The same will hold true for the other boundary conditions Eq. (4.1.46).

To drive home the importance of homogeneous boundary conditions, let us consider a replacement for one of the conditions in Eq. (4.1.46).

$$u(0,t) = A(t)\,,$$

where $A(t)$ is some function. Even if $u_1(x,t)$ and $u_2(x,t)$ satisfy this condition,

$$u_1(0,t) = A(t)\,, \qquad u_2(0,t) = A(t)\,,$$

their sum $u(x,t) = u_1(x,t) + u_2(x,t)$ will not!

$$u(0,t) = u_1(0,t) + u_2(0,t) = 2A(t)\,. \tag{4.1.48}$$

Only if $A(t) = 0$ will they all satisfy the same boundary condition!

---

### Principle 24

*Any sum of solutions to a homogeneous partial differential equation and homogeneous boundary conditions is a solution. If the sum is infinite, convergence of the sum must be checked.*

---

The importance of this Principle 24 is that it gives us the opportunity to compose a general solution, such as Eq. (4.1.36), that we can use to satisfy the remaining condition, the initial condition Eq. (4.1.37) or Eq. (4.1.39). The main purpose of the method of separation of variables, then, is to construct all the solutions to the homogeneous differential equation and homogeneous boundary conditions in order to form a general solution. Then use the general solution to satisfy any remaining condition that may contain some function, for example, the initial condition.

### 4.1.6    *Exercises*

(1) Confirm that the solutions Eq. (4.1.10) and Eq. (4.1.38) are the same.

**A fundamental property of the diffusion equation is revealed in the next exercise.**

(2) In the solutions Eq. (4.1.38) and Eq. (4.1.42), the constant in the Fourier series does not change in time. This means that the average value of $\rho$ remains constant in time, a natural consequence of the conservation of the amount of salt. There is another way this result can be seen. Integrate Eq. (4.1.6) over the length of the pipe, and use the boundary conditions Eq. (4.1.11) to show that

$$\frac{\mathrm{d}}{\mathrm{d}t} \int_0^L \rho(x,t)\,\mathrm{d}x = 0\,.$$

Since the integral is just $a_0 L$, you will have shown that $a_0$ does not change in time and that it must agree with the value determined by the initial profile.

**Convergence of the Fourier series**

(3) Note the form of the Fourier series in the solution Eq. (4.1.42). What is the nature of the decay of the Fourier coefficients at $t = 0$? Is this the expected result? How does the convergence change when $t > 0$? When $t = 0.001$ s, estimate the number of terms needed to obtain the graph in Fig. 4.4.

**The possibility of a point source for an initial condition is considered in the next exercise.**

(4) Suppose that the initial salt concentration in the pipe with sealed ends is given by

$$\rho(x,0) = \begin{cases} Q/(AH), & \text{for } (L - H)/2 < x < (L + H)/2\,, \\ 0, & \text{elsewhere}\,. \end{cases}$$

The salt is deposited in the middle over a region of length $H < L$. The cross-sectional area is $A$. Show that the total amount of salt deposited is $Q$. Find the Fourier coefficients in the solution Eq. (4.1.36). Plot the results for the choices $D/L^2 = 1$ s$^{-1}$, and $H/L = 1/2, 1/4, 1/8$ at times $t = 0.01, 0.1$ s: you need only consider the first three terms in the Fourier series for the solution. What do you observe happening? Can you take the limit as $H \to 0$; this describes the deposit of a point source at $L/2$. What does the solution look like at $t = 0.01, \ldots, 0.1$ s? Compare it to the sequence in $H$.

**Answer:** Results at $t = 0.01$ s are shown in Fig. 4.5. Observe that the profile for small $H$ looks somewhat like a Gaussian.

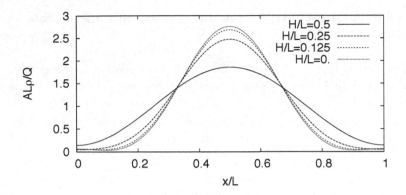

Fig. 4.5    Concentration profiles at $t = 0.01$ s for different values of $H/L$: $D/L^2 = 1\ \mathrm{s}^{-1}$.

**The next exercise explores the importance of the origin of the coordinate system.**

(5) Let us reconsider the salt concentration problem Eq. (4.1.6) and Eq. (4.1.11) but on a shifted interval $(-L/2 < x < L/2)$. In other words, the boundary conditions are

$$\frac{\partial \rho}{\partial x}(-L/2, t) = 0, \quad \text{and} \quad \frac{\partial \rho}{\partial x}(L/2, t) = 0.$$

The initial condition is

$$\rho(x, 0) = \frac{\rho_m}{2}\left\{1 - \cos\left[\frac{2\pi}{L}\left(x + \frac{L}{2}\right)\right]\right\}.$$

Construct the solution by the method of separation of variables. Plot the solution for the same circumstances as in Fig. 4.2. You should expect the same profiles just shifted over. Is there a connection between the Fourier series for the shifted region and the Fourier series in Eq. (4.1.38)? Which choice of coordinates proves easier to treat during separation of variables?

**Different boundary conditions**

(6) Suppose the contamination has filled the sealed pipe so that the density is uniform $\rho_m$. Now study the consequence of removing the contamination at the ends. Imagine there are exchange pumps that have the ability to remove the contaminated water, a filtering process removes the contaminant, and the exchange pump returns clean water so that the volume of water is unchanged. Assume the process can reduce the density to zero at the ends and maintain zero density for the rest of time. How long will it take to reduce the maximum density in the pipe to below a tolerance level $\rho_c$?

**Answer:** For the choice $D/L^2 = 1\ \mathrm{s}^{-1}$ and $\rho_c = 0.01\rho_m$, the time will be 0.5 s.

## 4.2   Diffusion: Part II

There are other ways the diffusion equation Eq. (4.1.16) arises but with the same main idea, namely, that the flux is proportional to the gradient but in the opposite direction. Another important example is the transport of energy. For solids, the internal energy is a measure of the agitation of the molecules making up the solid: The more agitated the molecules the higher the internal energy. Unfortunately, the internal energy is difficult to measure: Instead the temperature is used. For most solids, a slight increase in the temperature, produced by adding heat, increases the internal energy slightly. Mathematically, if $T$ (K) is the temperature and $e(T)$ (J/kg) is the internal energy per unit mass, then

$$\frac{de}{dT}(T) = c(T),$$

where the slope $c(T)$ (J/(kg·K)) is called the heat capacity per unit mass. Provided the changes in $T$ are not too large, $c(T)$ is a constant, dependent only on the material properties of the solid.

The temperature in a solid, for example a rod, may vary along the rod and change in time. The concentration of internal energy is then $\rho e(T(x,t))$ (J/m$^3$) where $\rho$ (kg/m$^3$) is the density of the solid, assumed a constant. Consequently, the rate of change of the energy concentration will be

$$\frac{\partial}{\partial t}\left[\rho e(T(x,t))\right] = \rho \frac{de}{dT}(T)\frac{\partial T}{\partial t}(x,t) = \rho c \frac{\partial T}{\partial t}(x,t). \tag{4.2.1}$$

Fourier (1822) introduced the idea that the flux of internal energy $q(x,t)$ (J/(m$^2 \cdot$ s)) is proportional to the negative of temperature gradient.

$$q(x,t) = -k\frac{\partial T}{\partial x}(x,t), \tag{4.2.2}$$

where $k$ (J/(m·K·s)) is called the thermal conductivity. The insertion of (4.2.1) and (4.2.2) into (4.1.2) gives the equation for the conduction of heat in a solid.

$$\frac{\partial T}{\partial t}(x,t) = \alpha \frac{\partial^2 T}{\partial x^2}(x,t), \tag{4.2.3}$$

where $\alpha = k/(\rho c)$ (m$^2$/s) is called the diffusivity. Note the obvious similarity of Eq. (4.2.3) and Eq. (4.1.6).

### 4.2.1   *Inhomogeneous boundary conditions*

Let us consider as the next example the heat diffusion equation Eq. (4.2.3) applied to a cylindrical rod encased in a thermal insulator[7] except at the ends where we maintain some constant temperatures.

$$T(0,t) = T_0, \quad \text{and} \quad T(L,t) = T_1. \tag{4.2.4}$$

These boundary conditions are not homogeneous. As demonstrated in Eq. (4.1.48), inhomogeneous boundary conditions do not allow linear combinations of solutions, and so the solution cannot be constructed by the method of separation of variables.

---

[7]There is no loss of energy through the cylindrical surface.

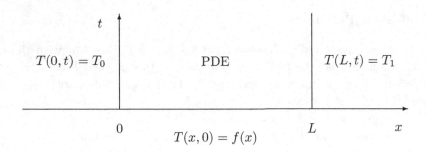

It is helpful to draw a mathematical diagram of what is going on. The region in the $(x, t)$ plane where the solution is desired is shown in the diagram with vertical boundaries drawn at $x = 0$ and $x = L$ for $t > 0$, and with the bottom along $0 < x < L$ for $t = 0$. The top is at $t = \infty$ between 0 and $L$. The partial differential equation determines the solution in the interior and the boundary conditions specify the solution on the sides. The initial condition starts the behavior of the solution and at each subsequent moment in time, a profile of the solution is found along a horizontal line in the region.

This diagram helps us appreciate that the boundary conditions lie along fixed vertical lines at $x = 0$ and $x = L$, and that they can depend on time. For the example we are considering, the functions of time are just constants, but note that there are different constants on the left and right. If these constants were both zero then we could apply the method of separation of variables as done in the previous section. So we follow the standard strategy by determining a *particular* solution that takes care of the forcing effects at the boundaries and then adding a *homogeneous* solution using homogeneous boundary conditions. Our expectation then is that the *general* solution we construct this way can be used to satisfy the initial condition, and then we are done.

Since the boundary conditions can depend on time, the approach to the choice of particular solution focuses on the time behavior. Since the temperatures at the end appear as forcing effects that are constants in $t$, we should try a guess that is a constant in $t$. On the other hand, there is no reason why the particular solution cannot vary with $x$. Indeed, it must vary with $x$ otherwise the particular solution could not match different constants at each boundary. So let us try $T_p(x, t) = f(x)$ as the guess for the particular solution.

For the particular solution to be successful, it must satisfy the differential equation and the two boundary conditions, but it does not have to satisfy the initial condition. Substitute the guess for the particular solution into Eq. (4.2.3),

$$\alpha \frac{d^2 f}{dx^2}(x) = 0 \,.$$

This equation is easily integrated.

$$f(x) = c_0 + c_1 x \,.$$

Now apply the boundary conditions Eq. (4.2.4) to obtain

$$T_{\mathrm{p}}(x,t) = f(x) = T_0 + \frac{T_1 - T_0}{L}\, x\,. \tag{4.2.5}$$

This then is the particular solution!

Next, we seek a general solution solution by adding a homogeneous solution to the particular solution.

$$T(x,t) = T_{\mathrm{p}}(x,t) + T_{\mathrm{h}}(x,t) = T_0 + \frac{T_1 - T_0}{L}\, x + T_{\mathrm{h}}(x,t)\,. \tag{4.2.6}$$

First, let us confirm that $T_{\mathrm{h}}(x,t)$ must satisfy the homogeneous partial differential equation. By substituting Eq. (4.2.6) into Eq. (4.2.3), we find

$$\frac{\partial (T_{\mathrm{p}} + T_{\mathrm{h}})}{\partial t}(x,t) - \alpha\, \frac{\partial^2 (T_{\mathrm{p}} + T_{\mathrm{h}})}{\partial x^2}(x,t) = \frac{\partial T_{\mathrm{h}}}{\partial t}(x,t) - \alpha\, \frac{\partial^2 T_{\mathrm{h}}}{\partial^2 x}(x,t) = 0\,.$$

The result should be obvious because $T_{\mathrm{p}}(x,t)$ is chosen to satisfy Eq. (4.2.3). Next, let us check the boundary conditions that $T_{\mathrm{h}}(x,t)$ must satisfy. Substitute Eq. (4.2.6) into Eq. (4.2.4) to obtain

$$T_{\mathrm{p}}(0,t) + T_{\mathrm{h}}(0,t) = T_0 + T_{\mathrm{h}}(0,t) = T_0\,, \quad \text{so} \quad T_{\mathrm{h}}(0,t) = 0\,, \tag{4.2.7}$$

$$T_{\mathrm{p}}(L,t) + T_{\mathrm{h}}(L,t) = T_1 + T_{\mathrm{h}}(L,t) = T_1\,, \quad \text{so} \quad T_{\mathrm{h}}(L,t) = 0\,. \tag{4.2.8}$$

The intent of the particular solution is to take care of the inhomogeneous boundary conditions and it does, leaving only homogeneous boundary conditions for $T_{\mathrm{h}}(x,t)$.

### 4.2.2  *Homogeneous solution*

Well, the construction of the homogeneous solution proceeds by the method of separation of variables.

**1. Separate the variables.**
Set $T_{\mathrm{h}}(x,t) = F(x)\, G(t)$ and substitute into Eq. (4.2.3).

$$F(x)\, \frac{\mathrm{d}G}{\mathrm{d}t}(t) = \alpha G(t)\, \frac{\mathrm{d}^2 F}{\mathrm{d}x^2}(x)\,.$$

**2. Separate the equation.**
Divide by $\alpha F G$ and make sure that only terms with each independent variable appear on different sides of the equation. Arranging for $\alpha$ to be with the term in $t$ makes the algebra easier later on.

$$\frac{1}{\alpha G(t)}\, \frac{\mathrm{d}G}{\mathrm{d}t}(t) = \frac{1}{F(x)}\, \frac{\mathrm{d}^2 F}{\mathrm{d}x^2}(x)\,. \tag{4.2.9}$$

The only way Eq. (4.2.9) makes sense is that the left and right hand sides are the same constant, $c$ say. Thus we have two ordinary differential equations to solve.

$$\frac{\mathrm{d}G}{\mathrm{d}t}(t) = \alpha\, c\, G(t)\,, \tag{4.2.10}$$

$$\frac{\mathrm{d}^2 F}{\mathrm{d}x^2}(x) = c\, F(x)\,. \tag{4.2.11}$$

## 3. Separate the boundary conditions.

When we substitute $T_h(x,t) = F(x)\,G(t)$ into the boundary conditions Eq. (4.2.7) and Eq. (4.2.8), we find

$$F(0) = F(L) = 0. \tag{4.2.12}$$

These are the boundary conditions to be used with Eq. (4.2.11), and together we have a homogeneous boundary value problem.

## 4. Solve the homogeneous boundary value problem.

Homogeneous boundary value problems always have the trivial solution, but we are interested in the special values for $c$ that will allow non-trivial solutions. We start with an exponential as the guess for the solution,

$$F(x) = C\,e^{rx}, \quad \text{where } r^2 = c.$$

Clearly, the nature of the solution depends on the sign of $c$, so we must consider three possible cases: $c > 0, = 0, < 0$.

## 5. Consider all possible choices for the separation constant.

(a) For the first case, assume $c > 0$, then $r = \pm\sqrt{c}$. The general solution is

$$F(x) = b_1\,e^{\sqrt{c}\,x} + b_2\,e^{-\sqrt{c}\,x},$$

and the application of the boundary conditions Eq. (4.2.12) leads to

$$b_1 + b_2 = 0,$$
$$b_1\,e^{\sqrt{c}\,L} + b_2\,e^{-\sqrt{c}\,L} = 0.$$

The only solution is $b_1 = b_2 = 0$, which leads to the trivial solution, and that is no good.

(b) For the second case, assume $c = 0$. The solution to Eq. (4.2.11) is

$$F(x) = b_1 + b_2\,x.$$

When Eq. (4.2.12) is applied, the only solution is $b_1 = b_2 = 0$, which is also no good.

(c) For the last case, assume $c < 0$. It is more convenient to write $c = -d^2$ where $d$ is real and $d > 0$. Clearly this choice forces $c < 0$. Further, $r = \sqrt{c} = \sqrt{-d^2} = \pm i\,d$. The general solution to Eq. (4.2.11) can be written as

$$F(x) = b_1 \cos(dx) + b_2 \sin(dx).$$

Apply the boundary conditions Eq. (4.2.12). The first condition gives $b_1 = 0$. The second condition then gives

$$b_2 \sin(dL) = 0.$$

This is the only case where we are not forced to set $b_1 = b_2 = 0$. Instead, there is another option, $\sin(dL) = 0$, or $dL = n\pi$ for some positive integer $n$.[8] Consequently, we find non-trivial solutions,

$$F(x) = b_2 \sin\left(\frac{n\pi}{L}x\right) \quad \text{with } c = -\frac{n^2\pi^2}{L^2}. \qquad (4.2.13)$$

In summary, we have found the family of solutions Eq. (4.2.13) as the only candidates for $F(x)$.

## 6. Solve the remaining differential equation.
Now that we have determined $F(x)$, we turn to Eq. (4.2.10) for $G(t)$ and solve it with the values of $c$ determined in Eq. (4.2.13).

$$G(t) = a\,e^{-\alpha n^2\pi^2 t/L^2}. \qquad (4.2.14)$$

## 7. Compose a homogeneous solution.
For each choice $n$, there is a specific separation constant $c$, and solutions for $G(t)$ Eq. (4.2.14) and $F(x)$ Eq. (4.2.13). By multiplying them together, we obtain the separated solution

$$T_h(x,t) = a\,e^{-\alpha n^2\pi^2 t/L^2}\, b_2 \sin\left(\frac{n\pi}{L}x\right).$$

This solution contains two unknown coefficients, $a$ and $b_2$, multiplied together. By Principle 6, we may replace this product by another single coefficient. Our choice should reflect the fact that we did this for a choice of $n$; the coefficient will be different for each choice of $n$. Let us pick the coefficient to be $b_n$.

Finally, we compose a homogeneous solution by summing over all choices for $n$.

$$T_h(x,t) = \sum_{n=1}^{\infty} b_n e^{-\alpha n^2\pi^2 t/L^2} \sin\left(\frac{n\pi}{L}x\right).$$

The general solution is simply the addition of the particular solution Eq. (4.2.5) and the homogeneous solution.

$$T(x,t) = T_0 + \frac{T_1 - T_0}{L}x + \sum_{n=1}^{\infty} b_n e^{-\alpha n^2\pi^2 t/L^2} \sin\left(\frac{n\pi}{L}x\right). \qquad (4.2.15)$$

## 8. Satisfy the initial condition.
To complete the solution, we need some initial condition. Suppose the temperature was at $T_0$ initially and then we suddenly raise the temperature at $x = L$ to $T_1$. In other words,

$$T(x,0) = T_0. \qquad (4.2.16)$$

---

[8]Why only positive integers? Obviously $n = 0$ is not allowed since $d \neq 0$, and if we pick a negative $n$, then we find nothing new since $\sin(n\pi x/L) = -\sin(-n\pi x/L)$. We have merely obtained a copy of the solution with $n$ positive.

Notice that the initial condition does not satisfy the boundary condition $T((L, 0) = T_1$. Instead, we interpret the temperature at the end as $T(L, 0) = T_0$, but $T(L, t) = T_1$ for $t > 0$. The temperature has a jump discontinuity in time at the end $x = L$.[9]

By applying Eq. (4.2.16) to Eq. (4.2.15),

$$T_0 + \frac{T_1 - T_0}{L} x + \sum_{n=1}^{\infty} b_n \sin\left(\frac{n\pi}{L} x\right) = T_0 \,,$$

or

$$\sum_{n=1}^{\infty} b_n \sin\left(\frac{n\pi}{L} x\right) = -\frac{T_1 - T_0}{L} x \,. \tag{4.2.17}$$

A Fourier sine series appears on the left and a linear function, defined only in $(0, L)$, appears on the right. The way to balance these two functions is to express the linear function on the right in a Fourier sine series as well. In effect, the function is extended by odd symmetry to $(-L, L)$ and then periodically as a $2L$-periodic function. Its Fourier coefficients are

$$B_n = -\frac{2}{L} \int_0^L \frac{T_1 - T_0}{L} x \sin\left(\frac{n\pi}{L} x\right) dx$$

$$= 2 (T_1 - T_0) \frac{(-1)^n}{n\pi} \,.$$

Then Eq. (4.2.24) can be written as

$$\sum_{n=1}^{\infty} b_n \sin\left(\frac{n\pi}{L} x\right) = -\frac{T_1 - T_0}{L} x = \sum_{n=0}^{\infty} B_n \sin\left(\frac{n\pi}{L} x\right).$$

Now it is now easy to balance the two Fourier series by requiring $b_n = B_n$.

Altogether, the solution may be written as

$$\frac{T(x, t) - T_0}{T_1 - T_0} = \frac{x}{L} + 2 \sum_{n=1}^{\infty} \frac{(-1)^n}{n\pi} e^{-\alpha n^2 \pi^2 t / L^2} \sin\left(\frac{n\pi x}{L}\right). \tag{4.2.18}$$

Profiles of the solution are plotted in Fig. 4.6 at two different times. The parameters have been set so that $\alpha/L^2 = 1 \text{ s}^{-1}$. Here we see the temperature move in slowly and fill in until the profile becomes linear. The temperature diffuses into the interior. By looking at the solution in Eq. (4.2.18), we see that all the Fourier coefficients decay in time, the larger $n$, the faster the decay. The slowest decay occurs with $n = 1$ and the dominant time scale is $L^2/(\alpha\pi^2)$ (s). Eventually, the solution simply becomes the particular solution $T_p(x, t) = f(x)$. In this example, diffusion has the effect of relaxing the transient solution until the steady state solution is reached.

---

[9]Clearly, this is a situation where there is no hope of constructing a Taylor series expansion in time.

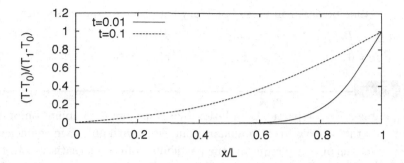

Fig. 4.6   Profiles of the temperature $T$ shown at the times indicated in the legend in seconds.

### 4.2.3   *Time-dependent boundary conditions*

That was easy so let us try something a little more complicated. Let us allow an exponential decay in one of the boundary conditions. In particular, suppose the temperature at $x = L$ is ramped up exponentially.

$$T(0,t) = T_0 \quad \text{and} \quad T(L,t) = T_1 + (T_0 - T_1)\,e^{-\beta t}. \tag{4.2.19}$$

We may consider this boundary condition as a more realistic model of the sudden initial jump in temperature at $x = L$ that was assumed in the previous example.

In order for the particular solution to match the behavior in time at the boundaries, it must contain terms that are constant in time and an exponential to match the exponential at $x = L$. We do not know yet how the particular solution must vary in $x$, so we guess that the particular solution must be of the form,

$$T_{\mathrm{p}}(x,t) = a(x) + b(x)\,e^{-\beta t}. \tag{4.2.20}$$

We need a constant in time to balance the constants in Eq. (4.2.19) and an exponential with the same argument to balance the exponential in Eq. (4.2.19). The undetermined coefficients, however, can be functions of $x$; thus $a(x)$ and $b(x)$.

Substitute the guess Eq. (4.2.20) into Eq. (4.2.3).

$$-\beta\,b(x)\,e^{-\beta t} - \alpha\left(\frac{\mathrm{d}^2 a}{\mathrm{d}x^2}(x) + \frac{\mathrm{d}^2 b}{\mathrm{d}x^2}(x)\,e^{-\beta t}\right) = 0.$$

Balance the terms that do not depend on $t$:

$$\frac{\mathrm{d}^2 a}{\mathrm{d}x^2}(x) = 0. \tag{4.2.21}$$

Balance the terms with the exponential:

$$\alpha\frac{\mathrm{d}^2 b}{\mathrm{d}x^2}(x) + \beta\,b(x) = 0. \tag{4.2.22}$$

As a result we have two differential equations for $a(x)$ and $b(x)$.

Now substitute the guess Eq. (4.2.20) into the boundary conditions Eq. (4.2.19). Thus

$$a(0) + b(0)\,e^{-\beta t} = T_0 \quad \text{and} \quad a(L) + b(L)\,e^{-\beta t} = T_1 + (T_0 - T_1)\,e^{-\beta t}.$$

Balancing the time behavior in these boundary conditions requires setting

$$a(0) = T_0, \qquad \text{and} \qquad a(L) = T_1, \qquad (4.2.23)$$
$$b(0) = 0, \qquad \text{and} \qquad b(L) = T_0 - T_1. \qquad (4.2.24)$$

> **Reflection**
>
> We pick a guess for the particular solution by composing a combination of the functions in $t$ that appear in the boundary conditions with unknown coefficients that are functions of $x$. By requiring the particular solution to satisfy both the partial differential equation and the boundary conditions we arrive at boundary value problems for the undetermined coefficients.

We now have the necessary boundary conditions Eq. (4.2.23) and Eq. (4.2.24) to solve Eq. (4.2.21) and Eq. (4.2.22), respectively. The solution to Eq. (4.2.21) is $a(x) = c_0 + c_1 x$ and the boundary conditions Eq. (4.2.23) then give $c_0 = T_0$ and $c_1 = (T_1 - T_0)/L$. So

$$a(x) = T_0 + \frac{T_1 - T_0}{L} x$$

which is the same as Eq. (4.2.5). It is the expected long term behavior.

Next, the general solution for (4.2.22) is

$$b(x) = c_0 \cos(\omega x) + c_1 \sin(\omega x),$$

where $\omega^2 = \beta/\alpha$. Applying the boundary conditions Eq. (4.2.24) gives $c_0 = 0$ and $c_1 = (T_0 - T_1)/\sin(\omega L)$. Note that we must require $\omega L \neq n\pi$! Altogether, the particular solution is

$$T_{\mathrm{p}}(x,t) = T_0 + \frac{T_1 - T_0}{L} x + \frac{T_0 - T_1}{\sin(\omega L)} \sin(\omega x) \, e^{-\beta t}. \qquad (4.2.25)$$

The particular solution has two parts, a linear profile that becomes the expected long term behavior and an exponentially decaying part that accommodates the transient behavior of the switch in temperature at $x = L$.

To complete the construction of the general solution, we must add the homogeneous solution $T_{\mathrm{h}}(x,t)$. The calculation of the homogeneous solution is exactly the same as in Sec. 4.2.2. Thus the general solution is

$$T(x,t) = T_0 + \frac{T_1 - T_0}{L} x + \frac{T_0 - T_1}{\sin(\omega L)} \sin(\omega x) \, e^{-\beta t}$$

$$+ \sum_{n=1}^{\infty} b_n e^{-\alpha n^2 \pi^2 t/L^2} \sin\left(\frac{n\pi}{L} x\right). \qquad (4.2.26)$$

What remains unknown are the Fourier coefficients $b_n$. These are determined by applying the initial condition Eq. (4.2.16).

$$T_0 + \frac{T_1 - T_0}{L} x + \frac{(T_0 - T_1)}{\sin(\omega L)} \sin(\omega x) + \sum_{n=1}^{\infty} b_n \sin\left(\frac{n\pi}{L} x\right) = T_0,$$

which can be restated as

$$\sum_{n=1}^{\infty} b_n \sin\left(\frac{n\pi}{L}x\right) = -\frac{T_1 - T_0}{L}x - \frac{(T_0 - T_1)}{\sin(\omega L)}\sin(\omega x). \qquad (4.2.27)$$

The next step, as done before, is to express the function on the right hand side of Eq. (4.2.27) as a Fourier sine series with the Fourier coefficients determined by

$$B_n = -\frac{2}{L}\int_0^L \left[\frac{T_1 - T_0}{L}x + \frac{(T_0 - T_1)}{\sin(\omega L)}\sin(\omega x)\right]\sin\left(\frac{n\pi}{L}x\right)dx.$$

The integral can be split into two parts: the first can be performed by integration by parts and the second can be simplified by the addition angle formula Eq. (D.2.6). The result is

$$B_n = 2\left(T_1 - T_0\right)\frac{(-1)^n}{n\pi} + 2\left(T_1 - T_0\right)\frac{(-1)^n n\pi}{\omega^2 L^2 - n^2\pi^2}. \qquad (4.2.28)$$

Replacing the function on the right hand side of Eq. (4.2.27) by its Fourier sine series means that

$$\sum_{n=1}^{\infty} b_n \sin\left(\frac{n\pi}{L}x\right) = \sum_{n=1}^{\infty} B_n \sin\left(\frac{n\pi}{L}x\right)$$

and the initial condition is satisfied with the balance $b_n = B_n$.

Finally, substitute Eq. (4.2.28) into Eq. (4.2.26) to obtain the specific solution,

$$\frac{T(x,t) - T_0}{T_1 - T_0} = \frac{x}{L} - \frac{\sin(\omega x)}{\sin(\omega L)}e^{-\beta t}$$

$$+ 2\sum_{n=1}^{\infty}(-1)^n \left[\frac{1}{n\pi} + \frac{n\pi}{\omega^2 L^2 - n^2\pi^2}\right]e^{-\alpha n^2\pi^2 t/L^2}\sin\left(\frac{n\pi x}{L}\right). \qquad (4.2.29)$$

The behavior of the solution in Eq. (4.2.29) is illustrated in Fig. 4.7 at the same times as in Fig. 4.6 and the same settings for the parameters with the additional choice $\beta = 36\,\text{s}^{-1}$ and $\omega L = 6$. The increasing value for $T(x,t)$ at the right boundary $x = L$ is clearly observed, with the corresponding diffusion into the interior. By

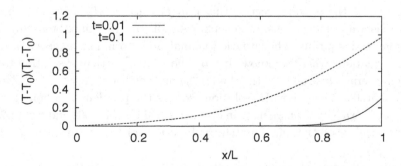

Fig. 4.7   Profiles of the temperature $T$ shown at the times indicated in the legend in seconds.

changing the choice for $\beta$, the pace of the diffusion into the interior can be controlled. The final result, a linear profile, will always occur.

We should not overlook that we require $\omega L \neq m\pi$ for the solution Eq. (4.2.29) to be valid. Why is this requirement necessary from a physical point of view? Well, it just means we must look more carefully at the solution. Obviously, the denominator $\sin(\omega L) = 0$ causes problems but there is also a term in the sum that has a denominator that becomes zero, namely, the term with $n = m$. Ah, now we see the light! Separating out the terms that cause problems,

$$-\frac{\sin(\omega x)}{\sin(\omega L)}\, \mathrm{e}^{-\beta t} + 2(-1)^m \frac{m\pi}{\omega^2 L^2 - m^2\pi^2}\, \mathrm{e}^{-\alpha m^2 \pi^2 t / L^2} \sin\left(\frac{m\pi x}{L}\right),$$

we ask whether the result remains sensible if $\omega L \to m\pi$. It does! The limit is

$$(-1)^m \mathrm{e}^{-\alpha m^2 \pi^2 t / L^2}\left[\frac{x}{L}\cos\left(\frac{n\pi x}{L}\right) - 2\frac{\alpha}{L^2}\, n\pi t \sin\left(\frac{n\pi x}{L}\right)\right].$$

While the solution looks different, if we graph the results with $\omega L$ near $m\pi$ and $\omega L = m\pi$, we will see hardly any difference. In other words, the limit is just a mathematical oddity – we never know quantities such as $\omega$ with more than a few digits of accuracy, so the exact limit $\omega L = m\pi$ never really happens.

### 4.2.4   *Abstract view*

The new feature in this section is the appearance of inhomogeneous (forcing) terms in the boundary conditions. Assume that the partial differential equation $u(x, t)$ is homogeneous,

$$\frac{\partial u}{\partial t}(x, t) - \alpha \frac{\partial^2 u}{\partial x^2}(x, t) = 0\,, \tag{4.2.30}$$

but that the boundary conditions are inhomogeneous, for example,

$$u(0, t) = g_1(t) \qquad \text{and} \qquad u(L, t) = g_2(t)\,. \tag{4.2.31}$$

The approach to treat these inhomogeneous boundary conditions does not depend on the type they are, but it is helpful to make a specific choice to illustrate the approach.

We know from the previous section that if $g_1(t) = g_2(t) = 0$, that is, the boundary conditions are homogeneous, then separation of variables may be used to construct the general solution. The situation is analogous to inhomogeneous ordinary differential equations where we know that by choosing a particular solution to care of the forcing term, we are left with a homogeneous equation that can be solved by using exponentials. The particular solution reduces the problem to a homogeneous one, and that is exactly what we need in order to solve Eq. (4.2.30) and Eq. (4.2.31). Here the particular solution must match the functions of $t$ that appear in $g_1(t)$ and $g_2(t)$.

Unfortunately, there is no general approach that guarantees success, but there are some special cases where we can guess the form of the particular solution. In particular, whenever $g_1(t)$ and $g_2(t)$ are of the form found in Table 1.4, we try to extend the method of undetermined coefficients.

First, we must identify all the different functions in $g_1(t)$ and $g_2(t)$ that we will need to match. Thus we are led to select the appropriate functions of $t$ that will need to be part of the particular solution. What we do not know is what behavior in $x$ that should be in the particular solution. The way forward is to multiply each contributing function of time in the particular solution by an unknown function of $x$ (for ordinary differential equations, we just multiplied each contribution by an unknown coefficient – hence the method of undetermined coefficients) and all contributions are added together.

## Example

To illustrate the idea, suppose

$$g_1(t) = 1 + t \quad \text{and} \quad g_2(t) = \exp(-t).$$

The appropriate functions in the particular solution that we need to match $g_1(t)$ are $1$ and $t$. For $g_2(t)$ we need $\exp(-t)$. Thus we add them altogether with unknown functions in place of unknown coefficients,

$$u_\mathrm{p}(t) = a_1(x) + a_2(x)\,t + a_3(x)\,\mathrm{e}^{-t}. \tag{4.2.32}$$

Once the guess for the particular solution is made, we proceed as we always do. We substitute our guess into the differential equation and the boundary conditions and demand that all functions of time match. We anticipate that the consequences will identify the requirements the unknown functions of $x$ must satisfy. When we substitute the guess into the partial differential equation, we should expect the requirements to contain derivatives of the unknown functions of $x$ and that it is likely that ordinary differential equations for the unknown functions will result.

## Example continued

By substituting Eq. (4.2.32) into Eq. (4.2.30) we obtain

$$a_2(x) - a_3(x)\,\mathrm{e}^{-t} - \alpha\,\frac{\mathrm{d}^2 a_1}{\mathrm{d}x^2}(x) - \alpha\,\frac{\mathrm{d}^2 a_2}{\mathrm{d}x^2}(x)\,t - \alpha\,\frac{\mathrm{d}^2 a_3}{\mathrm{d}x^2}(x)\,\mathrm{e}^{-t} = 0,$$

and by balancing terms with the same dependency in $t$, we obtain

$$\alpha\,\frac{\mathrm{d}^2 a_1}{\mathrm{d}x^2}(x) - a_2(x) = 0, \tag{4.2.33}$$

$$\alpha\,\frac{\mathrm{d}^2 a_2}{\mathrm{d}x^2}(x) = 0, \tag{4.2.34}$$

$$\alpha\,\frac{\mathrm{d}^2 a_3}{\mathrm{d}x^2}(x) + a_3(x) = 0. \tag{4.2.35}$$

Clearly Eq. (4.2.35) is a differential equation for $a_3(x)$, but Eq. (4.2.34) can be integrated directly to determine $a_2(x)$, and subsequently Eq. (4.2.33) can be integrated to determine $a_1(x)$.

By matching the time dependency in the boundary condition, we will obtain boundary conditions for the unknown functions in $x$ that supplement the differential equations and lead to specific boundary value problems.

## Example continued

By substituting Eq. (4.2.32) into Eq. (4.2.31), we obtain

$$a_1(0) + a_2(0)\,t + a_3(0)\,e^{-t} = 1 + t\,,$$
$$a_1(L) + a_2(L)\,t + a_3(L)\,e^{-t} = e^{-t}\,.$$

It is at this moment that we can confirm our guess for the particular solution is valid. We can set

$$a_1(0) = 1\,, \qquad\qquad a_1(L) = 0\,, \qquad\qquad (4.2.36)$$
$$a_2(0) = 1\,, \qquad\qquad a_2(L) = 0\,, \qquad\qquad (4.2.37)$$
$$a_3(0) = 0\,, \qquad\qquad a_3(L) = 1\,, \qquad\qquad (4.2.38)$$

and we have balanced all the time behaviors in the boundary conditions perfectly.

The final step then is to determine the functions of $x$ that satisfy the resulting differential equations or derivatives and their associated boundary conditions. As the example illustrates, we have an inhomogeneous boundary value problem for $a_3$ given by Eq. (4.2.35) and Eq. (4.2.38) which can be solved in the standard way as described in Chapter 3. Fortunately, Eq. (4.2.34) and Eq. (4.2.33) are not really differential equations and can be integrated directly. The constants of integration are determined by the boundary values Eq. (4.2.36) and Eq. (4.2.37). By integrating Eq. (4.2.34), $a_2(x) = c_1 + c_2\,x$, and by applying Eq. (4.2.37), $a_2(x) = 1 - x/L$. With $a_2(x)$ determined, Eq. (4.2.33) can be integrated and determined completely with the application of Eq. (4.2.36).

If we have been successful, then we should find that $u_p(x,t)$ satisfies Eq. (4.2.30) and Eq. (4.2.31). But we have not yet satisfied the initial condition. Thus we write the general solution as a combination of the particular solution $u_p(x,t)$ and a homogeneous solution $u_h(x,t)$:

$$u(x,t) = u_p(x,t) + u_h(x,t)\,.$$

Now substitute into the partial differential equation.

$$
\begin{aligned}
\frac{\partial(u_p + u_h)}{\partial t}(x,t) - \alpha\,\frac{\partial^2(u_p + u_h)}{\partial x^2}(x,t) &= \left[\frac{\partial u_p}{\partial t}(x,t) - \alpha\,\frac{\partial^2 u_p}{\partial x^2}(x,t)\right] \\
&\quad + \left[\frac{\partial u_h}{\partial t}(x,t) - \alpha\,\frac{\partial^2 u_h}{\partial x^2}(x,t)\right] \\
&= \frac{\partial u_h}{\partial t}(x,t) - \alpha\,\frac{\partial^2 u_h}{\partial x^2}(x,t)\,,
\end{aligned}
$$

where we have used the fact that $u_p(x, t)$ satisfies Eq. (4.2.30) by construction. For the general solution to satisfy Eq. (4.2.30), we require the homogeneous solution to also satisfy

$$\frac{\partial u_h}{\partial t}(x, t) - \alpha \frac{\partial^2 u_h}{\partial x^2}(x, t) = 0.$$

Finally, we check the boundary conditions Eq. (4.2.31)

$$u(0, t) = u_p(0, t) + u_h(0, t) = g_1(t) + u_h(0, t) = g_1(t) \quad \Rightarrow \quad u_h(0, t) = 0,$$
$$u(L, t) = u_p(L, t) + u_h(L, t) = g_2(t) + u_h(L, t) = g_2(t) \quad \Rightarrow \quad u_h(L, t) = 0.$$

Major success! The homogeneous solution can now be constructed by separation of variables, the initial condition can be applied and we are done.

## Principle 25

*Find a particular solution so that the homogeneous solution can be constructed by separation of variables.*

### 4.2.5 Exercises

**This exercise gives some practice in solving the diffusion equation with inhomogeneous boundary condition.**

(1) Consider a metal rod encased in a thermal insulator. Heat is supplied at an exponentially decaying rate at the left end, while the other end is thermally insulated. Mathematically,

$$q(0, t) = Q \, e^{-\beta t} \qquad q(L, t) = 0.$$

Suppose the initial temperature is just a constant, $T(x, 0) = T_0$. What is the long term temperature profile?

Face this question in two ways:

(a) Determine the specific solution and consider the result when $t \to \infty$.
(b) Adopt the approach in Exercise 1 of Sec. 4.1.6.

**Answer:** The temperature is raised by $\alpha Q / (\beta k L)$.

**This exercise points out the difficulty in propagating heat into a solid.**

(2) Diffusion tends to smooth profiles until they become flat or linear. What happens if we drive the system with an oscillatory forcing? To explore this question, set

$$T(0, t) = T_0 + A \cos(\omega t) \quad \text{and} \quad T(L, t) = T_0.$$

Initially, let $T(x,0) = T_0$ and determine the temperature later, especially for long times when the initial transients have decayed away. To simplify a study of the results, pick $L = 1$, $\omega = 2\pi$ and $\alpha = 1/\pi$.

**Answer:** For long times, only the particular solution survives and we may study the effective penetration of the oscillatory nature of the solution by simply plotting the temperature at selected locations along the rod, $x/L = 0, 0.25, 0.5, 0.75$. The variation of the temperature in time is shown in Fig. 4.8. The amplitude of the oscillation in the temperature is decaying rapidly as the location along the rod is increased.

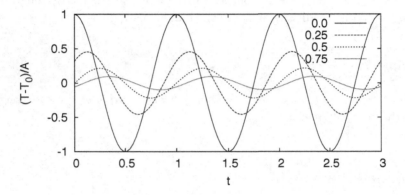

Fig. 4.8   Variation of the temperature in time at select locations, $x/L = 0.0, 0.25, 0.5, 0.75$.

**The next exercises explore the nature of the Fourier series when the boundary conditions are of mixed type.**

(3) Consider a long slab of stainless steel with a uniform rectangular cross section. It is long enough that the temperature does not vary along its length. The top and bottom of the cross section is insulated, and no vertical variation in the temperature is expected. The temperature does vary along the horizontal direction as a consequence of different boundary conditions at its ends. At the right edge of the cross section, located at $x = L$, the temperature is maintained at a constant value, $T(L,t) = T_1$, while the other edge, located at $x = 0$ is insulated. Insulated means that there is no flow of heat through the surface, or

$$\frac{\partial T}{\partial x}(0,t) = 0.$$

Suppose the initial temperature is $T_0$. Find the temperature profile at later times. You will need to think carefully about the form of the Fourier series; Exercise 5 in Sec. 2.3.4 should help.

**Answer:** An engineer has made measurements for a stainless steel bar whose thermal diffusivity is $\alpha = 0.05$ cm$^2$/s and whose length is $L = 10$ cm. The

Table 4.1 Measured
temperatures.

| Time in min | $R(t)$ |
| --- | --- |
| 1.0 | 1.000 |
| 2.0 | 0.992 |
| 3.0 | 0.963 |
| 4.0 | 0.918 |
| 5.0 | 0.864 |

temperature is measured at $x = 0$ for different times and given in the Table 4.1 in terms of $R(t) = (T(0,t) - T_1)/(T_0 - T_1)$.

(4) Repeat the previous calculation but switch the insulation to the other end. In other words, the boundary conditions are now

$$\frac{\partial T}{\partial x}(L,t) = 0 \quad \text{and} \quad T(0,t) = T_1.$$

Find the temperature profile at later times. Is there a connection between the Fourier series in this problem and that of the previous problem?

**Answer:** Consider a change in coordinate that switches the locations at $x = 0$ and $x = L$. The results should agree perfectly with the previous result. Which calculation seemed easier to perform?

**These exercises highlight the way to deal with external forcing in the differential equation.**

(5) Suppose there is no salt initially in a sealed pipe, and then salt concentration is added at the rate

$$\dot{\rho}_i(t) = Q_m \frac{\lambda}{AL} e^{-\lambda t},$$

where $A$ is the cross-sectional area of the pipe. Show that the total amount of salt that is added is $Q_m$. The equation for the conservation of the amount of salt now must include the additional salt added.

$$\frac{\partial \rho}{\partial t}(x,t) = D \frac{\partial^2 \rho}{\partial x^2}(x,t) + \dot{\rho}_i(t).$$

Construct a particular solution which also satisfies the homogeneous boundary conditions, and then add to it the homogeneous solution Eq. (4.1.33) to satisfy the initial condition. Plot the concentration in the middle of the pipe, in particular, $AL\rho(L/2,t)/Q_m$ as a function of $\lambda t$.

**Answer:** The value of plotting $AL\rho/Q_m$ as a function of $\lambda t$ is that the result is universal. In other words, the result does not depend on the choice for the parameters. The graph is shown in Fig. 4.9.

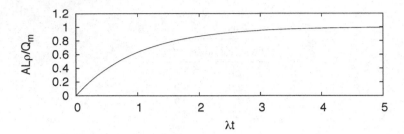

Fig. 4.9   Concentration in the middle of the pipe.

(6) Repeat the calculation in the previous exercise, but pick

$$\dot{\rho}_i(x,t) = Q_m \frac{\lambda}{AL} \left[ 1 - \cos\left( \frac{2\pi}{L} x \right) \right] e^{-\lambda t}.$$

Plot the concentration in the middle of the pipe in the form $AL\rho/Q_m$ as a function of $\lambda t$ for three choices of the parameter $R = 4\pi^2 D/(\lambda L^2) = 10.0, 1.0, 0.1$. Can you explain the results?

**Answer:** This time the results depend on a single dimensionless parameter $R$ that measures the ratio of the rate of diffusion with the rate of disposition. A graph of the concentration in the middle of the pipe is shown in Fig. 4.10 for three choices for $R$. When $R$ is large, the salt quickly diffuses; when $R$ is small, the disposition accumulates rapidly in the middle before slowly diffusing away. You may also wish to compare your results with those in Fig. 4.2 to assess the difference between instantaneous disposition (stated as an initial condition) and a more gradual disposition.

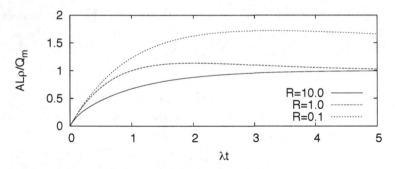

Fig. 4.10   Concentration in the middle of the pipe.

## 4.3  Propagation

Other situations besides the transport of quantities also lead to partial differential equations. For example, there is a simple partial differential equation that describes the propagation of information in space: sound waves, ocean waves and electromagnetic waves are important examples of wave propagation that dominate our lives.

But the one I'll pick is the propagation of signals along a co-axial cable; it describes our connections to the internet among other applications.

### 4.3.1  *Transmission line*

A co-axial cable has a core of copper at its center, surrounded by a dielectric material encased in a shield. The standard mathematical model for describing signals transmitting along a co-axial cable is to imagine it is a little electric circuit in a small segment of the cable. This idea comes about in the following way. If we apply a voltage difference $V$ (V) across the ends of the cable, then a current $I$ (A) will flow along it and it will appear to have a resistance to the flow in the standard form $V = RI$. The longer the cable, the larger the resistance. Indeed, the resistance is proportional to the length, and the usual way it is measured is the resistance per unit length of the cable, $\mathcal{R}$ ($\Omega$/m). In a small segment of length $\triangle x$ (m) then, the resistance will be $\mathcal{R} \triangle x$. For copper, the resistance is extremely low, which means the voltage to drive a current does not have to be too large.

There are other electric effects that occur in the cable. The dielectric material surrounding the copper wire acts as a capacitor between the copper wire and the outer shield. The effective capacitance will depend on the length of the cable, and, just like the resistance, it can be measured as a capacitance per unit length, $\mathcal{C}$ (F/m). There is also an effective inductance from the current flow along the wire, measured as $\mathcal{L}$ (H/m). All this means is that we can make a circuit diagram of a small segment of length $\triangle x$ as shown in Fig. 4.11.

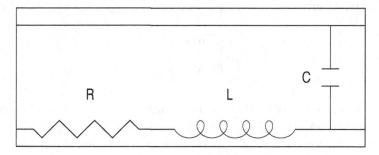

Fig. 4.11  A schematic of a small segment of a co-axial cable: the bottom line represents the copper wire, and the top line represents the shield.

We must use Kirchoff's laws to establish equations for the drop in potential $\triangle V(x,t)$ across a small segment $\triangle x$ of the copper wire and the change in the current $\triangle I(x,t)$ along the copper wire as some of it leaks across the capacitor.

$$\triangle V(x,t) = -\mathcal{L}\,\triangle x\,\frac{\partial I}{\partial t}(x,t) - \mathcal{R}\,\triangle x\,I(x,t)\,,$$

$$\triangle I(x,t) = -\mathcal{C}\,\triangle x\,\frac{\partial V}{\partial t}(x,t)\,.$$

After dividing by $\triangle x$, we take the limit and obtain two coupled, first-order partial differential equations that are also linear with constant coefficients. After rearranging,

$$\frac{\partial I}{\partial t}(x,t) = -\frac{1}{\mathcal{L}}\frac{\partial V}{\partial x}(x,t) - \frac{\mathcal{R}}{\mathcal{L}}I(x,t)\,, \tag{4.3.1}$$

$$\frac{\partial V}{\partial t}(x,t) = -\frac{1}{\mathcal{C}}\frac{\partial I}{\partial x}(x,t)\,. \tag{4.3.2}$$

We can obtain a single partial differential equation for the voltage $V(x,t)$ by differentiating Eq. (4.3.1) with respect to $x$ and then substituting the partial derivative of the current $I(x,t)$ with respect to $x$ from Eq. (4.3.2).

$$\frac{\partial^2 V}{\partial t^2}(x,t) + \frac{\mathcal{R}}{\mathcal{L}}\frac{\partial V}{\partial t}(x,t) - \frac{1}{\mathcal{L}\mathcal{C}}\frac{\partial^2 V}{\partial x^2}(x,t) = 0\,. \tag{4.3.3}$$

### 4.3.2   *Initial and boundary conditions*

The partial differential equation Eq. (4.3.3) is homogeneous with two independent variables $x$ and $t$. It is second-order in both the time and spatial derivatives. It also contains two parameters $\mathcal{R}/\mathcal{L}$ and $1/(\mathcal{L}\mathcal{C})$. Since there are two time derivatives, two initial conditions must be given to completely specify the potential later in time – see Principle 9. The difference with ordinary differential equations is that the potential and its rate of change must be known the whole length of the cable. Stated mathematically,

$$V(x,0) = f(x)\,, \tag{4.3.4}$$

$$\frac{\partial V}{\partial t}(x,0) = g(x)\,. \tag{4.3.5}$$

It certainly seems reasonable that we may know $f(x)$, the initial profile of the potential, but why would we know $g(x)$, the initial rate of change of the potential. The answer is that we are more likely to know the initial current $I(x,0) = h(x)$ and we can then use Eq. (4.3.2) evaluated at $t=0$ to determine

$$g(x) = \frac{\partial V}{\partial t}(x,0) = -\frac{1}{\mathcal{C}}\frac{\partial I}{\partial x}(x,0) = -\frac{1}{\mathcal{C}}\frac{\mathrm{d}h}{\mathrm{d}x}(x)\,.$$

Further, what happens at the ends of the cable will affect the potential along the cable. That information provides two additional conditions, boundary conditions, that will describe some external forcing to drive a signal along the cable. For example, there might be some time variation in the voltages at the ends of the cable that would drive the voltage along the cable. The situation is captured by a mathematical diagram.

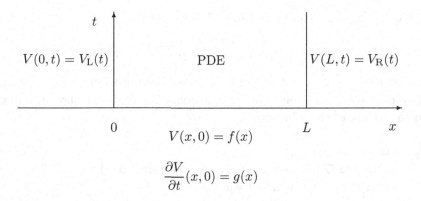

Mathematical diagrams such as the one shown helps us check the formulation of the problem. The region where the solution is desired is clearly identified. The number of partial derivatives in each independent variable dictates the additional information needed at the boundaries and initially. The partial differential equation Eq. (4.3.3) has two spatial derivatives, and two boundary conditions are supplied. It also has two partial time derivatives, and we have two initial conditions. All is well.

The presence of two inhomogeneous boundary conditions means a particular solution will be needed to account for this specific forcing, and to the particular solution we will need to add a homogeneous solution that captures the transient adjustment to the initial state of the voltage and current along the cable. The homogeneous solution is independent of the particular solution and it is a good place to start.

### 4.3.3 *The homogeneous solution*

The homogeneous problem is composed of the partial differential equation Eq. (4.3.3), and the boundary conditions

$$V(0,t) = V(L,t) = 0.  \qquad (4.3.6)$$

This problem also arises in important applications because a zero voltage may represent a grounded circuit. So we can imagine the cable is grounded at its ends but is affected initially by some external voltage and current as expressed by the initial conditions Eq. (4.3.4) and Eq. (4.3.5). This problem does not require a particular solution and we begin solving the homogeneous problem by separating the variables.

**1. Separate the variables.**
Separation of variables assumes the solution can be found in the form,

$$V(x,t) = X(x)\,T(t).  \qquad (4.3.7)$$

After substituting Eq. (4.3.7) into Eq. (4.3.3) we obtain

$$X(x)\,\frac{\mathrm{d}^2 T}{\mathrm{d}t^2}(t) + r\,X(x)\,\frac{\mathrm{d}T}{\mathrm{d}t}(t) - \omega^2\,T(t)\,\frac{\mathrm{d}^2 X}{\mathrm{d}x^2}(x) = 0\,.$$

The parameters $r = \mathcal{R}/\mathcal{L}$ and $\omega^2 = 1/(\mathcal{L}\mathcal{C})$ have been introduced for convenience. Clearly $r$ is a scaled resistance, and $\omega$ is a frequency that will play an important part in the nature of the solutions.

## 2. Separate the equation.
By dividing by $X(x)\,T(t)$, we may rearrange the terms so that each side contains only a function of each independent variable.

$$\frac{\omega^2}{X(x)}\,\frac{\mathrm{d}^2 X}{\mathrm{d}x^2}(x) = \frac{r}{T(t)}\,\frac{\mathrm{d}T}{\mathrm{d}t}(t) + \frac{1}{T(t)}\,\frac{\mathrm{d}^2 T}{\mathrm{d}t^2}(t)\,. \tag{4.3.8}$$

We are in the same position as before, trying to balance two functions of different independent variables. The only possibility is for each side of Eq. (4.3.8) to be some constant $c$ (the separation constant). Thus we have two ordinary differential equations for $X(x)$ and $T(t)$ separately.

$$\frac{\mathrm{d}^2 T}{\mathrm{d}t^2}(t) + r\,\frac{\mathrm{d}T}{\mathrm{d}t}(t) = c\,T(t)\,, \tag{4.3.9}$$

$$\omega^2\,\frac{\mathrm{d}^2 X}{\mathrm{d}x^2}(x) = c\,X(x)\,. \tag{4.3.10}$$

## 3. Separate the boundary conditions.
Substitute Eq. (4.3.7) into Eq. (4.3.6).

$$T(t)\,X(0) = 0\,, \quad \text{or} \quad X(0) = 0\,, \tag{4.3.11}$$

$$T(t)\,X(L) = 0\,, \quad \text{or} \quad X(L) = 0\,. \tag{4.3.12}$$

In both cases, we have a choice of factors to set to zero, but we do not want to make the choice $T(t) = 0$, since we would then have set $V(x,t) = X(x)\,T(t) = 0$, the useless trivial solution. Instead we have homogeneous boundary conditions, Eq. (4.3.11) and Eq. (4.3.12), to add to the homogeneous differential equation Eq. (4.3.10) – we have a homogeneous boundary value problem.

## 4. Solve the homogeneous boundary value problem.
As before, we seek non-trivial solutions to the homogeneous boundary value problem by placing restrictions on the separation constant. We try the obvious exponential trial solution $X(x) = \exp(\beta x)$. As a result, the characteristic equation is

$$\omega^2 \beta^2 = c\,. \tag{4.3.13}$$

## 5. Consider all possible choices for the separation constant.
Since we do not know $c$, we must consider all three possible cases: $c > 0$, $c = 0$, and $c < 0$. Each case leads to a different type of solution for Eq. (4.3.3).

(a) Assume $c > 0$. Then $\beta = \sqrt{c}/\omega$. The general solution to (4.3.10) is

$$X(x) = a_1 e^{\sqrt{c}\, x/\omega} + a_2 e^{-\sqrt{c}\, x/\omega}.$$

By applying the boundary conditions, Eq. (4.3.11) and Eq. (4.3.12), we end up with two equations to determine $a_1$ and $a_2$.

$$a_1 + a_2 = 0,$$
$$a_1 e^{\sqrt{c}\, L/\omega} + a_2 e^{-\sqrt{c}\, L/\omega} = 0.$$

The first equation gives $a_2 = -a_1$, and after we substitute the result into the second equation, we obtain

$$a_1 \left( e^{\sqrt{c}\, L/\omega} - e^{-\sqrt{c}\, L/\omega} \right) = 0.$$

Since $\exp(\sqrt{c}\, L/\omega) \neq \exp(-\sqrt{c}\, L/\omega)$, the only solution is $a_1 = 0$ and $a_2 = 0$, and this means $X(x) = 0$, which is useless!

(b) Assume $c = 0$. The solution to Eq. (4.3.10) is

$$X(x) = a_1 + a_2\, x,$$

and the boundary conditions $X(0) = X(L) = 0$ become

$$a_1 = 0,$$
$$a_1 + a_2 L = 0.$$

Once again, $a_1 = a_2 = 0$, and still no useful solution.

(c) Assume $c < 0$. For convenience let $c = -\alpha^2 \omega^2$. Then Eq. (4.3.13) gives $\beta = \pm i\alpha$, and the general solution to Eq. (4.3.10) can be written as

$$X(x) = a_1 \cos(\alpha x) + a_2 \sin(\alpha x).$$

Now the boundary conditions $X(0) = X(L) = 0$ give

$$a_1 = 0,$$
$$a_1 \cos(bL) + a_2 \sin(bL) = 0.$$

Since $a_1 = 0$, the second condition reads

$$a_2 \sin(\alpha x) = 0.$$

We are faced with two choices: either $a_2 = 0$, but this would be useless since $X(x) = 0$, or $\sin(\alpha L) = 0$ and all we have to do is set $\alpha L = n\pi$ for any positive integer $n$. Thus we have many possible solutions,

$$X(x) = a_2 \sin\left( \frac{n\pi x}{L} \right), \qquad (4.3.14)$$

associated with different choices for the separation constant,

$$c = -\alpha^2 \omega^2 = -\frac{n^2 \pi^2 \omega^2}{L^2} \qquad (4.3.15)$$

for any positive integer $n$.

## 6. Solve the remaining equation.

With a fixed choice of $n$ in mind, we now solve Eq. (4.3.9). Let $T(t) = \exp(\lambda t)$, and substitute into Eq. (4.3.9) to obtain the characteristic equation,

$$\lambda^2 + r\lambda + \frac{n^2\pi^2\omega^2}{L^2} = 0.$$

The roots of this quadratic are

$$2\lambda = -r \pm \sqrt{r^2 - \frac{4n^2\pi^2\omega^2}{L^2}}.$$

The nature of the roots, whether real or imaginary, depends on the sign of the discriminant

$$D(n) = r^2 - \frac{4n^2\pi^2\omega^2}{L^2}.$$

Whether this quantity is positive or negative depends on our choice for $n$. Obviously, if the choice of $n$ is large enough, $D(n)$ will be negative. If there are choices for $n$ that make $D$ positive, then the roots will be real and $T(t)$ will have decaying exponentials. To keep the example simple, let us assume that $D(1) < 0$. Consequently, $D(n) < 0$ for any choice of $n > 1$. The roots are always complex, and it is convenient to introduce a parameter $\Omega$ to represent the imaginary part.

$$\lambda = -\frac{r}{2} \pm i\Omega_n, \quad \text{where} \quad \Omega_n = \frac{1}{2}\sqrt{\frac{4n^2\pi^2\omega^2}{L^2} - r^2}. \tag{4.3.16}$$

The solution to (4.3.9) is

$$T(t) = e^{-rt/2}\left[c_1 \cos(\Omega_n t) + c_2 \sin(\Omega_n t)\right]. \tag{4.3.17}$$

## 7. Compose a general solution.

We have both solutions for $X(x)$ and $T(t)$, so we have found a solution of the form

$$V(x,t) = T(t)\,X(x)$$

$$= e^{-rt/2}\left[c_1 \cos(\Omega_n t) + c_2 \sin(\Omega_n t)\right] a_2 \sin\left(\frac{n\pi}{L}x\right)$$

$$= e^{-rt/2}\left[a_n \cos(\Omega_n t) + b_n \sin(\Omega_n t)\right] \sin\left(\frac{n\pi}{L}x\right)$$

for each choice of $n$. Note that the products of unknown coefficients $c_1 a_2$ and $c_2 a_2$ have been replaced by new coefficients $a_n$ and $b_n$ respectively – recall Principle 6. The choice of subscript is to help us remember these new coefficients depend on the choice for $n$.

As Principle 24 suggests, we are now in the position to add up all the possible solutions, and it is this combination that gives us hope of finishing the problem by satisfying the initial conditions.

$$V(x,t) = e^{-rt/2} \sum_{n=1}^{\infty}\left[a_n \cos(\Omega_n t) + b_n \sin(\Omega_n t)\right] \sin\left(\frac{n\pi x}{L}\right). \tag{4.3.18}$$

A crucial test of this solution is that it must satisfy the homogeneous boundary conditions $V(0,t) = V(L,t) = 0$. Since each solution satisfies these conditions, the sum will satisfy the boundary conditions.

> **Reflection**
>
> Two time derivatives in the partial differential equation lead to a second-order differential equation in time after separation of variables is invoked. Consequently, the solution contains two sets of unknown coefficients that require two initial conditions to determine them specifically.

## 8. Satisfy the initial condition.

Normally we would add the particular solution to the homogeneous solution to create a general solution before applying the initial conditions. But for the moment we are considering a homogeneous problem with the boundary conditions Eq. (4.3.6) and this gives us the opportunity to understand how the initial conditions must be applied.

The two initial conditions, Eq. (4.3.4) and Eq. (4.3.5), will be used to determine the coefficients $a_n$ and $b_n$ in the general solution Eq. (4.3.18). First, Eq. (4.3.4) gives

$$\sum_{n=1}^{\infty} a_n \sin\left(\frac{n\pi x}{L}\right) = f(x). \tag{4.3.19}$$

This match has possibilities because it looks like a Fourier series.

If we are to regard the left hand side of Eq. (4.3.19) as a Fourier series, then we must take note of two oddities about it. First, there are no cosine terms. Second, the periodicity of the series is $2L$, but $f(x)$ applies to only $[0, L]$. In other words, we do not know $f(x)$ completely in a $2L$ period.

We wiggle our way out of these difficulties by exploiting an important property of sine functions. They are odd functions. If only $f(x)$ were an odd function, then it would have only a sine series. So we make it so since it does not matter what $f(x)$ is outside the range $[0, L]$. We regard $f(x)$ as odd and extend to be a $2L$-periodic function. Then, we express $f(x)$ as a Fourier series,

$$f(x) = \sum_{n=0}^{\infty} F_n \sin\left(\frac{n\pi x}{L}\right)$$

where the coefficients are calculated by

$$F_n = \frac{2}{L} \int_0^L f(x) \sin\left(\frac{n\pi x}{L}\right) dx. \tag{4.3.20}$$

The initial condition Eq. (4.3.19) is now restated as

$$\sum_{n=1}^{\infty} a_n \sin\left(\frac{n\pi x}{L}\right) = \sum_{n=1}^{\infty} F_n \sin\left(\frac{n\pi x}{L}\right)$$

with the obvious conclusion that $a_n = F_n$.

We are not quite finished since we also must determine $b_n$ in Eq. (4.3.18) from the other initial condition Eq. (4.3.5). After differentiating $V(x,t)$ in Eq. (4.3.18) and applying Eq. (4.3.5), we find

$$-\frac{r}{2}\sum_{n=1}^{\infty} a_n \sin\left(\frac{n\pi x}{L}\right) + \sum_{1}^{\infty}\Omega_n b_n \sin\left(\frac{n\pi x}{L}\right) = g(x).\qquad(4.3.21)$$

Clearly, the way forward is to express $g(x)$ as a Fourier sine series. We interpret $g(x)$ as an odd function and extend it periodically with period $2L$. Then it has a sine series

$$g(x) = \sum_{n=1}^{\infty} G_n \sin\left(\frac{n\pi x}{L}\right),$$

where

$$G_n = \frac{2}{L}\int_0^L g(x)\sin\left(\frac{n\pi x}{L}\right)dx.\qquad(4.3.22)$$

Finally, we match the Fourier coefficients in Eq. (4.3.21).

$$\Omega_n b_n = \frac{r}{2} a_n + G_n,\qquad(4.3.23)$$

which determines $b_n$. With both $a_n$ and $b_n$ determined, the solution Eq. (4.3.18) is determined. Time to make the process concrete with a specific example.

### 4.3.3.1  *Example*

Suppose some voltage is induced in the middle of the cable. The mathematical representation for this initial voltage is given by

$$V(x,0) = f(x) = \begin{cases} \dfrac{A}{2} + \dfrac{A}{2}\cos\left[\dfrac{2\pi}{W}\left(x - \dfrac{L}{2}\right)\right], & \text{for } \left|x - \dfrac{L}{2}\right| \leq \dfrac{W}{2}, \\[2ex] 0, & \text{otherwise}. \end{cases}\qquad(4.3.24)$$

Its Fourier coefficients are calculated by Eq. (4.3.20);

$$F_n = I_1 + I_2,$$

where

$$I_1 = \frac{A}{L}\int_{(L-W)/2}^{(L+W)/2} \sin\left(\frac{n\pi x}{L}\right)dx$$

$$= -\frac{A}{n\pi}\left\{\cos\left[\frac{n\pi}{L}\left(\frac{L+W}{2}\right)\right] - \cos\left[\frac{n\pi}{L}\left(\frac{L-W}{2}\right)\right]\right\},$$

and

$$
I_2 = \frac{A}{L} \int_{(L-W)/2}^{(L+W)/2} \cos\left[\frac{2\pi}{W}\left(x - \frac{L}{2}\right)\right] \sin\left(\frac{n\pi x}{L}\right) dx
$$

$$
= \frac{AW}{2\pi}\left\{\cos\left[\frac{n\pi}{L}\left(\frac{L+W}{2}\right)\right] - \cos\left[\frac{n\pi}{L}\left(\frac{L+W}{2}\right)\right]\right\}\Big/(nW + 2L)
$$

$$
+ \frac{AW}{2\pi}\left\{\cos\left[\frac{n\pi}{L}\left(\frac{L+W}{2}\right)\right] - \cos\left[\frac{n\pi}{L}\left(\frac{L+W}{2}\right)\right]\right\}\Big/(nW - 2L).
$$

The integral of a product of a sine and a cosine is converted into a sum of integrals of sines through the trigonometric identity Eq. (D.2.7). The result is valid provided there is no value of $n$ for which $nW = 2L$. If so, a modification is needed that follows the reasoning in Eq. (2.2.18) or Eq. (2.2.19).

By adding $I_1$ and $I_2$ together, and performing some algebra, we obtain

$$
F_n = a_n = -\frac{8AL^2}{n\pi(n^2 W^2 - 4L^2)} \sin\left(\frac{n\pi}{2}\right) \sin\left(\frac{n\pi W}{2L}\right). \tag{4.3.25}
$$

**Reflection**

When calculating Fourier coefficients, we should always check the decay in their magnitudes and make sure it agrees with the prediction based on the nature of the function. Here, we have a function Eq. (4.3.24) that is continuous with a continuous derivative: the second derivatives has a jump discontinuity. The prediction is that $|a_n| \sim 1/n^3$ and the result Eq. (4.3.25) is consistent with this prediction. So all is well.

Suppose that the voltage is initially induce in the cable without generating any current. In other words, $g(x) = 0$ and $G_n = 0$. Then Eq. (4.3.23) determines

$$
b_n = \frac{r}{2\Omega_n} a_n. \tag{4.3.26}
$$

By substituting the results Eq. (4.3.25) and Eq. (4.3.26) into Eq. (4.3.18), the voltage is completely determined. As scientists and engineers, we prefer to express the result in dimensionless variables. Introduce,

$$
X = \frac{x}{L}, \quad \tau = \frac{\omega t}{L}, \quad \alpha = \frac{W}{L}, \quad \beta = \frac{rL}{2\omega}, \quad B_n = \sqrt{1 - \frac{\beta^2}{n^2\pi^2}}, \tag{4.3.27}
$$

then

$$
\frac{V(x,t)}{A} = e^{-\beta\tau} \sum_{n=1}^{\infty} \left[\cos(n\pi B_n \tau) + \frac{\beta}{n\pi B_n} \sin(n\pi B_n \tau)\right] A_n \sin(n\pi X), \tag{4.3.28}
$$

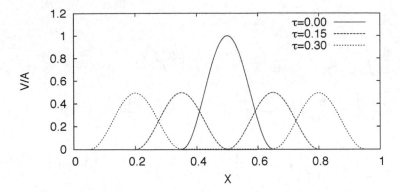

Fig. 4.12   Profiles of the voltage along the cable at various times.

where

$$A_n = -\frac{8}{n\pi(n^2\alpha^2 - 4)}\sin\left(\frac{n\pi}{2}\right)\sin\left(\frac{n\pi\alpha}{2}\right).$$

Profiles of $V(x,t)/A$ are shown in Fig. 4.12 with the choices $\alpha = 0.3$ and $\beta = 0.05$. The initial profile shows a single peak that subsequently splits into two that moves away from the center. The initial profile is producing two signals that move in opposite directions. While the geometry is different, the phenomenon is similar to a pebble causing ripples after it has struck a water surface.

### 4.3.4   *Inhomogeneous boundary conditions*

The homogeneous solution shows the presence of propagating signals that are generated by the initial conditions. What is the influence of the boundary conditions? As an example, consider the co-axial cable to have no initial signal, but then the voltage at the end of the cable $x = L$ is suddenly increased to $V_0$ (V). In other words, $V(L,t) = V_0$ while the other end remains grounded, $V(0,t) = 0$.

Of course the standard approach is to seek a particular solution to take care of the inhomogeneous boundary conditions, just as we did before when treating the boundary conditions Eq. (4.2.4). The plan is to identify the functions of time in the boundary conditions and then compose the particular solution with the same time behavior but with coefficients that depend on $x$. The boundary conditions here do not vary with time, and so we take a constant in time as our guess for the particular solution. This constant in time can vary with $x$.

$$V_{\mathrm{p}}(x,t) = a(x).\qquad\qquad(4.3.29)$$

Substitute Eq. (4.3.29) into Eq. (4.3.3).

$$\frac{\mathrm{d}^2 a}{\mathrm{d}x^2}(x) = 0\,,$$

with the obvious result $a(x) = c_1 + c_2 x$. By applying the conditions, $V_p(0,t) = 0$ and $V_p(L,t) = V_0$, the result is easily found,

$$V_p(x) = \frac{V_0}{L} x \,.$$

Time to create the general solution, a combination of the particular solution and the homogeneous solution Eq. (4.3.18),

$$V(x,t) = V_p(x,t) + V_h(x,t)$$
$$= \frac{V_0}{L} x + e^{-rt/2} \sum_{n=1}^{\infty} \left[ a_n \cos(\Omega_n t) + b_n \sin(\Omega_n t) \right] \sin\left(\frac{n\pi x}{L}\right). \quad (4.3.30)$$

Finally, we apply the initial conditions,

$$V(x,0) = 0 \,, \quad \frac{\partial V}{\partial t}(x,0) = 0 \,,$$

to determine the coefficients $a_n$ and $b_n$.

The first initial condition becomes

$$V(x,0) = \frac{V_0}{L} x + \sum_{n=1}^{\infty} a_n \sin\left(\frac{n\pi x}{L}\right) = 0 \,,$$

or

$$\sum_{n=1}^{\infty} a_n \sin\left(\frac{n\pi x}{L}\right) = -\frac{V_0}{L} x \,.$$

By extending the right hand side as an odd function that is $2L$-periodic and computing the coefficients of the Fourier sine series, the match of the two sine series gives

$$a_n = -\frac{2V_0}{L^2} \int_0^L x \sin\left(\frac{n\pi x}{L}\right) dx$$
$$= \frac{2V_0}{\pi n} (-1)^n \,. \quad (4.3.31)$$

The second initial condition becomes

$$\frac{\partial V}{\partial t}(x,0) = -\frac{r}{2} \sum_{n=1}^{\infty} a_n \sin\left(\frac{n\pi x}{L}\right) + \sum_{n=1}^{\infty} \Omega_n b_n \sin\left(\frac{n\pi x}{L}\right) = 0 \,,$$

or

$$b_n = \frac{rV_0}{\Omega_n \pi n} (-1)^n \,. \quad (4.3.32)$$

By substituting Eq. (4.3.31) and Eq. (4.3.32) into Eq. (4.3.30) and using the dimensionless variables Eq. (4.3.27), we may express the final result as

$$\frac{V}{V_0} = X + 2e^{-\beta\tau} \sum_{n=1}^{\infty} \left[ \cos(n\pi B_n \tau) + \frac{\beta}{n\pi B_n} \sin(n\pi B_n \tau) \right] \frac{(-1)^n}{n\pi} \sin(n\pi X) \,. \quad (4.3.33)$$

This result unfortunately does not help us see what is going on and the best way forward is to look at profiles of the voltage for a specific case: pick the same

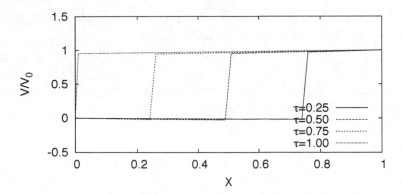

Fig. 4.13   Profiles of the voltage at various times during $0 < \tau \leq 1$.

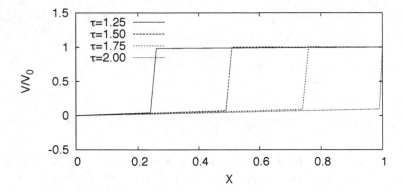

Fig. 4.14   Profiles of the voltage at various times during $1 < \tau \leq 2$.

value $\beta = 0.05$ that is used in Fig. 4.12. There is a big difference in evaluating the sums in Eq. (4.3.28) – the magnitude of the Fourier coefficients decreases as $1/n^3$ – and the sum in Eq. (4.3.33) – the decrease is only $1/n$. As a consequence the sum in Eq. (4.3.33) is evaluated on a computer with 10,000 terms to obtain plotting accuracy. The resulting profiles of the voltage are shown for a sequence of times from $t = 0$ until $t = 1$ in Fig. 4.13. What is striking about the profiles during this time is that they show an apparent jump discontinuity propagating to the left until it reaches $x = 0$. After that, shown in Fig. 4.14, the discontinuity propagates to the right. The signal has bounced from the grounded end.

Is it plausible that the solution has a propagating jump discontinuity? Fix a time $t$ and consider how the Fourier coefficients decrease in magnitude with increasing $n$. The coefficients $a_n$ given in Eq. (4.3.33) decrease the slowest and since the dependence of the decrease in $n$ is only $1/n$, we expect a jump discontinuity in $V(x,t)$.

As the jump discontinuity propagates back and forth on the time scale $\tau$, the contribution from the homogeneous solution decreases because of the exponential

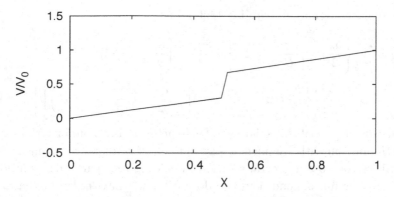

Fig. 4.15  Profile of the voltage at $\tau = 20.5$.

decay $\exp(-\beta\tau)$. The remaining part of the signal starts to look like a straight line with an increasing slope – as illustrated in Fig. 4.15 when $\tau = 20.5$. Eventually, on the time scale of $1/\beta$, the signal approaches a steady state $V(x,t)/V_0 = X$.

> **Reflection**
>
> The homogeneous solution describes the propagation of signals generated by the initial and/or boundary conditions. The presence of small resistance $r \ll 1$ causes these signals to slowly die away.

### 4.3.5  *No resistance*

For most transmission lines, including co-axial cable, $r$ is very small, and for intermediate times when the decay in signal strength is negligible, we can set $r = 0$. Thus

$$\Omega_n = \frac{n\pi\omega}{L}$$

and the homogeneous solution Eq. (4.3.18) becomes

$$V_{\mathrm{h}}(x,t) = \sum_{n=1}^{\infty} \left[ a_n \cos\left(\frac{n\pi\omega t}{L}\right) + b_n \sin\left(\frac{n\pi\omega t}{L}\right) \right] \sin\left(\frac{n\pi x}{L}\right). \qquad (4.3.34)$$

This special solution provides an explanation for the propagation of signals to the left and right. The trigonometric products in the sum in Eq. (4.3.34) can be rewritten as

$$2\cos\left(\frac{n\pi\omega t}{L}\right)\sin\left(\frac{n\pi x}{L}\right) = \sin\left[\frac{n\pi}{L}(x+\omega t)\right] + \sin\left[\frac{n\pi}{L}(x-\omega t)\right],$$

$$2\sin\left(\frac{n\pi\omega t}{L}\right)\sin\left(\frac{n\pi x}{L}\right) = \cos\left[\frac{n\pi}{L}(x-\omega t)\right] - \cos\left[\frac{n\pi}{L}(x+\omega t)\right].$$

As a consequence, we may rewrite Eq. (4.3.34) as

$$V_{\mathrm{h}}(x,t) = \sum_{n=1}^{\infty} \left\{ \frac{b_n}{2} \cos\left[\frac{n\pi}{L}(x - \omega t)\right] + \frac{a_n}{2} \sin\left[\frac{n\pi}{L}(x - \omega t)\right] \right\}$$

$$- \sum_{n=1}^{\infty} \left\{ \frac{b_n}{2} \cos\left[\frac{n\pi}{L}(x + \omega t)\right] - \frac{a_n}{2} \sin\left[\frac{n\pi}{L}(x + \omega t)\right)\right] \right\}. \qquad (4.3.35)$$

The remarkable aspect of this solution is that each sum is a Fourier series and thus represents a function with argument $x - \omega t$ in the first sum and $x + \omega t$ in the second sum. Until we compose a general solution by adding in a particular solution and then applying the initial conditions, we do not know what these functions are, so we simply write $h_{\mathrm{r}}(x - \omega t)$ and $h_{\mathrm{l}}(x + \omega t)$ as the result of the sum in each case. Thus Eq. (4.3.35) has the form

$$V_{\mathrm{h}}(x,t) = h_{\mathrm{r}}(x - \omega t) + h_{\mathrm{l}}(x + \omega t). \qquad (4.3.36)$$

How can such a simple representation of the solution be possible? We must consider the differential equation that the solution satisfies. With $r = 0$, Eq. (4.3.3) becomes

$$\frac{\partial^2 V}{\partial t^2}(x,t) - \omega^2 \frac{\partial^2 V}{\partial x^2}(x,t) = 0. \qquad (4.3.37)$$

Let us verify that Eq. (4.3.36) is a solution by direct substitution. Since

$$\frac{\partial^2 V_{\mathrm{h}}}{\partial t^2}(x,t) = \omega^2 h_{\mathrm{r}}''(x - \omega t) + \omega^2 h_{\mathrm{l}}''(x + \omega t),$$

$$\frac{\partial^2 V_{\mathrm{h}}}{\partial x^2}(x,t) = h_{\mathrm{r}}''(x - \omega t) + h_{\mathrm{l}}''(x + \omega t),$$

where the double primes mean two derivatives of the function with respect to its argument, $x - \omega t$ in the first case and $x + \omega t$ in the second. It is now easy to see that Eq. (4.3.37) is satisfied.

What does the solution Eq. (4.3.36) mean? The function $h_{\mathrm{l}}(x - \omega t)$ describes a signal propagating to the right since the argument $x - \omega t$ means that the function has been shifted to the right by a distance $\omega t$, and this shift in distance is equivalent to a shift by a uniform speed $\omega$. The other function, $h_{\mathrm{l}}(x + \omega t)$ represents a signal propagating to the left. The solution is composed of two signals traveling in opposite directions, a wonderful phenomenon!

The partial differential equation is one of the classic equations; it is called the *wave equation*.

### 4.3.6    *Abstract view*

We have seen a few examples of partial differential equations, the diffusion equation (4.1.6), the transmission line equation (4.3.3) and the wave equation (4.3.37). They all apply in a restricted region as shown in the diagram below; the partial differential equations are linear. The examples require a pair of boundary conditions to describe what is happening at the ends of the region. They differ in the specification of the initial conditions; the diffusion equation requires just the initial profile, while the wave equation requires both the initial profile and the initial rate of change. The solutions also differ fundamentally in behavior; diffusion smooths the initial profile and leaks information from the ends into the interior, while the wave equation generates propagating signals.

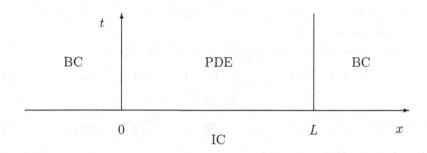

Each partial differential equation has two spatial derivatives implying the need for two boundary conditions that reflect what occurs at the edges of the device. Two popular choices are knowledge of the solution – it is being controlled – or its spatial derivative – usually associated with insulation or a barrier. If these boundary conditions are inhomogeneous, then a particular solution is required as stated in Principle 25. The forcing terms in the boundary conditions may depend on $t$. If the forcing term is expressed in terms of polynomial, exponentials and/or trigonometric functions, then a particular solution can be guessed. The homogeneous problem is solved by separation of variables, leading to a Fourier series with a form that is dictated by the homogeneous boundary conditions. The initial condition is usually treated by matching with this Fourier series.

These ideas can be expressed abstractly in mathematical terms by introducing some new terminology. Let $\mathcal{L}\{u\}$ represent any of the following partial differential equations:

(1) If the partial differential equation has only one derivative in one of the independent variables, it is called a *parabolic equation* – the classic example is the diffusion equation,

$$\mathcal{L}\{u\}(x,t) \equiv \frac{\partial u}{\partial t}(x,t) - D\frac{\partial^2 u}{\partial x^2}(x,t) = 0. \qquad (4.3.38)$$

(2) *Hyperbolic equations* allow wave propagation and one example is the transmission line equation,

$$\mathcal{L}\{u\}(x,t) \equiv \frac{\partial^2 u}{\partial t^2}(x,t) + r\frac{\partial u}{\partial t}(x,t) - \omega^2\frac{\partial^2 u}{\partial x^2}(x,t) = 0. \qquad (4.3.39)$$

(3) Another example is the wave equation,

$$\mathcal{L}\{u\}(x,t) \equiv \frac{\partial^2 u}{\partial t^2}(x,t) - \omega^2 \frac{\partial^2 u}{\partial x^2}(x,t) = 0\,. \qquad (4.3.40)$$

The symbol $\mathcal{L}$ is called an operator because it involves differentiation "operating" on a function $u$. A key property of the operator is that it is linear. That means

$$\mathcal{L}\{u_1 + u_2\}(x,t) = \mathcal{L}\{u_1\}(x,t) + \mathcal{L}\{u_2\}(x,t)\,, \qquad (4.3.41)$$

a simple statement of the property of differentiation, namely, derivatives of a sum are the sum of the derivatives.

The boundary conditions can also be stated abstractly;

$$\mathcal{B}_{\mathrm{L}}\{u\} \equiv u(0,t) = f(t) \qquad \text{or} \qquad \mathcal{B}_{\mathrm{L}}\{u\} \equiv \frac{\partial u}{\partial x}(0,t) = f(t)\,, \qquad (4.3.42)$$

and

$$\mathcal{B}_{\mathrm{R}}\{u\} \equiv u(L,t) = g(t) \qquad \text{or} \qquad \mathcal{B}_{\mathrm{R}}\{u\} \equiv \frac{\partial u}{\partial x}(L,t) = g(t)\,. \qquad (4.3.43)$$

Both operators $\mathcal{B}_{\mathrm{L}}$ and $\mathcal{B}_{\mathrm{R}}$ are linear.

Abstractly, all the problems that we have considered may be written as

$$\mathcal{L}\{u\}(x,t) = 0\,, \quad \text{subject to } \mathcal{B}_{\mathrm{L}}\{u\} = f(t) \text{ and } \mathcal{B}_{\mathrm{R}}\{u\} = g(t) \qquad (4.3.44)$$

together with appropriate initial conditions. The solution is constructed with two parts, $u(x,t) = u_{\mathrm{p}}(x,t) + u_{\mathrm{h}}(x,t)$.

### 4.3.6.1   *The particular solution*

$$\mathcal{L}\{u_{\mathrm{p}}\}(x,t) = 0\,, \quad \text{subject to } \mathcal{B}_{\mathrm{L}}\{u_{\mathrm{p}}\} = f(t) \text{ and } \mathcal{B}_{\mathrm{R}}\{u_{\mathrm{p}}\} = g(t)\,. \qquad (4.3.45)$$

Provided the functions $f(t)$ and $g(t)$ contain polynomials, exponentials and/or sines and cosines, the particular solution can be constructed by the method of undetermined coefficients where the coefficients are unknown functions of $x$.

### 4.3.6.2   *The homogeneous solution*

$$\mathcal{L}\{u_{\mathrm{h}}\}(x,t) = 0\,, \quad \text{subject to } \mathcal{B}_{\mathrm{L}}\{u_{\mathrm{h}}\} = 0 \text{ and } \mathcal{B}_{\mathrm{R}}\{u_{\mathrm{h}}\} = 0\,. \qquad (4.3.46)$$

The method of separation of variables leads to some Fourier series dictated by the nature of the homogeneous boundary conditions. There are unknown coefficients in the Fourier series and they are determined by requiring the general solution $u(x,t) = u_{\mathrm{p}}(x,t) + u_{\mathrm{h}}(x,t)$ to satisfy the initial conditions.

### 4.3.6.3   *Verification*

Substitute $u(x,t) = u_{\mathrm{p}}(x,t) + u_{\mathrm{h}}(x,t)$ directly into Eq. (4.3.44) and use the property of linearity Eq. (4.3.41).

$$\mathcal{L}\{u_{\mathrm{p}} + u_{\mathrm{h}}\}(x,t) = \mathcal{L}\{u_{\mathrm{p}}\}(x,t) + \mathcal{L}\{u_{\mathrm{h}}\}(x,t) = 0\,,$$

by using the properties in Eq. (4.3.45) and Eq. (4.3.46). Similarly the boundary conditions are satisfied.

$$\mathcal{B}_L\{u_p + u_h\}(0,t) = \mathcal{B}_L\{u_p\}(0,t) + \mathcal{B}_L\{u_h\}(0,t) = f(t),$$
$$\mathcal{B}_R\{u_p + u_h\}(L,t) = \mathcal{B}_R\{u_p\}(L,t) + \mathcal{B}_R\{u_h\}(L,t) = g(t).$$

### 4.3.7 *Exercises*

**These examples explore the method of separation of variables for the wave equation.**

(1) Find the solution to the wave equation Eq. (4.3.37) with the boundary conditions

$$\frac{\partial V}{\partial x}(0,t) = 0,$$
$$\frac{\partial V}{\partial x}(L,t) = 0,$$

and the initial conditions,

$$V(x,0) = \begin{cases} 1 + \cos(4\pi x/L), & \text{for } |x - L/2| < L/4, \\ 0, & \text{otherwise}. \end{cases}$$

$$\frac{\partial V}{\partial t}(x,0) = 0.$$

Plot the profile at a few moments in time to assess what happens to the initial profile.

**Answer:** Profiles are shown in Fig. 4.16 at $t = 0$, $L/(8\omega)$, $L/(4\omega)$.

(2) Repeat the previous problem but with the initial conditions replaced by

$$V(x,0) = \begin{cases} 1 + \cos(4\pi x/L), & \text{for } |x - L/2| < L/4, \\ 0, & \text{otherwise}. \end{cases}$$

$$\frac{\partial V}{\partial t}(x,0) = \frac{4\pi\omega}{L} \begin{cases} \sin(4\pi x/L), & \text{for } |x - L/2| < L/4, \\ 0, & \text{otherwise}. \end{cases}$$

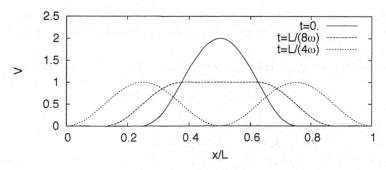

Fig. 4.16  Profiles at selected times.

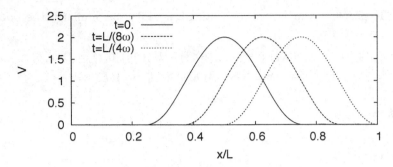

Fig. 4.17 Profiles at selected times.

**Answer:** In contrast to the previous exercise, the signal propagates to the right as shown in Fig. 4.17 for the same selected times.

(3) If instead of just raising the voltage at the end of the cable to some fixed value, we apply an alternating voltage,

$$V(L, t) = V_0 \sin(\sigma t).$$

Keep the circuit closed at the other end,

$$V(0, t) = 0$$

and assume $r = 0$.

(a) Find the particular solution in the form,

$$V_p(x, t) = A(x) \cos(\sigma t) + B(x) \sin(\sigma t).$$

(b) Construct the general solution and assume that the cable has no voltage nor current initially.

(c) Express the solution $V/V_0$ in terms of dimensionless variable $X = x/L$ and $\tau = \omega t/L$ and the dimensionless parameter $\gamma = \sigma L/\omega$.

(d) Discuss the implications of the solution you find.

**Answer:** Figure 4.18 show the temporal variation of the voltage at three spatial locations for the choice $\gamma = \pi/4$. Resonance occurs whenever $\sigma = m\pi\omega/L$ for some integer $m$.

(4) Repeat the previous exercise but now ramp up the voltage at the end $x = L$,

$$V(L, 0) = V_0 \left(1 - e^{-\alpha t}\right).$$

Express the solution in term of the dimensionless variables $X = x/L$ and $\tau = \omega t/L$ and the dimensionless parameter $\gamma = \alpha L/\omega$.

**Answer:** Figure 4.19 show the temporal variation of the voltage at three spatial locations for the choice $\gamma = \pi/4$.

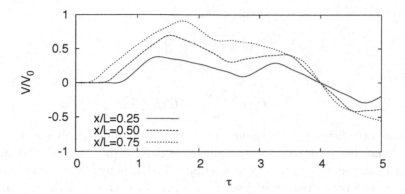

Fig. 4.18 Temporal variation of the voltage at three selected locations.

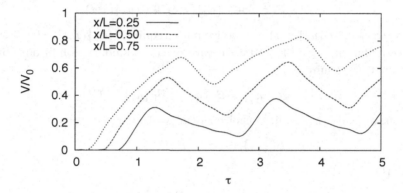

Fig. 4.19 Temporal variation of the voltage at three selected locations.

### 4.3.8 *Inhomogeneous partial differential equations*

There are occasions when the partial differential equation has a forcing term, for example, Exercises 5 and 6 in Sec. 4.2.5. The particular solution must now take care of the forcing in the equation and the forcing in the boundary conditions. Usually, it is best to split the particular solution into two parts $u_p(x, t) = u_1(x, t) + u_2(x, t)$ where $u_1(x, t)$ takes care of the forcing in the differential equation and $u_2(x, t)$ takes care of the forcing in the boundary conditions.

The abstract version of the problem replaces Eq. (4.3.44) with

$$\mathcal{L}\{u\}(x, t) = F(x, t), \quad \text{subject to } \mathcal{B}_L\{u\} = f(t) \text{ and } \mathcal{B}_R\{u\} = g(t). \quad (4.3.47)$$

The construction of $u_1(x, t)$ depends on the nature of the forcing function $F(x, t)$. Provided $F(x, t)$ is a combination of functions in $x$ and $t$ that contain only polynomials, exponentials and/or sines and cosines, the solution can be determined by the method of undetermined coefficients. For example, the guess for the particular solution to the presence of the forcing term,

$$Q_m \frac{\lambda}{AL}\left[1 - \cos\left(\frac{2\pi x}{L}\right)\right]e^{-\lambda t},$$

should be

$$u_1(x,t) = \left[A + B\sin\left(\frac{2\pi x}{L}\right) + C\cos\left(\frac{2\pi x}{L}\right)\right]e^{-\lambda t}.$$

The unknown coefficients are determined by substitution into the differential equation and balancing terms separately. Thus this part of the particular solution is determined completely and has no connection with the boundary conditions.

After we substitute the general solution $u(x,t) = u_1(x,t) + u_2(x,t) + u_h(x,t)$ into the differential equation, we are left with

$$\mathcal{L}\{u\}(x,t) = \mathcal{L}\{u_1\}(x,t) + \mathcal{L}\{u_2\}(x,t) + \mathcal{L}\{u_h\}(x,t)$$
$$= F(x,t) + \mathcal{L}\{u_2\}(x,t) + \mathcal{L}\{u_h\}(x,t).$$

Obviously, we want $\mathcal{L}\{u_h\}(x,t) = 0$, so we must require $\mathcal{L}\{u_2\}(x,t) = 0$.

The contribution $u_2(x,t)$ must take care of the boundary conditions, but be careful. What we require is

$$\mathcal{B}_L\{u_1 + u_2 + u_h\} = \mathcal{B}_L\{u_1\} + \mathcal{B}_L\{u_2\} = f(t),$$

where we have made the required choice $\mathcal{B}_L\{u_h\} = 0$. Consequently,

$$\mathcal{B}_L\{u_2\} = f(t) - \mathcal{B}_L\{u_1\},$$

and similarly

$$\mathcal{B}_R\{u_2\} = g(t) - \mathcal{B}_r\{u_1\}.$$

The boundary conditions for $u_2(x,t)$ are modified so that the general solution satisfies the given boundary conditions.

We are left with the calculation of the homogeneous solution by separation of variables.

## 4.4 Laplace's equation

In the previous two sections of this chapter, our attention has been on time dependent processes that lead to either diffusion or propagation. We consider spatial variation in only one direction. While this is reasonable when we consider thin pipes or co-axial cable, there are many other circumstances where we must allow spatial variations in more directions. We should anticipate, then, that there will be more spatial derivatives in the partial differential equations, and the need for more boundary conditions.

The challenges we face in this section are the adaption of the method of separation of variables to partial differential equations with more than two independent variables and the presence of additional boundary conditions. The first challenge is

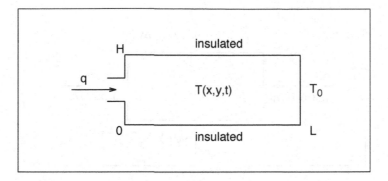

Fig. 4.20   Heat conduction in a rectangular slab of metal. The view is of the cross section.

resolved by "peeling an onion;" as we construct the solution with attention placed on a single independent variable, we find a new problem containing the remaining independent variables. The second challenge is dealt with by recognizing that time is not a special variable mathematically, and that initial conditions can be replaced by boundary conditions in the method of separation of variables.

Let us consider a specific example. Suppose a long metallic beam has a rectangular cross section. Assume that there is no variation in the temperature along this beam, variation occurring only in the cross section. A schematic is presented in Fig. 4.20 that shows the cross section. The top and bottom sides of the beam are insulated, as well as the left side except for a part that is exposed to a heat source, a furnace, for instance. Assume that this heat source produces a steady uniform flux of heat. The right side is cooled by a refrigerant that maintains a temperature of $T_0$. The challenge is to determine the steady temperature distribution inside the rectangle and to locate the maximum temperature. One can imagine the metal beam is a support for the furnace and we want to be sure that the maximum temperature does not exceed the melting temperature of the metal.

### 4.4.1   *Heat transport in two dimensions*

The derivation of the diffusion equation in two dimensions relies on the derivation of the transport equation in two dimensions which in turn is an extension of the derivation of the transport equation in one dimension (4.1.2). It is based on evaluating the increase in energy inside a small volume from a net flux of heat across surfaces surrounding the volume. Imagine a small length $\triangle z$ of the beam, and consider a small rectangular cross section as shown in Fig. 4.21. We assume that there is no flux of heat in the $z$ direction and that the temperature does not vary along $z$, but there can be heat fluxes through the surfaces on the right and left, and the top and bottom. These fluxes can be, of course, different but they are measured in the same way. For example, $q_x(x, y, t)$ measures the amount per unit area per unit time that passes a surface located at $(x, y)$ with the normal to the surface pointing

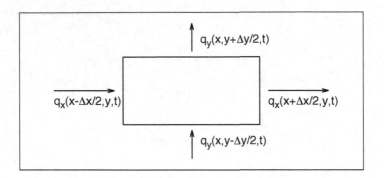

Fig. 4.21    An illustration of the conservation of heat in a small cross section.

in the $x$ direction. Not only might the flux measurement be different at different times – hence the dependency on $t$, but it can also depend on where the small area $\triangle y, \triangle z$ is placed in $(x, y)$ – hence the dependency on $x, y$. Similarly, $q_y(x, y, t)$ measures the heat flux crossing a surface located at $(x, y)$ with the normal to the surface pointing in the $y$ direction. All the fluxes entering and leaving the small rectangle are displayed schematically in Fig. 4.21.

If $\triangle T(x, y, t)$ is the increase in the temperature, then the increase in the internal energy per unit mass is approximated by

$$\triangle e = \frac{\mathrm{d}e}{\mathrm{d}T} \triangle T = c \triangle T$$

where we have used the definition of the heat capacity per unit mass – see Eq. (4.2.1). The mass in the small volume is given by $\rho \triangle x \triangle y \triangle z$. So the change in amount of internal energy is

$$\rho c \triangle T(x, y, t) \triangle x \triangle y \triangle z \,,$$

and it must be the result of the net flow of heat through the surfaces. The flux of heat into the small volume in $\triangle t$ is

$$q_x(x - \triangle x/2, y, t) \triangle y \triangle z \triangle t + q_y(x, y - \triangle y/2, t) \triangle x \triangle z \triangle t$$

and the flux out is

$$q_x(x + \triangle x/2, y, t) \triangle y \triangle z \triangle t + q_y(x, y + \triangle y/2, t) \triangle x \triangle z \triangle t \,.$$

The conservation of energy then requires

$$\rho c \triangle T(x, y, t) \triangle x \triangle y \triangle z = q_x(x - \triangle x/2, y, t) \triangle y \triangle z \triangle t$$
$$+ q_y(x, y - \triangle y/2, t) \triangle x \triangle z \triangle t - q_x(x + \triangle x/2, y, t) \triangle y \triangle z \triangle t$$
$$- q_y(x, y + \triangle y/2, t) \triangle x \triangle z \triangle t \,.$$

Divide this expression by $\triangle x \, \triangle y \, \triangle z \, \triangle t$ and rearrange the terms.

$$\rho c \frac{\triangle T}{\triangle t}(x, y, t) = -\frac{q_x(x + \triangle x/2, y, t) - q_x(x - \triangle x/2, y, t)}{\triangle x}$$

$$-\frac{q_y(x, y + \triangle y/2, t) - q_y(x, y - \triangle y/2, t)}{\triangle y}.$$

Take the limit as $\triangle t, \triangle x, \triangle y \to 0$ and recall the definition of a partial derivative;

$$\rho c \frac{\partial T}{\partial t}(x, y, t) = -\frac{\partial q_x}{\partial x}(x, y, t) - \frac{\partial q_y}{\partial y}(x, y, t). \tag{4.4.1}$$

If we regard $q_x$ and $q_y$ as the components of a vector, the heat flux vector $\mathbf{q} = q_x \mathbf{i} + q_y \mathbf{j}$, then Eq. (4.3.1) can also be written as

$$\rho c \frac{\partial T}{\partial t} = -\nabla \cdot \mathbf{q}. \tag{4.4.2}$$

This result is the vector statement of the conservation of energy.

The final step in deriving an equation for the temperature is to invoke Fourier's law of conduction, the heat flux depends on the change in temperature in the direction of the flux. In other words, the heat flux passing through a surface depends on the change in temperature along the normal to the surface, but opposite in direction.

$$q_x(x, y, t) = -k \frac{\partial T}{\partial x}(x, y, t), \quad q_y(x, y, t) = -k \frac{\partial T}{\partial y}(x, y, t)$$

$$\text{or} \quad \mathbf{q} = -k \, \nabla T. \tag{4.4.3}$$

By substituting Eq. (4.4.3) into Eq. (4.4.2), we obtain the multi-dimensional diffusion equation,

$$\frac{\partial T}{\partial t}(x, y, t) = \alpha \left( \frac{\partial^2 T}{\partial x^2}(x, y, t) + \frac{\partial^2 T}{\partial y^2}(x, y, t) \right), \tag{4.4.4}$$

where $\alpha = k/(\rho c)$. This equation is linear and homogeneous (no forcing terms). It has one time derivative and two second-order spatial derivatives. Thus we need one initial condition and two pairs of boundary conditions, one pair at $x = 0$ (left) and $x = L$ (right) and another pair at $y = 0$ (bottom) and $y = H$ (top).

From the description of the situation, the insulated top and bottom require

$$\frac{\partial T}{\partial y}(x, 0, t) = \frac{\partial T}{\partial y}(x, H, t) = 0. \tag{4.4.5}$$

At the right boundary,

$$T(L, y, t) = T_0, \tag{4.4.6}$$

and at the left boundary,

$$\frac{\partial T}{\partial x}(0, y, t) = \begin{cases} 0 & \text{for } 0 < y < (H - h)/2 \text{ and } (H + h)/2 < y < H, \\ -q/k & \text{for } (H - h)/2 < y < (H + h)/2. \end{cases} \tag{4.4.7}$$

The decision has been made to place the exposed part of the beam in the middle of the left boundary. It has length $h$.

An appropriate initial condition is

$$T(x, y, 0) = T_0\,, \tag{4.4.8}$$

indicating that the furnace is turned on at $t = 0$.

### 4.4.2  *Steady state solution as a particular solution*

From a mathematical viewpoint, the unknown function $T(x, y, t)$ satisfies a partial differential equation in all of its independent variables, $x$, $y$, and $t$. Two of the boundary conditions are homogeneous – Eq. (4.4.5) – and two are not – Eq. (4.4.6) and Eq. (4.4.7). The obvious strategy to follow is the one that is developed in the construction of solutions to the diffusion equation and the wave equation. That strategy depends on separation of variables when all the boundary conditions are homogeneous. With the hope that we can use separation of variables when there are two pairs of homogeneous boundary conditions, the first step in constructing a solution is to find a particular solution that will take care of the inhomogeneous boundary conditions. Since the forcing terms (inhomogeneous terms) in Eq. (4.4.6) and Eq. (4.4.7) do not vary with $t$, our guess should be that $T_p(x, y, t)$ is a constant in time, but a constant in time does not exclude dependency on $x$ or $y$. Thus we guess

$$T_p(x, y, t) = u(x, y)\,. \tag{4.4.9}$$

Substitute Eq. (4.4.9) into Eq. (4.4.4).

$$\frac{\partial^2 u}{\partial x^2}(x, y) + \frac{\partial^2 u}{\partial y^2}(x, y) = 0\,. \tag{4.4.10}$$

Substitute Eq. (4.4.9) into Eq. (4.4.5).

$$\frac{\partial u}{\partial y}(x, 0) = \frac{\partial u}{\partial y}(x, H) = 0\,. \tag{4.4.11}$$

Substitute Eq. (4.4.9) into Eq. (4.4.6) and Eq. (4.4.7).

$$u(L, y) = T_0\,, \tag{4.4.12}$$

and

$$\frac{\partial u}{\partial x}(0, y) = \begin{cases} 0 & \text{for } 0 < y < (H - h)/2 \text{ and } (H + h)/2 < y < H\,, \\ -q/k & \text{for } (H - h)/2 < y < (H + h)/2\,. \end{cases} \tag{4.4.13}$$

We have obtained a new mathematical problem.[10] The new partial differential equation Eq. (4.4.10) is called Laplace's equation. It is linear and homogeneous and contains two spatial derivatives for each independent variable $x$ and $y$. There

---

[10]It is quite common in solving mathematical problems that we proceed by reducing a problem to a simpler one. For example, note how separation of variables reduces solving partial differential equations to solving ordinary differential equations. Here we have converted a partial differential equation in time and space into one just in space.

are four boundary conditions altogether, two are homogeneous – Eq. (4.4.11) – and two are not – Eq. (4.4.12) and Eq. (4.4.13). Now we set aside how this problem came about and ask ourselves what must we do to solve it.

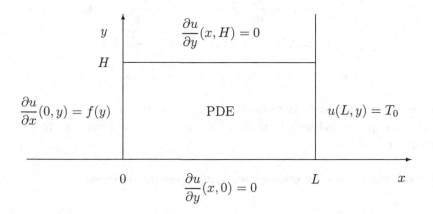

It is useful to draw a mathematical diagram as we did for the diffusion equation and the wave equation but now in the $x, y$ plane. The diagram provides the clue on how to approach construction of a solution. Recall how separation of variables applied to a homogeneous partial differential equation and homogeneous boundary conditions leads to a Fourier series that then can be used to match the initial conditions. Imagine that we separate the variables and obtain a Fourier series in $y$ because the homogeneous boundary conditions are at the top and bottom. Instead of matching initial conditions, we match the boundary conditions on the left and right. So let us implement this approach.

**1. Separate the variables.**
Set $u(x, y) = X(x) Y(y)$ and substitute into Eq. (4.4.10)
$$Y(y) \frac{\mathrm{d}^2 X}{\mathrm{d}x^2}(x) + X(x) \frac{\mathrm{d}^2 Y}{\mathrm{d}y^2}(y) = 0 \,.$$

**2. Separate the equation.**
Rewrite the equation as
$$\frac{1}{X(x)} \frac{\mathrm{d}^2 X}{\mathrm{d}x^2}(x) = -\frac{1}{Y(y)} \frac{\mathrm{d}^2 Y}{\mathrm{d}y^2}(y) \,.$$
The only way this statement can make sense is if each side is a constant, $c$ say. Thus
$$\frac{\mathrm{d}^2 X}{\mathrm{d}x^2}(x) - c\, X(x) = 0 \,, \tag{4.4.14}$$

$$\frac{\mathrm{d}^2 Y}{\mathrm{d}y^2}(y) + c\, Y(y) = 0 \,. \tag{4.4.15}$$
We have converted the problem into two homogeneous ordinary differential equations.

## 3. Separate the boundary conditions.
Since it is the homogeneous boundary conditions we are interested in for now, substitute $u(x, y) = X(x) Y(y)$ into (4.4.11).

$$\frac{\partial u}{\partial y}(x, 0) = X(x) \frac{dY}{dy}(0) = 0, \quad \Rightarrow \quad \frac{dY}{dy}(0) = 0, \tag{4.4.16}$$

$$\frac{\partial u}{\partial y}(x, H) = X(x) \frac{dY}{dy}(H) = 0, \quad \Rightarrow \quad \frac{dY}{dy}(H) = 0. \tag{4.4.17}$$

We now have homogeneous boundary conditions for $Y(y)$.

## 4. Solve the homogeneous boundary value problem.
Set $Y(y) = \exp(ry)$ and substitute into Eq. (4.4.15). The resulting characteristic equation is $r^2 + c = 0$.

## 5. Consider all possible choices for the separation constant.

(1) Assume $c > 0$. The general solution to Eq. (4.4.15) is

$$Y(y) = b_1 \cos(\sqrt{c} \, y) + b_2 \sin(\sqrt{c} \, y).$$

Substitute into Eq. (4.4.16).

$$b_2 \sqrt{c} = 0.$$

Substitute into Eq. (4.4.17) (using $b_2 = 0$ from the previous step).

$$-b_1 \sqrt{c} \, \sin(\sqrt{c} \, H) = 0.$$

To ensure $b_1 \neq 0$, we must choose $\sqrt{c} H = n\pi$, or

$$c = \frac{n^2 \pi^2}{H^2} \quad \text{for any positive integer } n. \tag{4.4.18}$$

We have found many solutions. For each choice for $n$,

$$Y(y) = b_1 \cos\left(\frac{n\pi y}{H}\right). \tag{4.4.19}$$

(2) Now try $c = 0$. The general solution to Eq. (4.4.15) is

$$Y(y) = b_1 + b_2 y.$$

By substituting into Eq. (4.4.16) and Eq. (4.4.17), we find $b_2 = 0$, but no restriction on $b_1$. So we have found another solution,

$$Y(y) = b_1. \tag{4.4.20}$$

(3) Finally, assume $c < 0$. For convenience, set $c = -d^2$. The general solution is

$$Y(y) = b_1 e^{dy} + b_2 e^{-dy}$$

and, when the boundary conditions are applied, the only possible solution is the trivial solution $b_1 = b_2 = 0$. This case fails.

As before, the homogeneous boundary value problem has non-trivial solutions only for restricted choices of the separation constant $c$.

## 6. Solve the remaining differential equation.

The remaining equation, Eq. (4.4.14), must be solved for the choices of $c$ that lead to non-trivial solutions in the previous step.

(1) For the choice Eq. (4.4.18), Eq. (4.4.14) becomes

$$\frac{d^2 X}{dx^2}(x) - \frac{n^2 \pi^2}{H^2} X(x) = 0,$$

which has the solution,

$$X(x) = a_1 e^{n\pi x/H} + a_2 e^{-n\pi x/H}. \tag{4.4.21}$$

(2) Next, consider $c = 0$. Then

$$X(x) = a_1 + a_2 x. \tag{4.4.22}$$

We do not try yet to satisfy the inhomogeneous boundary conditions Eq. (4.4.12) and Eq. (4.4.13) but first compose a general solution.

## 7. Compose a general solution.

Take all the solutions we have found for each different choice of $c$ and add them all together with an appropriate choice for the unknown coefficients – see Principle 6.

(1) $c = n^2 \pi^2 / H^2$,

$$X(x) Y(y) = \left[ a_1 e^{n\pi x/H} + a_2 e^{-n\pi x/H} \right] b_1 \cos\left( \frac{n\pi y}{H} \right)$$

$$= \left[ c_n e^{n\pi x/H} + d_n e^{-n\pi x/H} \right] \cos\left( \frac{n\pi}{H} y \right).$$

(2) $c = 0$,

$$X(x) Y(y) = (a_0 + a_1 x) b_1 = c_0 + d_0 x.$$

Adding together,

$$u(x,y) = c_0 + d_0 x + \sum_{n=1}^{\infty} \left[ c_n e^{n\pi x/H} + d_n e^{-n\pi x/H} \right] \cos\left( \frac{n\pi}{H} y \right) \tag{4.4.23}$$

which has the appearance of a cosine Fourier series in $y$ with rather complicated Fourier coefficients that depend on $x$.

---

**Reflection**

The general solution Eq. (4.4.23) is similar to Eq. (4.3.18). The difference is that the "Fourier coefficients" depend on $t$ in the first case and $x$ in the second case. Also the functions in the "Fourier coefficients" are different, sines and cosines in the first case and exponentials in the second case. Both have two sets of unknown coefficients, $a_n$ and $b_n$ in the first case and $c_n$ and $d_n$ in the second case. Initial conditions determined the unknown coefficients $a_n$ and $b_n$; we look to the remaining inhomogeneous boundary conditions to determine $c_n$ and $d_n$.

## 8. Satisfy the remaining boundary conditions.
The first condition Eq. (4.4.12) requires

$$u(L,y) = c_0 + d_0 L + \sum_{n=1}^{\infty} \left[ c_n\, e^{n\pi L/H} + d_n\, e^{-n\pi L/H} \right] \cos\left( \frac{n\pi}{H} y \right) = T_0. \quad (4.4.24)$$

Let us look carefully at what this requirement states. To help see the result clearly, let us replace

$$c_0 + d_0 L = a_0,$$

and

$$c_n\, e^{n\pi L/H} + d_n\, e^{-n\pi L/H} = a_n.$$

Then the statement Eq. (4.4.24) reads

$$a_0 + \sum_{n=1}^{\infty} a_n \cos\left( \frac{n\pi}{H} y \right) = T_0.$$

In this form, it is easy to see that

$$a_0 = c_0 + d_0 L = T_0, \quad (4.4.25)$$
$$a_n = c_n\, e^{n\pi L/H} + d_n\, e^{-n\pi L/H} = 0. \quad (4.4.26)$$

One last boundary condition Eq. (4.4.13) to apply!

$$d_0 + \sum_{n=1}^{\infty} \frac{n\pi}{H} (c_n - d_n) \cos\left( \frac{n\pi}{H} y \right) = f(y)$$

$$\equiv \begin{cases} 0 & \text{for } 0 < y < (H-h)/2 \text{ and } (H+h)/2 < y < H, \\ -q/k & \text{for } (H-h)/2 < y < (H+h)/2. \end{cases} \quad (4.4.27)$$

Once again we see the appearance of a Fourier cosine series on the left and on the right a discontinuous function. As before, the way forward is to cast the function on the right hand side in a Fourier cosine series,

$$f(y) = A_0 + \sum_{n=1}^{\infty} A_n \cos\left( \frac{n\pi y}{H} \right).$$

Obviously, we must extend the function as an even function and demand it is $2H$-periodic. Easily done! Next we determine the Fourier coefficients.

$$A_0 = -\frac{1}{H} \int_{(H-h)/2}^{(H+h)/2} \frac{q}{k} \, dy$$

$$= -\frac{qh}{kH} \, ,$$

$$A_n = -\frac{2}{H} \int_{(H-h)/2}^{(H+h)/2} \frac{q}{k} \cos\left(\frac{n\pi}{H} y\right) dy$$

$$= -\frac{2q}{n\pi k} \left[ \sin\left(\frac{n\pi}{2H}(H+h)\right) - \sin\left(\frac{n\pi}{2H}(H-h)\right) \right]$$

$$= -\frac{4q}{n\pi k} \cos\left(\frac{n\pi}{2}\right) \sin\left(\frac{n\pi h}{2H}\right).$$

The match of the Fourier cosine series in Eq. (4.4.27) gives

$$d_0 = A_0 = -\frac{qh}{kH} \, , \tag{4.4.28}$$

$$\frac{n\pi}{H}(c_n - d_n) = A_n = -\frac{4q}{n\pi k} \cos\left(\frac{n\pi}{2}\right) \sin\left(\frac{n\pi h}{2H}\right). \tag{4.4.29}$$

The boundary conditions have provided equations to determine $c_n$ and $d_n$. From Eq. (4.4.25) and Eq. (4.4.28),

$$c_0 = T_0 + \frac{qhL}{kH} \, , \qquad D_0 = -\frac{qh}{kH} \, .$$

From Eq. (4.4.26) and Eq. (4.4.29),

$$c_n = -\frac{4qh \, \exp(-n\pi L/H)}{n^2 \pi^2 k \left[\exp(n\pi L/H) + \exp(-n\pi L/H)\right]} \cos\left(\frac{n\pi}{2}\right) \sin\left(\frac{n\pi h}{2H}\right),$$

$$d_n = \frac{4qh \, \exp(n\pi L/H)}{n^2 \pi^2 k \left[\exp(n\pi L/H) + \exp(-n\pi L/H)\right]} \cos\left(\frac{n\pi}{2}\right) \sin\left(\frac{n\pi h}{2H}\right).$$

The result can be simplified further by observing the pattern displayed in Table 4.2. Clearly, only the even coefficients are non-zero and we may introduce a new

Table 4.2 Pattern in $\cos(n\pi/2)$.

| $n$ | $\cos(n\pi/2)$ |
|-----|----------------|
| 1 | 0 |
| 2 | −1 |
| 3 | 0 |
| 4 | 1 |
| 5 | 0 |
| 6 | −1 |

counter $m = 2n$. Then,

$$\cos\left(\frac{n\pi}{2}\right) = \cos(m\pi) = (-1)^m.$$

Gathering all the information together, the solution can be expressed as

$$u(x,y) = T_o + \frac{qh}{kH}(L-x) + \frac{qH}{k\pi^2}\sum_{m=1}^{\infty}\left[\frac{(-1)^m \sin(m\pi h/H)}{m^2}\right.$$

$$\left. \times \frac{\sinh\left[2m\pi(L-x)/H\right]}{\cosh(2m\pi L/H)}\cos\left(\frac{2m\pi}{H}y\right)\right]. \quad (4.4.30)$$

The result involves several parameters, $T_0$, $q$, $h$, $H$ and $L$. The information can be consolidated by introducing dimensionless variables,

$$X = \frac{x}{H}, \quad Y = \frac{y}{H}, \quad \rho = \frac{h}{H}, \quad r = \frac{L}{H}.$$

Then

$$F(x,y) = \frac{kH}{qh}(u - T_0) = r - X$$

$$+ \frac{1}{\rho\pi^2}\sum_{m=1}^{\infty}\frac{(-1)^m}{m^2}\sin(m\pi\rho)\frac{\sinh\left[2m\pi(r-X)\right]}{\cosh(2m\pi r)}\cos(2m\pi Y)$$

and the result depends on only two parameters, $r$ and $\rho$.

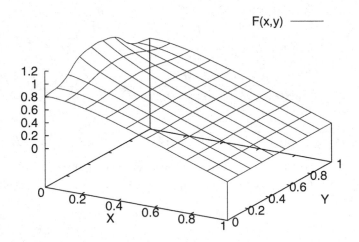

Fig. 4.22 Surface plot of the temperature field.

To gain some understanding of the behavior of the solution, let us set the parameters as follows: $r = 1$ and $\rho = 0.5$. The surface plot is shown in Fig. 4.22. The heat diffuses from the exposed part in the left boundary into the rest of the slab and quickly adjusts to a uniform conduction of heat (constant temperature gradient) towards the right boundary.

---

**Reflection**

Each time we have used separation of variables, it is the occurrence of a homogeneous boundary value with an unknown separation constant that has led to success. As long as the partial differential equation is homogeneous, we expect homogeneous ordinary differential equation to result from separation of variables. So the important part is the presence of homogeneous boundary conditions. Then the separation constant can only be certain values for non-trivial solutions.

---

### 4.4.3  *Electrostatic potential*

Laplace's equation also arises as the partial differential equation that determines potentials. For example, the electric field **E** is the gradient of the electrostatic potential $\phi$:

$$\mathbf{E} = \nabla\phi. \qquad (4.4.31)$$

In the absence of electric charges and magnetic field effects, the electric field must be divergence-free.

$$\nabla \cdot \mathbf{E} = 0,$$

and by using Eq. (4.4.31), we obtain Laplace's equation for the potential,

$$\nabla \cdot \nabla\phi = \frac{\partial^2 \phi}{\partial x^2}(x, y) + \frac{\partial^2 \phi}{\partial y^2}(x, y) = 0. \qquad (4.4.32)$$

Here we are considering the two-dimensional version.

Suppose we have a thin rectangular slice of a semi-conductor with a width $W$ and height $H$. Suppose we apply a potential distribution along the left side and along half of the top, but keep the potential on the remainder of the boundary at $\phi = 0$ (grounded). Specifically,

$$\phi(0, y) = \frac{y}{H}, \qquad (4.4.33)$$

$$\phi(W, y) = 0, \qquad (4.4.34)$$

$$\phi(x, 0) = 0, \qquad (4.4.35)$$

$$\phi(x, H) = f(x) \equiv \begin{cases} 1 - 2x/W & \text{for } 0 < x < W/2, \\ 0 & \text{for } W/2 < x < W. \end{cases} \qquad (4.4.36)$$

Our task, then, is to construct the solution for the potential $\phi(x, y)$ subject to these boundary conditions.

The starting point is to draw a mathematical diagram to show the boundary conditions. The boundary condition on top is written as $f(x)$ instead of Eq. (4.4.36) for convenience. While the partial differential equation Eq. (4.4.32) is homogeneous, there are two inhomogeneous boundary conditions, Eq. (4.4.33) and Eq. (4.4.36), and two homogeneous boundary conditions, Eq. (4.4.34) and Eq. (4.4.35). Clearly, there is no pair of homogeneous boundary conditions in either of the variations in $x$ or $y$. That means we will need a particular solution to take care of one set of inhomogeneous boundary conditions and the obvious choice is to take the pair Eq. (4.4.33) and Eq. (4.4.34) because the inhomogeneous term in Eq. (4.4.33) is just a linear function and a guess for the particular solution will be easy. In contrast, it is impossible to guess a particular solution for the pair Eq. (4.4.35) and Eq. (4.4.36) because $f(x)$ is a discontinuous function.

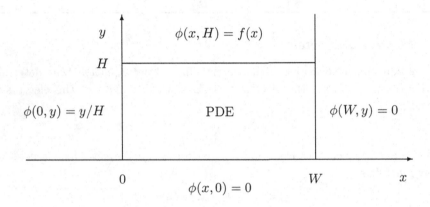

The particular solution $\phi_p(x, y)$ should be a linear function as far as the variation in $y$ goes, but can depend on $x$. Thus we try

$$\phi_p(x, y) = g(x) + h(x)y.$$

By substituting this guess into Eq. (4.4.32), we obtain

$$\frac{d^2 g}{dx^2}(x) + \frac{d^2 h}{dx^2}(x)\,y = 0,$$

with the consequence that $g(x) = a + bx$ and $h(x) = c + dx$. Now apply Eq. (4.4.33) and Eq. (4.4.34) with the result that

$$\phi_p(x, y) = \frac{y}{H}\left(1 - \frac{x}{W}\right). \tag{4.4.37}$$

We are left with a homogeneous problem: a homogeneous partial differential equation,

$$\frac{\partial^2 \phi_h}{\partial x^2}(x, y) + \frac{\partial^2 \phi_h}{\partial y^2}(x, y) = 0 \tag{4.4.38}$$

with homogeneous boundary conditions,

$$\phi_h(0, y) = 0, \tag{4.4.39}$$

$$\phi_h(W, y) = 0. \tag{4.4.40}$$

We set aside the remaining boundary conditions for now until we have obtain a solution by separation of variables. This time we flash through the procedure as quickly as possible.

## 1. Separate the variables.

Write $\phi_h(x, y) = X(x) Y(y)$ and substitute into Eq. (4.4.38).

$$Y(y) \frac{d^2 X}{dx^2}(x) + X(x) \frac{d^2 Y}{dy^2}(y) = 0.$$

## 2. Separate the equation.

Rewrite the equation as

$$\frac{1}{X(x)} \frac{d^2 X}{dx^2}(x) = -\frac{1}{Y(y)} \frac{d^2 Y}{dy^2}(y).$$

Set each side to be a constant.

$$\frac{d^2 X}{dx^2}(x) + c X(x) = 0, \qquad (4.4.41)$$

$$\frac{d^2 Y}{dy^2}(y) - c Y(y) = 0. \qquad (4.4.42)$$

## 3. Separate the boundary conditions.

Substitute $\phi_h(x, y) = X(x) Y(y)$ into Eq. (4.4.39) and Eq. (4.4.40).

$$\phi(0, y) = X(0) Y(y) = 0 \quad \Rightarrow \quad X(0) = 0, \qquad (4.4.43)$$
$$\phi(W, y) = X(W) Y(y) = 0 \quad \Rightarrow \quad X(W) = 0. \qquad (4.4.44)$$

## 4. Solve the homogeneous boundary value problem.

Solve Eq. (4.4.41) subject to Eq. (4.4.43) and Eq. (4.4.44). The characteristic equation is $\lambda^2 + c = 0$.

## 5. Consider all possible choices for the separation constant.

(1) $c > 0$. The general solution to Eq. (4.4.41) is

$$X(x) = a_1 \cos(\sqrt{c}\, x) + a_2 \sin(\sqrt{c}\, x).$$

Apply the boundary conditions Eq. (4.4.43) and Eq. (4.4.44):

$$a_1 = 0,$$
$$a_2 \sin(\sqrt{c}\, W) = 0.$$

Choose $\sqrt{c}\, W = n\pi$ for any positive integer $n$, or $c = n^2\pi^2/W^2$, and the solution is

$$X(x) = a_2 \sin\left(\frac{n\pi x}{W}\right). \qquad (4.4.45)$$

(2) $c = 0$. The general solution to Eq. (4.4.41) is $X(x) = a_1 + a_2 x$. Apply the boundary conditions Eq. (4.4.43) and Eq. (4.4.44). The result is $a_1 = a_2 = 0$ – no good.

(3) $c < 0$. Set $c = -\alpha^2$, then the general solution to Eq. (4.4.41) is

$$X(x) = a_1 e^{\alpha x} + a_2 e^{-\alpha x}.$$

Apply the boundary conditions Eq. (4.4.43) and Eq. (4.4.44). The result is $a_1 = a_2 = 0$ – no good.

## 6. Solve the remaining differential equation.
With the choice $c = n^2\pi^2/W^2$, Eq. (4.4.42) becomes

$$\frac{d^2 Y}{dy^2}(y) - \frac{n^2\pi^2}{W^2} Y(y) = 0,$$

which has the general solution

$$Y(y) = b_1 e^{n\pi y/W} + b_2 e^{-n\pi y/W}. \tag{4.4.46}$$

## 7. Compose a general solution.
For each choice of $n$, we have a solution that is a product $X(x)Y(y)$. By using Eq. (4.4.45) and Eq. (4.4.46),

$$\phi_{\mathrm{h}}(x, y) = \left[ b_1 e^{n\pi y/W} + b_2 e^{-n\pi y/W} \right] a_2 \sin\left( \frac{n\pi x}{W} \right).$$

Add up all the solutions with new coefficients that reflect values for different $n$. Thus,

$$\phi_{\mathrm{h}}(x, y) = \sum_{n=1}^{\infty} \left[ a_n e^{n\pi y/W} + b_n e^{-n\pi y/W} \right] \sin\left( \frac{n\pi x}{W} \right).$$

The general solution is

$$\begin{aligned}
\phi(x, y) &= \phi_{\mathrm{p}}(x, y) + \phi_{\mathrm{h}}(x, y) \\
&= \frac{y}{H}\left(1 - \frac{x}{W}\right) + \sum_{n=1}^{\infty} \left[ a_n e^{n\pi y/W} + b_n e^{-n\pi y/W} \right] \sin\left( \frac{n\pi x}{W} \right).
\end{aligned} \tag{4.4.47}$$

We are now ready to apply the remaining boundary conditions, Eq. (4.4.35) and Eq. (4.4.36). The hope is that these two conditions will be enough to determine the two families of coefficients, $a_n$ and $b_n$.

## 8. Satisfy the remaining conditions.
For Eq. (4.4.35),

$$\sum_{n=1}^{\infty} \left( a_n + b_n \right) \sin\left( \frac{n\pi x}{W} \right) = 0.$$

So we must have $b_n = -a_n$ and we may now write Eq. (4.4.47) in the form

$$\phi(x, y) = \frac{y}{H}\left(1 - \frac{x}{W}\right) + 2\sum_{n=1}^{\infty} a_n \sinh\left( \frac{n\pi y}{W} \right) \sin\left( \frac{n\pi x}{W} \right). \tag{4.4.48}$$

Apply the last condition Eq. (4.4.36).

$$1 - \frac{x}{W} + 2\sum_{n=1}^{\infty} a_n \sinh\left(\frac{n\pi H}{W}\right) \sin\left(\frac{n\pi x}{W}\right) = f(x)$$

$$= \begin{cases} 1 - 2x/W & \text{for } 0 < x < W/2, \\ 0 & \text{for } W/2 < x < W, \end{cases}$$

or rather

$$2\sum_{n=1}^{\infty} a_n \sinh\left(\frac{n\pi H}{W}\right) \sin\left(\frac{n\pi x}{W}\right) = f(x) - 1 + \frac{x}{W}. \qquad (4.4.49)$$

To proceed we must cast the right hand side of Eq. (4.4.49) into a Fourier sine series,

$$f(x) - 1 + \frac{x}{W} = \sum_{n=1}^{\infty} A_n \sin\left(\frac{n\pi x}{W}\right)$$

where the coefficients are given by

$$A_n = \frac{2}{W} \int_0^W \left(f(x) - 1 + \frac{x}{W}\right) \sin\left(\frac{n\pi x}{W}\right) dx$$

$$= -\frac{4}{n^2\pi^2} \sin\left(\frac{n\pi}{2}\right).$$

By balancing all the Fourier coefficients in Eq. (4.4.49),

$$2\, a_n \sinh\left(\frac{n\pi H}{W}\right) = A_n = -\frac{4}{n^2\pi^2} \sin\left(\frac{n\pi}{2}\right)$$

and $a_n$ has been determined.

When $a_n$ is substituted into Eq. (4.4.48), we have finally completed solution. In terms of dimensionless variables,

$$X = \frac{x}{W}, \quad Y = \frac{y}{W} \quad \text{and} \quad \rho = \frac{H}{W},$$

the solution is

$$\phi(x,y) = \frac{Y}{\rho}(1 - X) - \frac{4}{\pi^2} \sum_{n=1}^{\infty} \frac{\sin(n\pi/2)}{n^2 \sinh(n\pi\rho)} \sinh(n\pi Y) \sin(n\pi X). \qquad (4.4.50)$$

With the choice $\rho = 0.5$, the solution is displayed as a surface plot in Fig. 4.23.

### 4.4.4  *Abstract view*

Laplace's equation is an example of an *elliptic equation* and contributes another partial differential equation,

$$\mathcal{L}\{u\}(x,y) \equiv \frac{\partial^2 u}{\partial x^2}(x,y) + \frac{\partial^2 u}{\partial y^2}(x,y) \qquad (4.4.51)$$

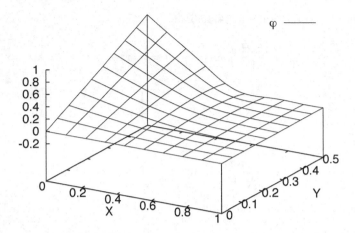

Fig. 4.23    Surface plot of the potential $\phi$.

to the collection Eq. (4.3.38), Eq. (4.3.39) and Eq. (4.3.40). But the abstract statement of the problem is different from Eq. (4.3.44). This time the problem is

$$\mathcal{L}\{u\}(x,y) = 0\,, \tag{4.4.52}$$

subject to two pairs of boundary conditions. One pair is the same as Eq. (4.3.42) and Eq. (4.3.43);[11] the new pair is

$$\mathcal{B}_{\mathrm{T}}\{u\} \equiv u(x,H) = h(x) \quad \text{or} \quad \mathcal{B}_{\mathrm{T}} \equiv \frac{\partial u}{\partial y}(x,H) = h(x)\,, \tag{4.4.53}$$

and

$$\mathcal{B}_{\mathrm{B}}\{u\} \equiv u(x,0) = k(x) \quad \text{or} \quad \mathcal{B}_{\mathrm{B}} \equiv \frac{\partial u}{\partial y}(x,0) = k(x)\,. \tag{4.4.54}$$

The strategy to solve the problem Eq. (4.3.44) is straightforward: determine a particular solution to take care of forcing terms in the boundary conditions Eq. (4.3.42) and Eq. (4.3.43), use separation of variables to solve the partial differential equation with homogeneous boundary conditions, add the two contributions to solve the initial conditions. The spirit of this strategy applies to the solution of the problem Eq. (4.4.52), Eq. (4.4.53) and Eq. (4.4.54).

The work horse in solving the problems is separation of variables. But separation of variables requires homogeneous boundary conditions. Not surprising then, we turn our attention to the nature of the boundary conditions first, and the best way to assess the situation is by drawing a mathematical diagram. If there is a pair of homogeneous boundary conditions, either Eq. (4.3.42), Eq. (4.3.43) or Eq. (4.4.53), Eq. (4.4.54), then separation of variables should lead to the construction of some type of Fourier series that can then be used to satisfy the remaining pair of boundary conditions.

---

[11]Obviously, the independent variable $t$ is replaced by $y$ in the arguments.

If there is no pair of homogeneous boundary conditions, then a particular solution is needed to treat the inhomogeneous boundary conditions with the intent to create homogeneous boundary conditions for the method of separation of variables. Check first whether there is a pair of boundary conditions that contain only polynomials, exponentials and/or trigonometric functions. If so, try a guess with the appropriate form, remembering that any coefficients may be functions of the other variable in which direction the boundary conditions are being applied. For example, if the unknown function is $u(x, y)$ and the boundary condition states $u(x, 0) = x + \exp(x)$, then the guess would be $a(y) + b(y)\, x + c(y) \exp(x)$; the same functions of $x$, but coefficients that depend on $y$. The consequence will be that the boundary conditions will be converted to homogeneous ones, subsequently enabling separation of variables to work.

---

**Principle 26**

*If the partial differential equation is homogeneous, then examine the nature of the boundary conditions. If they are homogeneous, then use separation of variables. If the boundary conditions are inhomogeneous, but contain functions that allow a guess for a particular solution, then use the particular solution to lead to homogeneous boundary conditions and then apply the method of separation of variables. Use the general solution created by separation of variables to satisfy any remaining boundary conditions or initial conditions.*

---

A word of warning; not all problems can be solved by particular solutions combined with homogeneous solutions determined by separation of variables. Often, an engineer will modify the problem so that this strategy will work, for example, by representing forcing effects at a boundary by an approximate function that allows a particular solution to be guessed. Both factors come into play; the choice of model and the choice of strategy for the solution.

### 4.4.5 *Exercises*

**Explore physical properties of the solution Eq. (4.4.30).**

(1) Since the temperature field constructed in Sec. 4.4.2 is stationary (does not depend on time), the amount of energy per unit time entering into the device must match the amount leaving it in unit time. Can you verify this expectation from the constructed solution?

The profile in Fig. 4.22 and symmetry suggests the maximum temperature occurs in the middle of the inlet on the left boundary. Derive an expression for the maximum temperature and find its value for the choice of parameters used in Fig. 4.22.

**Answer:** The maximum temperature is $T_0 + 1.19\, qh/k$.

**Exercises with a variety of boundary conditions.**

(2) Use Principle 26 on the following problem:

$$T(x,0) = 0\,,$$
$$T(x,H) = f(x)\,,$$
$$T(0,y) = y/H\,,$$
$$T(L,y) = 0\,.$$

Try a guess of the form $T = A(x) + B(x)\,y$ and make sure $A(x)$ and $B(x)$ satisfy the partial differential equation and the side boundary conditions. Then use separation of variable to find a series solution that can satisfy the top and bottom conditions. Complete the details when

(a) $f(x) = 1 - x/L$.
(b) $f(x) = 1 - x^2/L^2$.

**Answer:** For $H = 2L$, the temperature in the middle of the rectangle is (a) 0.25, (b) 0.7611.

(3) This problem requires the construction of a solution for Laplace's equation in a rectangle with arbitrary conditions on the sides. So imagine the steady solution to the heat diffusion equation with the temperature at the top and bottom held fixed with zero temperature. The temperature at the sides are unspecified. In other words,

$$T(x,0) = 0\,,$$
$$T(x,H) = 0\,,$$
$$T(0,y) = T_{\mathrm{L}}(y)\,,$$
$$T(L,y) = T_{\mathrm{R}}(y)\,.$$

Complete the details when

(a) $T_{\mathrm{L}}(y) = y(y - H)/H^2$.
(b) $T_{\mathrm{R}}(y) = -y(y - H)/H^2$.

**Answer:** The temperature is 0 in the middle of the rectangle.

(4) The next problem requires the construction of the solution for Laplace's equation in a rectangle with arbitrary conditions at the top and bottom. This time the boundary conditions are

$$T(x,0) = T_{\mathrm{B}}(x)\,,$$
$$T(x,H) = T_{\mathrm{T}}(x)\,,$$
$$T(0,y) = 0\,,$$
$$T(L,y) = 0\,.$$

Complete the details when

$$T_{\rm B}(x) = \sin(\pi x/L)\,,$$

$$T_{\rm T}(x) = \frac{2}{L}\begin{cases} x & \text{for } 0 < x < L/2\,, \\ (L-x) & \text{for } L/2 < x < L\,. \end{cases}$$

**Answer**: The temperature in the middle of a rectangle with $H = L$ is 0.52.

(5) Show that the general solution to Laplace's equation in a rectangle with boundary conditions,

$$T(x,0) = T_{\rm B}(x)\,,$$
$$T(x,H) = T_{\rm T}(x)\,,$$
$$T(0,y) = T_{\rm L}(y)\,,$$
$$T(L,y) = T_{\rm R}(y)\,,$$

can be constructed by adding the solutions from the previous two exercises.

### 4.4.6  *Heat transport in two dimensions – continued*

We started this section by considering heat diffusion in a rectangular cross section of a metal beam. The partial differential equation Eq. (4.4.4) applies to the time-dependent evolution of the temperature $T(x,y,t)$ inside the rectangular region. The boundary conditions are given by Eq. (4.4.5), Eq. (4.4.6). Equation (4.4.7) and only the conditions Eq. (4.4.5) are homogeneous. So we sought a particular solution $u(x,y)$ that solves the partial differential equation – it reduces to Laplace's equation – and satisfies the boundary conditions, in particular, the inhomogeneous boundary conditions Eq. (4.4.6) and Eq. (4.4.7).

The next step is to construct the homogeneous solution $T_{\rm h}(x,y,t)$ that satisfies

$$\frac{\partial T_{\rm h}}{\partial t}(x,y,t) = \alpha\left(\frac{\partial^2 T_{\rm h}}{\partial x^2}(x,y,t) + \frac{\partial^2 T_{\rm h}}{\partial y^2}(x,y,t)\right) \qquad (4.4.55)$$

and the two pairs of homogeneous boundary conditions

$$\frac{\partial T}{\partial y}(x,0,t) = \frac{\partial T}{\partial y}(x,H,t) = 0\,, \qquad (4.4.56)$$

$$T_{\rm h}(0,y,t) = T_{\rm h}(L,y,t) = 0\,. \qquad (4.4.57)$$

Separate the variables $T_{\rm h}(x,y,t) = X(x)\,Y(y)\,G(t)$ and substitute into Eq. (4.4.55). The result can be written as

$$\frac{1}{\alpha G(t)}\frac{{\rm d}G}{{\rm d}t}(t) = \frac{1}{X(x)}\frac{{\rm d}^2 X}{{\rm d}x^2}(x) + \frac{1}{Y(y)}\frac{{\rm d}^2 Y}{{\rm d}y^2}(y)\,. \qquad (4.4.58)$$

On the right there is a function of $x$ and $y$, but on the left a function of $t$ only. It must be a constant $c_1$. So,

$$\frac{{\rm d}G}{{\rm d}t}(t) - \alpha\,c_1\,G(t) = 0\,. \qquad (4.4.59)$$

Now rewrite Eq. (4.4.58) as

$$\frac{1}{X(x)}\frac{d^2X}{dx^2}(x) = c_1 - \frac{1}{Y(y)}\frac{d^2Y}{dy^2}(y).$$

On the left there is a function of $x$ only and on the right a function of $y$ only. They must be the same constant $c_2$. Thus,

$$\frac{d^2X}{dx^2}(x) - c_2\,X(x) = 0,\qquad(4.4.60)$$

$$\frac{d^2Y}{dy^2}(y) + (c_2 - c_1)\,Y(y) = 0.\qquad(4.4.61)$$

Clearly, we have two separation constants $c_1$ and $c_2$ to determine.

The boundary conditions also separate;

$$X(0) = X(L) = 0,\qquad(4.4.62)$$

$$\frac{dY}{dy}(0) = \frac{dY}{dy}(H) = 0.\qquad(4.4.63)$$

While there are two separation constants, $c_1$ and $c_2$, to be determined, there are also two homogeneous boundary value problems to solve. One of them contains the differential equation Eq. (4.4.60) with boundary conditions Eq. (4.4.62); the only non-trivial solutions are

$$X(x) = a\sin\!\left(\frac{n\pi x}{L}\right)\quad\text{with } c_2 = -\frac{n^2\pi^2}{L^2},\ n \geq 1.\qquad(4.4.64)$$

The other contains the differential equation Eq. (4.4.61) with boundary conditions Eq. (4.4.63); the only non-trivial solutions are

$$Y(y) = b\quad\text{with } c_2 - c_1 = 0,\qquad(4.4.65)$$

$$Y(y) = b\cos\!\left(\frac{m\pi y}{H}\right)\quad\text{with } c_2 - c_1 = \frac{m^2\pi^2}{H^2},\ m \geq 0.\qquad(4.4.66)$$

The results give two sets of values for the separation constants and we must solve the remaining differential equation Eq. (4.4.59) for all possible choices:

(1) For $c_1 = c_2$, we use Eq. (4.4.64) and Eq. (4.4.65).

$$c_1 = -\frac{n^2\pi^2}{L^2},\quad X(x) = a\sin\!\left(\frac{n\pi x}{L}\right),\quad Y(y) = b,$$

$$G(t) = c\,e^{-\alpha n^2\pi^2 t/L^2}.\quad(4.4.67)$$

(2) Otherwise we use Eq. (4.4.64) and Eq. (4.4.66).

$$c_1 = -\left(\frac{n^2\pi^2}{L^2} + \frac{m^2\pi^2}{H^2}\right),\quad X(x) = a\sin\!\left(\frac{n\pi x}{L}\right),\quad Y(y) = b\cos\!\left(\frac{m\pi y}{H}\right),$$

$$G(t) = c\exp\!\left[-\alpha\left(\frac{n^2\pi^2}{L^2} + \frac{m^2\pi^2}{H^2}\right)t\right].\qquad(4.4.68)$$

For each choice $n$, $m$, form the solution $X(x)Y(y)G(t)$ and then add up all possible contributions listed in Eq. (4.4.67) and Eq. (4.4.68). The solution is therefore

$$T_h(x, y, t) = \sum_{n=1} A_{n,0} \exp\left(-\frac{\alpha n^2 \pi^2 t}{L^2}\right) \sin\left(\frac{n\pi x}{L}\right)$$

$$+ \sum_{n=1}^{\infty} \sum_{m=1}^{\infty} A_{n,m} \exp\left[-\alpha\left(\frac{n^2\pi^2}{L^2} + \frac{m^2\pi^2}{H^2}\right)t\right] \sin\left(\frac{n\pi x}{L}\right) \cos\left(\frac{m\pi y}{H}\right). \quad (4.4.69)$$

The coefficient $A_{n,m}$ now reflects that there are two integer counters. The result has the appearance of two Fourier series, one a sine series in $x$ and the other a cosine series in $y$, multiplied together.

Before we can apply the last requirement, the initial condition, we must compose the full solution,

$$T(x, y, t) = T_p(x, y, t) + T_h(x, y, t).$$

The particular solution is calculated in Sec. 4.4.2 and is given by $T_p(x, y, t) = u(x, y)$ in Eq. (4.4.30).

Suppose the initial condition is $T(x, y, 0) = f(x, y)$. Then,

$$T_h(x, y, 0) = \sum_{n=1}^{\infty} A_{n,0} \sin\left(\frac{n\pi x}{L}\right)$$

$$+ \sum_{n=1}^{\infty} \sum_{m+1}^{\infty} A_{n,m} \sin\left(\frac{n\pi x}{L}\right) \cos\left(\frac{m\pi y}{H}\right) = f(x, y) - T_p(x, y). \quad (4.4.70)$$

Following the idea of matching Fourier series, we must find the double Fourier series representation for $f(x, y) - T_p(x, y)$,

$$f(x, y) - T_p(x, y) = \sum_{n=1}^{\infty} B_{n,0} \sin\left(\frac{n\pi x}{L}\right)$$

$$+ \sum_{n=1}^{\infty} \sum_{m=1}^{\infty} B_{n,m} \sin\left(\frac{n\pi x}{L}\right) \cos\left(\frac{m\pi y}{H}\right) \quad (4.4.71)$$

where $B_{n,m}$ is determined by the natural extension of the ideas in Chapter 2. It does not matter which way we proceed but the idea is to freeze one of the independent variables and calculate the series for the other one. To illustrate, keep $y$ fixed and rewrite Eq. (4.4.71) as

$$f(x, y) - T_p(x, y) = \sum_{n=1}^{\infty} F_n(y) \sin\left(\frac{n\pi x}{L}\right).$$

The "coefficients" are determined by

$$F_n(y) = \frac{2}{L} \int_0^L [f(x, y) - T_p(x, y)] \sin\left(\frac{n\pi x}{L}\right) dx.$$

Comparing with the expression in Eq. (4.4.71), we conclude

$$F_n(y) = B_{n,0} + \sum_{n-1}^{\infty} B_{n,m} \cos\left(\frac{m\pi y}{H}\right).$$

The coefficients in the cosine series can be determined by

$$B_{n,0} = \frac{2}{H} \int_0^H F_n(y)\,dy,$$

$$B_{n,m} = \frac{2}{H} \int_0^H F_n(y) \cos\left(\frac{m\pi y}{H}\right) dy.$$

Finally, by matching Eq. (4.4.70) and Eq. (4.4.71), $A_{n,m} = B_{n,m}$ and when we substitute the result into Eq. (4.4.69) we have calculated the specific solution. Unfortunately, the whole process is very labor intensive. Fortunately, these days we make the computer do the work!

While the details are missing, if you have followed the steps conceptually, you would understand how separation of variables can be applied in higher dimensions.

# Systems of Differential Equations

# 5

## 5.1 First-order equations

Systems of differential equations are common in science and engineering because they provide mathematical models for the motion of particles, bodies and structures through dynamics, for chemical reactions of different species, for electric circuits, and for population dynamics, to mention just a few of the important applications. They can be either linear or nonlinear equations. Systems of linear differential equations with constant coefficients are generally easy to solve because the homogeneous equation allows exponential solutions. That is not the case with nonlinear equations. There are certain tricks we can try but we have no guarantees with nonlinear equations, and they are usually solved by computer simulations based on numerical procedures.

Linear and nonlinear equations are often taught separately, because they are approached differently. But there is a view that provides a global perspective on the behavior of the solutions: we can sketch the solutions without knowing their detailed behavior and even when we can determine the details as in the case of linear equations with constant coefficients, it is often more useful to capture the important behavior through the global perspective.

We start by illustrating the global perspective on an important example from population dynamics. Even though this example contains only a single differential equation and really does not constitute a system, it is still possible to develop a global perspective that can be extended to systems.

### 5.1.1 *Population dynamics*

The simplest differential equation is

$$\frac{\mathrm{d}N}{\mathrm{d}t}(t) = \sigma N(t) \tag{5.1.1}$$

which arises, for example, when we consider the instantaneous percentage change in the growth of bacteria, an example we consider in the very first section of this book. The quantity $N(t)$ measures the population size, for example, the number of bacteria counted in a small container, or the number of people in the world, etc. The parameter $\sigma$ per unit time is the growth rate. The solution

$$N(t) = N_0 \, e^{\sigma t} \,,$$

to Eq. (5.1.1) captures very well the early growth of the bacteria population. The growth is exponential which means on a time scale of $1/\sigma$ the population will increase relatively by a factor $e \approx 2.72$. It is obvious, though, that growth cannot continue indefinitely. Sooner or later the food runs out.

More generally, we should regard the parameter $\sigma$ as the result of the difference between the percentage increase from births and the percentage loss from deaths. As the population rises and food becomes sparse, competition – and disease and war and so on – causes an increase in the death rate and a drop in the overall growth rate. A simple modification to the mathematical model that reflects this behavior is to assume that $\sigma$ should be replaced by $\sigma(1 - N/K)$, a linear decrease as the population increases. The parameter $K$ has units of the population size. As a consequence, Eq. (5.1.1) is replaced by

$$\frac{dN}{dt}(t) = \sigma N(t) \left( 1 - \frac{N(t)}{K} \right). \tag{5.1.2}$$

This equation is a phenomenological equation for resource-limited growth and is called the *logistic equation*. The equation is nonlinear since $N^2$ appears in the right hand side. An attempt to find an exponential solution will fail.

Fortunately, there is a clever way to solve this equation. It is based on the observation that the right hand side is a function of $N$ only; no explicit dependency on $t$. What springs to mind is the idea in separation of variables for partial differential equations. We should gather all the information about $N$ on one side of the equation. Specifically, write Eq. (5.1.2) as

$$\frac{K}{N(K-N)} \frac{dN}{dt} = \sigma \,. \tag{5.1.3}$$

Now the left hand side looks like the consequence of the chain rule for differentiation applied to a function $F(N)$,

$$\frac{d}{dt}F(N) = \frac{dF}{dN}(N) \frac{dN}{dt} \,. \tag{5.1.4}$$

By comparing with Eq. (5.1.3), we set

$$\frac{dF}{dN}(n) = \frac{K}{N(K-N)} \tag{5.1.5}$$

with the consequence that

$$F(N) = \int \frac{K}{N(K-N)} \, dN$$

$$= \int \left[ \frac{1}{N} + \frac{1}{K-N} \right] dN$$

$$= \ln \left[ \frac{N}{K-N} \right].$$

By gathering the information in Eq. (5.1.3), Eq. (5.1.4) and Eq. (5.1.5), we obtain

$$\frac{d}{dt} F(N) = \sigma$$

which can be integrated in time to give

$$F(N) = \ln \left[ \frac{N}{K-N} \right] = \sigma t + A, \qquad (5.1.6)$$

where $A$ is an integration constant. How is this constant determined? By some initial condition! Suppose $N(0) = N_0$, the starting population size. Then (5.1.6) evaluated at $t = 0$ becomes

$$\ln \left[ \frac{N_0}{K - N_0} \right] = A,$$

and $A$ is determined.

After rearranging (5.1.6), the solution is

$$N(t) = \frac{K N_0}{N_0 - (N_0 - K) \exp(-\sigma t)}. \qquad (5.1.7)$$

---

**Reflection**

If a nonlinear differential equation has a form that allows a separation of variables as in Eq. (5.1.4), then we may consider the derivative as the result of the application of the chain rule. Subsequently, the differential equation takes the form where direct integration is possible as in Eq. (5.1.6).

---

Let us look at a few specific examples of the solution to help understand the general behavior. The solutions are shown in Fig. 5.1 as a function of time for the choice $K = \sigma = 1$ and for various initial values. Quite clearly, the population size approaches a constant value $N(t) \to 1$ as time proceeds no matter what the initial value is. Now that we seen several examples of the behavior of the solution graphically, let us seek how that behavior is reflected in Eq. (5.1.7).

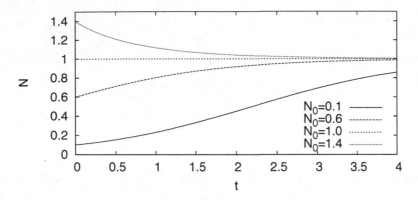

Fig. 5.1    Population size as a function of time for several different initial conditions.

By considering large times, the exponential in the denominator of Eq. (5.1.7) becomes extremely small and the solution approaches $N(t) \to K$; this result agrees with the solutions displayed in Fig. 5.1 since $K = 1$ there. Since

$$\frac{1}{1-x} \approx 1 + x$$

when $x$ is small, we can find the approximate behavior for the solution for large times as

$$N(t) \approx K \left[ 1 + \frac{N_0 - K}{N_0} e^{-\sigma t} \right]. \tag{5.1.8}$$

For either $N_0 - K$ positive or negative, the population size approaches $K$ from above or below respectively.

The only solution in Fig. 5.1 that does not appear to be initially approaching $N = K$ exponentially is the one with the initial condition $N_0 = 0.1$. Instead, the solution starts to increase rapidly before switching to an exponential approach to $K$. When $N_0$ is small, the denominator is approximately $K \exp(-\sigma t)$ and the solution Eq. (5.1.7) behaves approximately as

$$N(t) \approx N_0 \, e^{\sigma t}. \tag{5.1.9}$$

So the initial growth is exponential. This scenario is of particular importance because it describes how a population, small in size initially, grows until nonlinear effects restrict the growth rate and the population stalls at size $K$.

What is amazing is that all the crucial aspects of the behavior of the solution as revealed graphically in Fig. 5.1 and by the approximations Eq. (5.1.8) and Eq. (5.1.9) can be obtained without knowing the precise solution Eq. (5.1.7)! The device that provides this information is called a *phase diagram*; the phase diagram displays the rate of change of the solution as a function of the solution. For example, Eq. (5.1.3), specifies that the rate of change of the population size is a quadratic function of the population size; Fig. 5.2 provides the graph when $K = \sigma = 1$. From the information revealed by the phase diagram, we can construct a very good picture of how the solution must behave, and now let us see how that is done.

First we notice that the rate of change of the population size is zero for two values of $N$; $N = 0$ and $N = K$. One of the solutions shown in Fig. 5.1 has $N_0 = K = 1$, and the solution is clearly a constant in time with a zero rate of change. The other case, $N_0 = 0$ will also remain $N(t) = 0$ for all time. These two points $N = 0, K$ correspond to the special solutions in which there is no change in time. Accordingly, they are called *stationary points*.

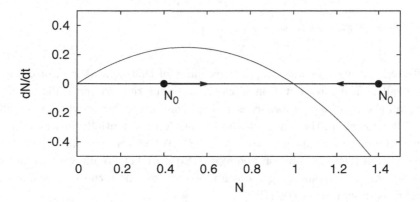

Fig. 5.2   The rate of change of the population as a function of the population.

So what will happen if we pick $N_0$ somewhere in between 0 and $K$? At that value, for example $N_0 = 0.4$, the graph in Fig. 5.2 indicates that the rate of change is positive and we draw an arrow on the axis pointing to the right to indicate that the population size will increase. Subsequently, the solution increases and moves to the right on the axis. The sign of the rate of change remains positive, so $N(t)$ will continue to increase, until it becomes very close to $N = K = 1$. The magnitude of the rate of change becomes small near $N = K = 1$, so $N(t)$ increases ever more slowly, eventually stalling when it reaches $K$, although it takes an infinite amount of time to reach it. On the other hand, if $N_0$ is to the right of $N = K = 1$, shown in Fig. 5.2 with $N_0 = 1.4$, then the graph indicates the rate of change is negative and we draw an arrow pointing to the left. Subsequently, the solution decreases and continues to do so as it approaches $N = K = 1$.

So far we have assessed the general behavior of the solution from the graphical information in the phase diagram, but we can be more specific about the behavior of the solution as it approaches the stationary point $N = K$ by a process called *linearization*. We pick an initial condition $N_0 = K + \varepsilon$, where $\varepsilon$ is considered small – it will be negative if we wish the initial point to be left of $N = K$. In other words, we start near the stationary point. Further, we assume that $N(t)$ remains near $N = K$ for at least some period of time. So we set $N(t) = K + \tilde{N}(t)$ and substitute this change of variable into Eq. (5.1.3). The result is

$$\frac{\mathrm{d}\tilde{N}}{\mathrm{d}t}(t) = \sigma\Big(K + \tilde{N}(t)\Big)\left[1 - \frac{1}{K}\Big(K + \tilde{N}(t)\Big)\right]$$

$$= -\sigma\tilde{N}(t) - \frac{\sigma}{K}\tilde{N}^2(t).$$

Now what we do is drop the $\tilde{N}^2(t)$ term on the grounds that it is too small to have any influence. Stated differently, we choose $\tilde{N}(t)$ to be small enough so that the $\tilde{N}^2$ term can be neglected. The result is a linear ordinary differential equation,

$$\frac{\mathrm{d}\tilde{N}}{\mathrm{d}t} = -\sigma\tilde{N}, \qquad\qquad (5.1.10)$$

which has the general solution,

$$\tilde{N} = \varepsilon e^{-\sigma t}. \qquad\qquad (5.1.11)$$

Whether $\varepsilon$ is negative or positive, $\tilde{N}(t)$ remains of that sign as it approaches zero exponentially. The interpretation is clear: a population to the right of $N = K$ remains on the right as it approaches $K$ exponentially in time, while a population on the left remains on the left as it approaches $K$ exponentially in time. We have recovered the behavior expressed in Eq. (5.1.8). Whenever the solution approaches $K$ close enough then the solution behaves as Eq. (5.1.11). While we may not know $\varepsilon$ precisely, we know something more important; we know the time scale $1/\sigma$ by which the solution approaches $K$.

There is a delightful geometric interpretation of linearization. Since $\tilde{N}(t) = N(t) - K$, we may rewrite the linear equation Eq. (5.1.10) as

$$\frac{\mathrm{d}N}{\mathrm{d}t}(t) = -\sigma\big(N(t) - K\big), \qquad\qquad (5.1.12)$$

and superimpose this approximate equation on the graph in Fig. 5.2 to produce Fig. 5.3. Now it is clear that the linear approximation Eq. (5.1.12) corresponds to the tangent line to the curve at $K$.

There is another stationary point $N = 0$ where the rate of change is zero. Linearization proceeds similarly. Set $N(t) = \tilde{N}(t)$ and assume it is small. Then Eq. (5.1.3) is approximately

$$\frac{\mathrm{d}\tilde{N}}{\mathrm{d}t}(t) = \sigma\tilde{N}(t),$$

and the solution is

$$\tilde{N}(t) = \varepsilon e^{\sigma t},$$

where we take the initial condition to be $\tilde{N}(0) = \varepsilon$. This time the behavior is different; the population grows exponentially in agreement with Eq. (5.1.9). Of course, as the population grows, the assumption that $\tilde{N}$ is small soon fails, but at least the behavior is clear; the solution grows and moves away from the point $N = 0$ as the arrows in Fig. 5.2 indicate. The situation is opposite from that at $N = K$ where the arrows point towards $N = K$.

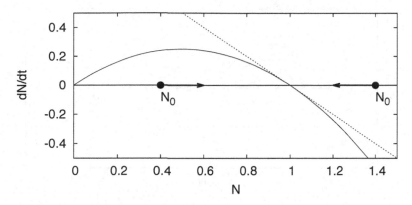

Fig. 5.3  The rate of change of the population size as a function of the population including the linear approximation.

In summary, if the initial population size is just to the right of $N = 0$, it will grow exponentially with the growth rate $\sigma$ as revealed by the linear approximation. The rate of change of the population size increases linearly with $N(t)$ until nonlinear effects cause the rate of change to slow down: this effect is evident as a decrease in the slope of the curve in Fig. 5.2. The behavior becomes linear again near $K$ and the population size approaches $K$ exponentially.

The example illustrates several important features of nonlinear differential equations. By graphing the rate of change of $N$ with $N$ we notice two points where the rate of change is zero; these points are called stationary or equilibrium points – called critical points by mathematicians – because the solution remains constant in time for these values. Next we notice that where the rate of change is positive, the solution increases – moves to the right. Where the rate of change is negative, the solution decreases – moves to the left. As a result of these observations, we can easily sketch the behavior of the solution. While the details would be inaccurate, the global behavior would be very good. The bottom line: we can safely predict that the population size will tend to $K$ for any initial condition.

## Reflection

Because the right hand side of Eq. (5.1.3) does not contain any explicit dependency on $t$, we can plot the rate of change of the population size as a function of the population size, valid for all time. The graph reveals the presence of two stationary (equilibrium) points, and by drawing arrows on the axis of the population size we can assess what the solution must do.

Another example illustrates how quickly we can assess the global nature of solutions without any analytic form for the solution. Certain species of animals or fish exhibit a survival pattern that demands a minimum level of population for survival.

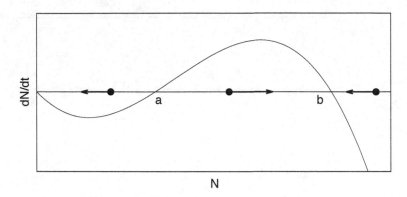

Fig. 5.4  The phase diagram showing the rate of change of the population size as a function of the population.

A simple mathematical equation that captures this scenario is

$$\frac{\mathrm{d}N}{\mathrm{d}t}(t) = f(N) \equiv \sigma N(t)\big(N(t) - a\big)\big(b - N(t)\big). \qquad (5.1.13)$$

The way the right hand side is written, it is clear that $N = 0, a, b$ are points where the rate of change is zero; they are stationary or equilibrium points, as displayed in Fig. 5.4. They play a crucial role because they separate regions of different behavior. Specifically, for $0 < N < a$, the rate of change is negative and the solution decreases until the population disappears. This is a good example of how some species can simply disappear if the population levels get too low. For $a < N < b$, the population increases until it reaches a stable equilibrium $N = b$. For $N > b$, the population decreases until it reaches the stable equilibrium $N = b$. Thus, $N = 0, b$ are stable equilibria and $N = a$ is an unstable equilibrium point.

Linearization round each of the equilibrium points will provide additional information about the time scales of instability or stability. Let $\overline{N} = 0, a, b$ be one of the equilibrium points, places where $f(\overline{N}) = 0$, and assume $N(t) = \overline{N} + \tilde{N}(t)$ with $\tilde{N}(t)$ a small perturbation away from the equilibrium point. We seek a linear approximation for $f(N)$ near the equilibrium points $\overline{N}$. So the assumption $N = \overline{N} + \tilde{N}(t)$ with $\tilde{N}(t)$ small must be substituted into (5.1.13).

$$\begin{aligned}
\frac{\mathrm{d}\tilde{N}}{\mathrm{d}t}(t) &= f\big(\overline{N} + \tilde{N}(t)\big) \\
&\approx f(\overline{N}) + \frac{\mathrm{d}f}{\mathrm{d}N}(\overline{N})\tilde{N}(t) + \cdots \\
&= \frac{\mathrm{d}f}{\mathrm{d}N}(\overline{N})\tilde{N}(t). \qquad (5.1.14)
\end{aligned}$$

In this approximation, the first term of the Taylor series expansion for $f(\overline{N} + \tilde{N})$ is retained but $\tilde{N}^2$ and all higher-order terms are dropped.[1] Despite appearances this is a linear differential equation with constant coefficients. In this case the derivative

---

[1] This explains the term linearization; only the linear term is retained.

of $N$ evaluated at an equilibrium point is just a parameter. Let us represent this parameter by

$$\lambda = \frac{\mathrm{d}f}{\mathrm{d}N}(\overline{N}).$$

Then the perturbation equation Eq. (5.1.14) becomes

$$\frac{\mathrm{d}\tilde{N}}{\mathrm{d}t}(t) = \lambda\tilde{N}(t),$$

which has the general solution

$$\tilde{N}(t) = \varepsilon e^{\lambda t}.$$

If $\lambda < 0$ then the perturbations die away and the solution approaches the equilibrium point exponentially; it is called stable because any perturbation, no matter how it is produced, will simply disappear in time. The time scale of the decay is $1/\lambda$. Alternatively, if $\lambda > 0$, the solution moves away from the equilibrium point no matter how small the perturbation is initially, and the equilibrium point is unstable. The time scale of the instability is $1/\lambda$.

For $f(N)$ given in (5.1.13),

$$\lambda = \frac{\mathrm{d}f}{\mathrm{d}N}(\overline{N}) = -\sigma ab + 2\sigma(a+b)\overline{N} - 3\sigma\overline{N}^2,$$

and at $\overline{N} = 0, a, b$, $\lambda = -\sigma ab, \sigma a(b-a), -\sigma b(b-a)$, respectively. The pattern is stable, unstable, stable respectively as we may have guessed from the graph in Fig. 5.4. By knowing the equilibrium points and their stability and the phase diagram Fig. 5.4, we can predict with certainty how the solution behaves. For example, if the initial population size is somewhere $a$ and $b$, then the solution increases in size until it exponentially approaches $b$ on a time scale that is the reciprocal of $\sigma b(b-a)$. If it starts near $a$, then we know the population size grows exponentially at first on a time scale that is the reciprocal of $\sigma a(b-a)$ before adjusting to the exponential approach to the equilibrium point at $b$.

### 5.1.2 *Abstract view*

A general form for a single, first-order differential equation is

$$\frac{\mathrm{d}y}{\mathrm{d}t}(t) = f(y,t). \tag{5.1.15}$$

Of course the nature of the solution depends entirely on the nature of $f(y,t)$. If the equation is linear, that is $f(y,t) = a(t) + b(t)\,y(t)$, then we should use the techniques in Chapter 1.

When $f(y,t)$ is nonlinear, there is no general formula for the solution to this differential equation. Only when $f(y,t)$ satisfies certain requirements can we make some progress. There are two important cases.

### 5.1.2.1   Separable equations

If $f(y,t)$ can be written as separated functions,

$$f(y,t) = F(y)\,G(t)\,, \tag{5.1.16}$$

then we can write Eq. (5.1.15) as

$$\frac{1}{F(y)}\frac{dy}{dt}(t) = G(t)\,, \tag{5.1.17}$$

and the equation has been separated. The next step is to identify the left hand side as the consequence of the use of the chain rule for differentiation. In order words, we seek a function $H(y)$ such that

$$\frac{dH}{dt} = \frac{dH}{dy}\frac{dy}{dt} = G(t)\,. \tag{5.1.18}$$

By setting

$$\frac{dH}{dy}(y) = \frac{1}{F(y)}\,,$$

we recover Eq. (5.1.17). Thus,

$$H(y) = \int \frac{1}{F(y)}\,dy\,. \tag{5.1.19}$$

Now we simply integrate Eq. (5.1.18), leading to

$$H(y) = \int G(t)\,dt\,. \tag{5.1.20}$$

The final step is to equate Eq. (5.1.19) and Eq. (5.1.20).

A different approach leading to the same conclusion is to integrate Eq. (5.1.17) and apply a change of integration variable.[2]

$$\int \frac{1}{F(y)}\frac{dy}{dt}\,dt = \int \frac{1}{F(y)}\,dy = H(y) = \int G(t)\,dt\,.$$

This looks wonderful until we realize that unless $F(y)$ and $G(t)$ are functions that can be integrated analytically, we have not gained much. Even if they are, all that results is an implicit equation for $y(t)$, and there is no guarantee we can write the answer in an explicit form where $y(t)$ equals some function of $t$.

A simple example drives home the point. Suppose

$$\frac{dN}{dt}(t) = \sigma\,N(t)\frac{\exp(-N(t) - \exp(-K)}{1 - \exp(-K)}\,. \tag{5.1.21}$$

The right hand side is chosen to contain the following properties: There are equilibrium points at $0$, $K$; the initial growth rate is $\sigma$; exponential decay replaces the

---

[2]One may view a change of integration variable as the reverse process of using the chain rule of differentiation.

linear decrease in the growth rate. To solve by separation of variables we must integrate

$$H(N) = \int \frac{1 - \exp(-K)}{N\left[\exp(-N) - \exp(-K)\right]} \, dN \, .$$

There is no known analytical expression for this integral.

### 5.1.2.2   *Autonomous equations*

When we design mathematical models for quantities changing in time, we usually expect that they do not depend on when we start the clock. In other words, $f(y,t)$ should not depend on time. Then Eq. (5.1.15) is of the form,

$$\frac{dy}{dt}(t) = f(y) \, . \tag{5.1.22}$$

The immediate consequence is that the equation is separable with $G(t) = 1$ – see Eq. (5.1.16). Since the equation is separable, a solution may be found by using Eq. (5.1.19) and Eq. (5.1.20), and if the integration can be performed we will obtain an implicit equation for $y(t)$; it has the form $H(y) = t$ – see Eq. (5.1.6) as an example. Most likely it will not be possible to write the solution $y(t)$ in an explicit form, but it may be possible to sketch the function $t = H(y)$ and then to transpose the axes. Alternatively, to use a graphics software package, consider the solution as given in parametric form, $t(p) = H(p), y(p) = p$.

There is no guarantee that the integration Eq. (5.1.19) can be performed to produce an analytic expression. Instead, we have at our disposal the power of a geometric interpretation[3] of the solution by graphing $f(y)$ with $y$ to produce a phase diagram, and the phase diagram will lead us to an understanding of the overall features of the solution.

Time to introduce some terminology that will prove useful when we consider systems of several nonlinear equations. When we consider the phase diagram, the graph of $f(y)$ versus $y$, the solution $y(t)$ appears as a moving point on the $y$-axis. We refer to this point as a *phase point* and the $y$-axis is called the *phase space*. In Fig. 5.2, the $N$-axis is the phase space and $N_0$ is the initial location of the phase point. The rate of change of the solution $dy/dt$ is interpreted as a "velocity" and is drawn as an arrow on the phase space. It has direction and its magnitude represents the speed; it is a vector! Some of these vectors are shown in Figs. 5.2 and 5.4, but we can easily imagine adding more vectors to the phase space as needed. The motion of the solution as a phase point in the phase space – the $y$-axis – is typically referred to as a *trajectory*. By now it should be obvious that much of this terminology reflects its origins, the description of the motion of particles governed by Newton's laws of motion, but it is freely used for any system of nonlinear differential equations.

The zeros of $f(y)$ give us equilibrium points, values of $y$ for which the solution is stationary – does not change in time; the velocity is zero. For example, the

---

[3]The adjective geometric arises because we draw arrows on the $y$-axis to indicate the direction of change in the solution.

equilibrium points for Eq. (5.1.21) occur at $N = 0, K$. Denote the location of an equilibrium point by $\overline{y}$, then the behavior of the solution near $\overline{y}$ can be assessed by linearization of $f(y)$; set $y = \overline{y} + \tilde{y}$ and approximate

$$f(\overline{y} + \tilde{y}) \approx f(\overline{y}) + \frac{\mathrm{d}f}{\mathrm{d}y}(\overline{y})\,\tilde{y}$$

$$\approx \frac{\mathrm{d}f}{\mathrm{d}y}(\overline{y})\,\tilde{y}, \tag{5.1.23}$$

where we use the fact that $f(\overline{y}) = 0$ because $\overline{y}$ is an equilibrium point. For convenience, set

$$\frac{\mathrm{d}f}{\mathrm{d}y}(\overline{y}) = \lambda. \tag{5.1.24}$$

Then approximate Eq. (5.1.22) by

$$\frac{\mathrm{d}\tilde{y}}{\mathrm{d}t} = \lambda\,\tilde{y} \tag{5.1.25}$$

which has the general solution,

$$\tilde{y}(t) = \varepsilon\,\mathrm{e}^{\lambda t}, \quad \text{or} \quad y(t) = \overline{y} + \varepsilon\,\mathrm{e}^{\lambda t}. \tag{5.1.26}$$

Of course, $|\varepsilon| \ll 1$ in order for Eq. (5.1.23) to be valid.

As an example, apply these ideas to Eq. (5.1.21). The equilibrium points are $\overline{N} = 0, K$ and

$$\lambda = \frac{\mathrm{d}f}{\mathrm{d}y}(\overline{N}) = \sigma\,\frac{(1 - \overline{N})\,\exp(-\overline{N}) - \exp(-K)}{1 - \exp(-K)}.$$

For $\overline{N} = 0, K$, $\lambda = \sigma$, $-K\exp(-K)/(1 - \exp(-K))$, respectively.

The result Eq. (5.1.26) provides an opportunity to introduce the mathematical concept of stability and instability. If $\lambda < 0$, then the perturbation $\tilde{y}(t)$ to the equilibrium solution $\overline{y}$ decays exponentially in time, and the solution eventually returns to the equilibrium point; we call this equilibrium point stable. For Eq. (5.1.21), $\overline{N} = K$ is a stable equilibrium point. On the other hand, if $\lambda > 0$, then the perturbation grows exponentially in magnitude and the solution moves away from the equilibrium point. This behavior is called instability in that even the smallest perturbation will cause the solution to change and leave the equilibrium point $\overline{y}$ behind. Of course, the solution is not valid as soon as the perturbation grows to be large enough in magnitude. Then the subsequent behavior of the solution must be assessed from the phase diagram. For Eq. (5.1.21), $\overline{N} = 0$ is an unstable equilibrium point.

Beware of the interpretation of mathematical statements such as $\lambda > 0$ means instability. Suppose $\lambda = 10^{-10}$ per year. By the mathematical definition, the solution is unstable, but we would have to wait $10^{10}$ years to notice any appreciable change from $\overline{y}$. Who cares! The point here is that $1/\lambda$ is the time scale of either the approach to $\overline{y}$ or the growth of instability and that is important information.

**Principle 27**

*The qualitative features of a solution to an autonomous nonlinear differential equation can be revealed by: locating all stationary (equilibrium) points; determining their stability; and by using the phase diagram to determine how the solution changes between stationary points.*

There are other possible features of the phase diagram, the graph of $f(y)$ with $y$, that affect the behavior of the solution. For example, what happens if $\lambda = 0$? How does $f(y)$ behave for $y \to \infty$ or $y \to -\infty$ and how does that affect the solution? What happens if there are vertical asymptotes in $f(y)$ at some finite value of $y$? What happens if $f(y)$ has singularities? Some of these questions are explored in the exercises.

### 5.1.3 *Exercises*

**Some exercises exploring separable equations.**

(1) Use separation to construct solutions to the following differential equations with the given initial conditions.

(a)

$$\frac{dy}{dt}(t) = 3\,y^2(t)\,t^2\,.$$

    i. $y(0) = 1$.
        **Answer:** The solution blows up at $t = 1$.
    ii. $y(0) = -1$.
        **Answer:** The solution is continuous for $t \geq 0$: $y(1) = -0.5$.

(b)

$$\frac{dx}{dt}(t) = e^{-x(t)}\,,$$

with $x(0) = 0$.
**Answer:** Even though the rate of change decreases exponentially in $x$, it decreases only as $1/(1+t)$ in $t$; the solution $\ln(1+t)$ increases indefinitely.

(c)

$$\frac{dy}{dx}(x) = y^2(x) + x^2$$

with $y(0) = 0$.
**Answer:** Hmmm.

(d)

$$\frac{dy}{dx}(x) = \cos(x)\, y(x),$$

with $y(0) = 1$.

Compare the answer with the answer that would be obtained if the procedure in Sec. 1.3 is followed.

(e) Sometimes a change of variable transforms a differential equation into separable form. For example, consider

$$\frac{dy}{dx}(x) = \big(y(x) - 4x\big)^2$$

with $y(0) = 2$. Introduce the new variable $v(x) = y(x) - 4x$. The new differential equation in $v(x)$ is separable. Complete the details to construct the solution for $y(x)$.

**Answer:**

$$y(x) = 4x + 2.$$

**Some applications.**

(2) For some species, the percentage growth rate is a maximum at an intermediate value of the population size $N$. It reflects the difficulty in finding mates when $N$ is small. A simple phenomenological model is given by the equation,

$$\frac{dN}{dt}(t) = \Big[r - a\,(N(t) - b)^2\Big]\,N(t).$$

Here $r$, $a$ and $b$ are positive parameters with $r > a\,b^2$.

(a) Draw the phase diagram.
(b) Locate all the equilibrium points.
(c) Determine the stability and the time scales at the equilibrium points.
(d) Graph the solution for a few different initial conditions.
(e) Compare the solutions to those found for the logistic equation. Are there any qualitative differences?

**Answer:** For $r = 9$, $a = 1$, $b = 2$, the growth rate at $N = 0$ is $1/\lambda = 1/5$ and at the stable equilibrium point it is $1/\lambda = -1/30$.

(3) The UN released a report giving estimates for the world population and its annual percentage increase; the data is recorded in Table 5.1. Assume that the percentage increase has the continuous limit,

$$\sigma(N) = \frac{1}{N}\frac{dN}{dt}.$$

Find a straight line approximation to $\sigma(N)$ from the data in the table, and use it to estimate the maximum world population. How close will the world population be to its maximum in 2040?

Table 5.1  World population.

| Year | Population (billions) | Percentage increase |
|------|----------------------|---------------------|
| 1970 | 3.74 | 2.08 |
| 1980 | 4.44 | 1.72 |
| 1990 | 5.30 | 1.63 |
| 2000 | 6.07 | 1.28 |
| 2010 | 6.85 | 1.14 |

According to the Food and Agriculture Organization, there is about $4.14 \times 10^3 \text{ km}^2$ available for food production. How big a plot of land does each person have available to live and grow their food when the world has reached its estimated maximum population?

**Answer**: Sobering! The maximum population is 10.8 billion, and in 2040, the population is about 8.8 billion. At maximum population, each person would have on average a square plot of land of size 36 m².

(4) Surely technology will slow the grow rate more gradually than the logistic model assumes. Let us consider the following model,

$$\frac{dN}{dt}(t) = \frac{aN(t)}{1 + bN^2(t)}.$$

(a) Draw the phase diagram and note that there are no stable equilibrium. The population will increase indefinitely.

(b) Plot $a/(1 + bN^2)$ for the choice $a = 0.033$ and $b = 0.04$ and compare it to the data in the table of Exercise 3.

(c) The equation is separable, so find an implicit representation of the solution.

(d) Plot the solution and compare it to a plot of the solution in Exercise 3. What conclusions can you draw from this comparison?

**Answer**: In 2040, there would be 9.1 billion people. In the near term, the prediction on overcrowding does not depend too sensitively on the mathematical model.

(5) Assume the deer population in some region follows the logistic equation, except that hunting is allowed. Assume a fixed annual rate $s$ of deer are killed; the rate is fixed due to government regulation. The population $N$ satisfies

$$\frac{dN}{dt}(t) = \sigma \left(1 - \frac{N(t)}{K}\right) N(t) - s.$$

For simplicity set $\sigma = 1$, $K = 1$. Draw the phase diagram for several choice of $s$: $s = 0.1, 0.2, 0.3$. What can you conclude about the possible impact on the deer population for different government regulations on hunting, that is, the different choices for $s$?

**Answer**: There are two equilibrium points, an unstable one at $\overline{N} = 0.5 \left[1 - \sqrt{1 - 4s}\right]$ and a stable one at $\overline{N} = 0.5 \left[1 + \sqrt{1 - 4s}\right]$.

(6) The model in the previous exercise suffers from a defect; when the deer population is low, the hunters still reduce the population at a rate higher than deer
present. The model can be adjusted to account for the success of the hunters
by reducing the amount killed as the population drops. Assume then that a
more appropriate model is given by the differential equation,

$$\frac{\mathrm{d}N}{\mathrm{d}t}(t) = \sigma\left(1 - \frac{N(t)}{K}\right)N(t) - \frac{sN(t)}{a + N(t)}.$$

To be specific, take $\sigma = 1$, $K = 1$ and $a = 0.04$ and draw the phase diagram
for the same values $s = 0.1, 0.2, 0.3$ as in Exercise 5. Calculate the equilibrium
points and determine their stability.

**Answer:** The equilibrium points and their stability are: $s = 0.1$, $\overline{N} = 0$ –
stable, $0.06$ – unstable, $0.89$ – stable; $s = 0.2$, $\overline{N} = 0$ – stable, $0.18$ – unstable,
$0.75$ – stable; $s = 0.3$, $\overline{N} = 0$ – stable. In the last case, the population will
disappear in time. This is exactly what happened in the overfishing of sardines
in San Francisco Bay.

(7) In Exercise 7 of Sec. 1.2.5, the motion of a falling body is described when the
air resistance is assumed proportional to the velocity. This model is unrealistic;
the air resistance is proportional to the square of the velocity.[4] The equation
of motion now takes the form,

$$m\frac{\mathrm{d}v}{\mathrm{d}t}(t) = \begin{cases} mg - kv^2(t) & \text{for } v > 0, \\ mg + kv^2(t) & \text{for } v < 0, \end{cases}$$

where $m$ is the mass of the body and $g$ is the gravitational constant. The
different cases arise because the resistance always acts in the direction opposite
to the motion – when $v$ is positive, the resistance is upwards, and when $v$ is
negative, the resistance is downwards. The parameter $k$ depends on the density
of the air and on an effective cross-sectional area that the body presents as it
falls. For a sky-diver of mass 200 lb, $k \approx 0.3$ lb/ft.

(1) Draw the phase diagram.
(2) Locate the equilibrium point and show that it is stable.
(3) What is the terminal velocity of the sky-diver?

(8) The growth of cancerous tumors can be described by

$$\frac{\mathrm{d}N}{\mathrm{d}t}(t) = -aN(t)\ln\big(bN(t)\big),$$

where $N(t)$ is a measure of the number of cells and $a, b > 0$ are two positive
parameters.

(a) Draw the phase diagram.
(b) Locate the equilibrium points and state their stability.

---

[4]When a body moves through a viscous velocity such as oil, its speed tends to be low and the
resistance is proportional to the velocity. When the motion of the body is faster then turbulence
sets in and the resistance is proportional to the square of the velocity.

(c) Interpret $a$ and $b$ biologically.

**Answer**: For the choice $a = 0.5$ and $b = 1.5 \times 10^{-4}$, there is a stable equilibrium point at $N = 6667$.

**Exercises to explore the behavior of solutions associated with some possible, non-standard geometric features of the phase diagram.**

(9) In the following exercises,

$$\frac{dy}{dt}(t) = f\big(y(t)\big),$$

but $f(y)$ has some different properties.
Be sure to draw the phase diagram and locate any equilibrium points. Determine general solutions and check their behavior against the phase diagram.

(a) A double zero:

$$f(y) = (y - a)^2.$$

(b) Singularity at the equilibrium point:

$$f(y) = a\, y^{1/3}.$$

(c) Horizontal asymptotes:

$$f(y) = a \tanh(y) = a\frac{\sinh(y)}{\cosh(y)}.$$

(d) Vertical asymptote:

$$f(y) = \frac{b}{y - a}.$$

## 5.2  Homogeneous linear equations

The value in considering a single differential equation in Sec. 5.1 is that new ideas could be introduced in a context as simple as possible. Now we face the challenge of incorporating those ideas into a system with more than one equation. A good first step is to consider two equations and the general form for two equations in two unknown functions is

$$\frac{dy_1}{dt}(t) = f_1\big(y_1(t), y_2(t), t\big), \tag{5.2.1}$$

$$\frac{dy_2}{dt}(t) = f_2\big(y_1(t), y_2(t), t\big). \tag{5.2.2}$$

We expect that two initial conditions, one each for $y_1(0)$ and $y_2(0)$, will be needed to determine a specific solution.

We proceed by retracing our steps in Sec. 5.1.2: we examine the possibility of separable solutions; we extend the ideas of the phase diagram to a system of two autonomous equations.

### 5.2.1 *Basic concepts*

As a start, we must acknowledge that there is no general formula for the solution
to Eq. (5.2.1) and Eq. (5.2.2), just as there is no general formula for the solution
to Eq. (5.1.15). Of course, there may be special cases where some progress can be
made; it is always advisable to study each case in its own right in the hope there
may be some tricks that allow progress in constructing solutions; some of them will
emerge later in this section.

Perhaps we will have some success if $f_1$ and $f_2$ are separable, that is, we can
write them as

$$f_1(y_1, y_2, t) = F_1(y_1) \, G_1(y_2) \, H_1(t) \,, \quad f_2(y_1, y_2, t) = F_2(y_1) \, G_2(y_2) \, H_2(t) \,.$$

But there is no way to separate the terms in the resulting equations as we do
when considering a single separable differential equation – see Eq. (5.1.19) and
Eq. (5.1.20).

Otherwise, we turn to the case of an autonomous system where $f_1$ and $f_2$ have
no explicit dependency on $t$.

$$\frac{dy_1}{dt}(t) = f_1\big(y_1(t), y_2(t)\big)\,, \tag{5.2.3}$$

$$\frac{dy_2}{dt}(t) = f_2\big(y_1(t), y_2(t)\big)\,. \tag{5.2.4}$$

What we now hope for is to draw a phase diagram that replaces Fig. 5.2 and gives
us the ability to infer the general feature of the solutions.

First, we must recognize that the phase space contains two independent variables, $y_1$ and $y_2$; so we have a "phase plane," illustrated in Fig. 5.5. At each
moment in time, the solution $\big(y_1(t), y_2(t)\big)$ is a point in this plane, called a phase
point; two such points are shown in Fig. 5.5. As the solution changes in time, the
phase point marks out a curve in the phase plane, called a trajectory. This curve
is defined through a parametric representation with $t$ as the independent variable.
The tangent to the trajectory is given by

$$\left(\frac{dy_1}{dt}(t), \, \frac{dy_2}{dt}(t)\right) = \Big(f_1\big(y_1(t), y(t)\big), f_2\big(y_1(t), y_2(t)\big)\Big)\,.$$

In other words, the functions $f_1$ and $f_2$ are the components of the tangent to the
trajectory as illustrated in Fig. 5.5. From a different perspective, the initial conditions determine the initial phase point $(y_1, y_2)_0$, and the tangent vector predicts
where the solution curve will head. At a moment later it will be at $(y_1, y_2)_1$ with a
new tangent vector. By linking these points together, we mark out the trajectory
in time.

Time to put these ideas into practice. The simplest examples where details can
be obtained are those where the equations are linear and homogeneous. The system

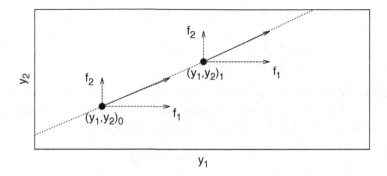

Fig. 5.5   Trajectories near the equilibrium point $(0, 0)$.

takes the special form

$$\frac{dy_1}{dt}(t) = a\, y_1(t) + b\, y_2(t),$$

$$\frac{dy_2}{dt}(t) = c\, y_1(t) + d\, y_2(t).$$

We have seen several examples in this book already and we will use them again to expose the features of the phase diagram.

### 5.2.2   *Chemical reactions*

An example of a chained chemical process is presented in 1.4.1. Three differential equations, Eq. (1.4.1), Eq. (1.4.2) and Eq. (1.4.3), describe the conversion of a chemical substance A to B and then to C. The last equation determines the concentration $[C](t)$ as an integration in time of $[B](t)$ and we may ignore it in favor of the first two equations, repeated here,

$$\frac{d[A]}{dt}(t) = -k_1\, [A],\qquad\qquad(5.2.5)$$

$$\frac{d[B]}{dt}(t) = k_1\, [A] - k_2\, [B].\qquad\qquad(5.2.6)$$

Of course, the standard approach to solving homogeneous linear equations with constant coefficients is to seek solutions in the form of an exponential. So we are ready to try

$$[A](t) = c_1\, e^{\lambda t},\quad [B](t) = c_2\, e^{\lambda t},\qquad\qquad(5.2.7)$$

as solutions to Eq. (5.2.5) and Eq. (5.2.6). Note that the exponential must be the same in both unknowns otherwise we will not be able to match the exponentials. Also, we now need different coefficients $c_1$ and $c_2$ because $[A](t)$ and $[B](t)$ will not be identically the same.

By substituting the guess Eq. (5.2.17) into Eq. (5.2.5) and Eq. (5.2.6), we obtain

$$\lambda\, c_1\, e^{\lambda t} = -k_1\, c_1\, e^{\lambda t}\,,$$
$$\lambda\, c_2\, e^{\lambda t} = k_1\, c_1\, e^{\lambda t} - k_2\, c_2\, e^{\lambda t}\,.$$

The wisdom in choosing guesses for the unknowns as exponentials with the same argument is now abundantly clear; all exponentials can be dropped, leaving a system of algebraic equations for the unknown coefficients $c_1$ and $c_2$.

$$(\lambda + k_1)\, c_1 = 0\,, \tag{5.2.8}$$
$$-k_1\, c_1 + (\lambda + k_2)\, c_2 = 0\,. \tag{5.2.9}$$

We are on familiar territory. This homogeneous algebraic system will only have the trivial solution unless we can find values for $\lambda$ that permit non-trivial solutions. Taking the equations one at a time, we see that Eq. (5.2.8) implies $c_1 = 0$, unless $\lambda = -k_1$ in which case $c_1$ is undetermined – it can have many values. The consequences on the other equation depend on which choice we make:

(1) If we set $\lambda = -k_1$, then Eq. (5.2.9) leads to $c_2 = k_1 c_1/(k_2 - k_1)$ and $c_1$ is undetermined.
(2) If $\lambda \neq -k_1$, then $c_1 = 0$ and Eq. (5.2.9) becomes

$$(\lambda + k_2)\, c_2 = 0\,.$$

Now $c_2 = 0$ unless $\lambda = -k_2$, in which case $c_2$ is undetermined.

In summary, we have found two different, non-trivial solutions: $\lambda = -k_1$, with $c_1$ undetermined and $c_2 = k_1 c_1/(k_2 - k_1)$; $\lambda = -k_2$ with $c_1 = 0$ and $c_2$ undetermined. By combining these solutions we obtain a general solution,

$$[\mathrm{A}](t) = c_1\, e^{-k_1 t}\,, \qquad [\mathrm{B}](t) = \frac{k_1 c_1}{k_2 - k_1}\, e^{-k_1 t} + c_2\, e^{-k_2 t}\,. \tag{5.2.10}$$

Of course, we are delighted at the presence of two unknown coefficients $c_1$ and $c_2$ in Eq. (5.2.10) because we need them to satisfy the initial conditions. The choice of initial conditions made in Sec. 1.4.1 is $[\mathrm{A}](0) = [\mathrm{A}_0]$ and $[\mathrm{B}](0) = 0$. By applying these initial conditions, we obtain two equations,

$$[\mathrm{A}_0] = c_1\,,$$
$$0 = \frac{k_1 c_1}{k_2 - k_1} + c_2\,.$$

The values for $c_1$ and $c_2$ are easily found, leading to the specific solution,

$$[\mathrm{A}](t) = [\mathrm{A}_0]\, e^{-k_1 t}\,, \qquad [\mathrm{B}](t) = \frac{k_1 [\mathrm{A}_0]}{k_2 - k_1} \left( e^{-k_1 t} - e^{-k_2 t} \right)\,,$$

a result that agrees with Eq. (1.4.17).

Three different methods have been used to solve the system Eqs. (1.4.1)–(1.4.3). Method one solves each equation separately – see Exercise 6 in Sec. 1.4.4. Method two combines the equation into a single equation Eq. (1.4.6). These first two methods cannot be applied to all systems, but work in this case because of the nature of the chain of equations. Method three, the method used in this section, always works for systems of linear equations with constant coefficients.

Now that we have the general solution Eq. (5.2.10) to Eq. (5.2.5) and Eq. (5.2.6), we can draw the phase diagram by selecting different choices for $c_1$ and $c_2$ and graphing the resulting curves. At the same time, by making the choice $k_1 = 1$ and $k_2 = 0.2$ we can establish connections between the phase diagram and the solutions shown in Fig. 1.14. The choices made for the phase diagram shown in Fig. 5.6 are: $c_1 = \pm 1, c_2 = 0$; $c_1 = 0, c_2 = \pm 1$; $c_1 = \pm 0.1, c_2 = \pm 0.9$; $c_1 = \pm 0.2, c_2 = \pm 0.8$; $c_1 = \pm 0.5, c_2 = \pm 0.5$. The diagram shows several interesting features, and it is worth commenting on them prior to relating them to the properties of the general solution. While direction arrows have not been drawn on all the trajectories, they all lead to the equilibrium point at $[A] = 0$, $[B] = 0$. The trajectories are shown for both positive and negative values of $[A]$ and $[B]$ to give a complete picture and also highlight some of the symmetries in the pattern. The physically relevant part of the diagram is just the first quadrant where $[A] \geq 0$ and $[B] \geq 0$.

There are four trajectories, marked with the arrows, that approach the equilibrium point as straight lines. They occur when we set either $c_1 = 0$ or $c_2 = 0$.

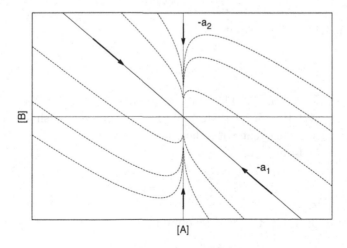

Fig. 5.6   Trajectories near the equilibrium point (0, 0).

For example, with $c_1 = 0$, the general solution becomes just $[A] = 0$ and $[B](t) = c_2 \exp(-k_2 t)$. The phase point moves down or up the $[B]$-axis towards the origin when $c_2$ is positive or negative, respectively. Next, with $c_2 = 0$, the solution becomes $[A](t) = c_1 \exp(-k_1 t)$, $[B](t) = k_1 c_1/(k_2 - k_1) \exp(-k_1 t)$. Both components to the solution decay with the same exponential. We may eliminate this exponential by substitution to obtain $[B] = k_1/(k_2 - k_1)\,[A]$, a straight line as shown in Fig. 5.6. In summary, the straight line solutions arise when we set either $c_1 = 0$ or $c_2 = 0$.

What about the other trajectories? Let us imagine what the solution looks like as $t$ becomes large. Remember we are considering the case $k_1 > k_2$. So $\exp(-k_1 t)$ decreases more rapidly than $\exp(-k_2 t)$. After long enough time, the terms containing $\exp(-k_1 t)$ will be negligible leaving only the solution that corresponds to the vertical trajectory $[A] = 0$. In other words, the first motion of the phase point representing the solution is to move towards the vertical axis.

The picture becomes more transparent when we introduce vector notation for the solution. Consider $[A](t)$ and $[B](t)$ as the components of a vector $\mathbf{y}(t)$, that is,

$$\mathbf{y}(t) = \begin{pmatrix} [A](t) \\ [B](t) \end{pmatrix}$$

with "velocity"

$$\frac{d\mathbf{y}}{dt}(t) = \frac{d}{dt}\begin{pmatrix} [A](t) \\ [B](t) \end{pmatrix} = \begin{pmatrix} \dfrac{d[A]}{dt}(t) \\ \dfrac{d[B]}{dt}(t) \end{pmatrix}.$$

With this introduction of vectors, we may write Eq. (5.2.5) and Eq. (5.2.6) as

$$\frac{d\mathbf{y}}{dt}(t) = \mathcal{A}\mathbf{y}(t), \qquad (5.2.11)$$

where $\mathcal{A}$ is a matrix with constant entries,

$$\mathcal{A} = \begin{bmatrix} -k_1 & 0 \\ k_1 & -k_2 \end{bmatrix}.$$

This compact form Eq. (5.2.11) invites comparison with a first-order differential equation Eq. (1.1.19). It suggests that the solution vector should contain an exponential in time, consistent with the choice made in Eq. (5.2.7). Expressing Eq. (5.2.7) in vector form gives

$$\mathbf{y}(t) = \begin{pmatrix} c_1 \\ c_2 \end{pmatrix} e^{\lambda t} = \mathbf{c}\,e^{\lambda t}. \qquad (5.2.12)$$

Upon substitution into Eq. (5.2.11), we obtain

$$\lambda \mathbf{c}\,e^{\lambda t} = \mathcal{A}\mathbf{c}\,e^{\lambda t},$$

and after dropping the exponentials,

$$\mathcal{A}\mathbf{c} = \lambda \mathbf{c}. \tag{5.2.13}$$

We end up with what is called an *eigenvalue problem* – see Appendix C.4.

Written out as separate equations,

$$-(k_1 + \lambda)\, c_1 = 0,$$
$$k_1\, c_1 - (k_2 + \lambda)\, c_2 = 0,$$

the same algebraic equations we obtained in (5.2.8) and (5.2.9). We found two different solutions:

(1) $\lambda = -k_1$, $c_2 = k_1 c_1/(k_2 - k_1)$ with $c_1$ undetermined. Thus there is a solution

$$\frac{c_1}{k_2 - k_1}\, \mathbf{a}_1\, e^{-k_1 t} \quad \text{with} \quad \mathbf{a}_1 = \begin{pmatrix} k_2 - k_1 \\ k_1 \end{pmatrix}. \tag{5.2.14}$$

The solution lies along the vector $a_1$. This solution corresponds to a trajectory on the straight line $[B] = k_1/(k_2 - k_1)\,[A]$.

(2) $\lambda = -k_2$, $c_1 = 0$ and $c_2$ undetermined. Thus there is another solution

$$c_2\, \mathbf{a}_2 e^{-k_2 t} \quad \text{with} \quad \mathbf{a}_2 = \begin{pmatrix} 0 \\ 1 \end{pmatrix}. \tag{5.2.15}$$

This solution corresponds to a trajectory on the vertical axis.

The general solution is made up of a linear combination of the two solutions (5.2.14) and (5.2.15),

$$\mathbf{y}(t) = \frac{c_1}{k_2 - k_1}\, \mathbf{a}_1\, e^{-k_1 t} + c_2\, \mathbf{a}_2\, e^{-k_2 t}. \tag{5.2.16}$$

The way to interpret this result is that the phase point is a vector that is the sum of different multiples of two vectors $\mathbf{a}_1$ and $\mathbf{a}_2$, which are eigenvectors to $\mathcal{A}$. To determine a specific solution, we must apply initial conditions to determine $c_1$ and $c_2$, but to draw the phase diagram we want to consider several different choices for $c_1$ and $c_2$. Whatever the choices for these coefficients, the exponential $\exp(-k_1 t)$ decreases more rapidly than the exponential $\exp(-k_2 t)$ when $k_1 > k_2$, the case we have assumed in this example. The means the contribution $\mathbf{a}_1$ to the sum of vectors in (5.2.16) decreases more rapidly than the contribution $\mathbf{a}_2$; the phase point moves parallel to the straight line defined by $\mathbf{a}_1$ as it approaches the line defined by $\mathbf{a}_2$. And that is exactly what we see in Fig. 5.6.

How does the behavior of the solution shown in Fig. 1.14 correspond to a trajectory in Fig. 5.6? Start with a point on the [A]-axis and note that the phase point moves initially mostly parallel to $\mathbf{a}_1$. On a time scale of $1/k_1 = 1$, the phase point approaches closely to $\mathbf{a}_2$; during this time the concentration $[B](t)$ grows and $[A](t)$ decays. Over the longer time scale $1/k_2 = 5$, $[B](t)$ decays exponentially as the phase point moves down the vertical axis – opposite to the direction given by $\mathbf{a}_2$.

### 5.2.3   *The LCR circuit*

A different example is afforded by the system of equations for an LCR-circuit. The equations are derived in Sec. 1.5.3 – see Eq. (1.5.23) and Eq. (1.5.24) – and restated here.

$$\frac{\mathrm{d}Q}{\mathrm{d}t}(t) = I(t) , \tag{5.2.17}$$

$$\frac{\mathrm{d}I}{\mathrm{d}t}(t) = -2\,\alpha\, I(t) - \omega_0^2\, Q(t) , \tag{5.2.18}$$

where the circuit parameters $\alpha$ and $\omega_0$ are defined in Eq. (1.5.30). We also ignore the external voltage $E(t)$ for the moment.

Equations (5.2.17) and (5.2.18) can be written as a system:

$$\frac{\mathrm{d}\mathbf{y}}{\mathrm{d}t}(t) = \mathcal{B}\,\mathbf{y}(t) , \tag{5.2.19}$$

where

$$\mathbf{y}(t) = \begin{pmatrix} Q(t) \\ I(t) \end{pmatrix} \quad \text{and} \quad \mathcal{B} = \begin{bmatrix} 0 & 1 \\ -\omega_0^2 & -2\,\alpha \end{bmatrix} .$$

Clearly $\mathbf{y} = 0$ is an equilibrium point. To construct a general solution we make the same guess Eq. (5.2.12) and substitute into Eq. (5.2.19). Thus we need to determine the eigenvalues and eigenvectors of $\mathcal{B}$, just as we do in Eq. (5.2.13). By following the procedures in Sec. C.4, the eigenvalues are given as the roots of the quadratic Eq. (C.4.4),

$$\lambda\,(\lambda + 2\alpha) + \omega_0^2 = 0 .$$

Under the assumption that $\alpha$ is small, specifically $\alpha^2 < \omega_0^2$, the eigenvalues are complex conjugates,

$$\lambda = -\alpha \pm \mathrm{i}\,\omega_\mathrm{d} \quad \text{where} \quad \omega_\mathrm{d} = \sqrt{\omega_0^2 - \alpha^2} .$$

The associated complex eigenvectors are given by Eq. (C.4.6).

$$\mathbf{e} = \mathbf{a} \pm \mathrm{i}\,\mathbf{b} = \begin{pmatrix} 1 \\ -\alpha \end{pmatrix} \pm \mathrm{i} \begin{pmatrix} 0 \\ \omega_\mathrm{d} \end{pmatrix} .$$

A general solution is composed by the addition of the two complex conjugate solutions.

$$\begin{aligned}
\mathbf{y}(t) &= c_1\,(\mathbf{a} + \mathrm{i}\,\mathbf{b})\,\mathrm{e}^{-\alpha t + \mathrm{i}\,\omega_\mathrm{d} t} + c_2\,(\mathbf{a} - \mathrm{i}\,\mathbf{b})\,\mathrm{e}^{-\alpha t - \mathrm{i}\,\omega_\mathrm{d} t} \\
&= \mathrm{e}^{-\alpha t}\left[c_1\,(\mathbf{a} + \mathrm{i}\,\mathbf{b})\,\mathrm{e}^{\mathrm{i}\,\omega_\mathrm{d} t} + c_2\,(\mathbf{a} - \mathrm{i}\,\mathbf{b})\,\mathrm{e}^{-\mathrm{i}\,\omega_\mathrm{d} t}\right] .
\end{aligned} \tag{5.2.20}$$

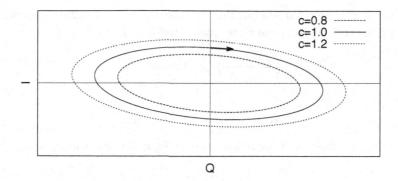

Fig. 5.7   Trajectories near the equilibrium point (0, 0).

Just as we discovered in Eq. (1.5.4), we must set $c_2 = \bar{c}_1$ to ensure solutions with real values. And as before in Eq. (1.5.10), the exponentials with complex arguments may be cast into sines and cosines. The easiest way to do so here is to introduce a polar form for $c_1 = (c/2)\exp(-\mathrm{i}\phi)$; then $c_2 = \bar{c}_1 = (c/2)\exp(\mathrm{i}\phi)$ and Eq. (5.2.20) becomes

$$\mathbf{y}(t) = \frac{c}{2}\,\mathrm{e}^{-\alpha t}\left[(\mathbf{a}+\mathrm{i}\,\mathbf{b})\,\mathrm{e}^{\mathrm{i}(\omega_\mathrm{d}t-\phi)} + (\mathbf{a}-\mathrm{i}\,\mathbf{b})\,\mathrm{e}^{-\mathrm{i}(\omega_\mathrm{d}t-\phi)}\right]$$

$$= c\,\mathrm{e}^{-\alpha t}\left[\mathbf{a}\cos(\omega_\mathrm{d}t-\phi) - \mathbf{b}\sin(\omega_\mathrm{d}t-\phi)\right]. \tag{5.2.21}$$

The coefficients $c_1$ and $c_2$ are replaced by new coefficients $c$ and $\phi$ and these new coefficients must be determined by initial conditions in the standard way.

Time to translate the general solution Eq. (5.2.21) into trajectories in the phase plane. The exponential $\exp(-\alpha t)$ indicates that the trajectories approach the equilibrium point at the origin, but how? Ignore the exponential part for the moment and focus on the other contribution to the general solution,

$$\mathbf{z}(t) = c\left[\mathbf{a}\cos(\omega_\mathrm{d}t-\phi) - \mathbf{b}\sin(\omega_\mathrm{d}t-\phi)\right]. \tag{5.2.22}$$

Observe that the solution is $2\pi/\omega_\mathrm{d}$-periodic and the coefficient $\phi$ is just a phase shift. What does a periodic solution mean? The vector $\mathbf{z}$ must return to the same value after a period. That means it must be a closed trajectory. The phase shift $\phi$ simply tells us where we start on this closed trajectory. The coefficient $c$ determines the length of the vector $\mathbf{z}$ which in turn determines how large the closed trajectory is. Figure 5.7 shows three of the closed trajectories for the choice $\alpha = 0.1$ and $\omega_\mathrm{d} = \sqrt{6}\,/5$, the same choice for the solution shown in Fig. 1.18.

The trajectories in Fig. 5.7 appear to be ellipses slightly rotated. Indeed, they are! We can confirm this by looking at the behavior of the components of $\mathbf{z}(t)$,

$$Q(t) = c\cos(\omega_\mathrm{d}t-\phi)\,, \tag{5.2.23}$$

$$I(t) = -c\,\alpha\cos(\omega_\mathrm{d}t-\phi) - c\,\omega_\mathrm{d}\sin(\omega_\mathrm{d}t-\phi)\,. \tag{5.2.24}$$

By eliminating the time variable in Eq. (5.2.23) and Eq. (5.2.24), we obtain an equation for the trajectory in the phase plane, the $Q - I$ plane.

$$\left(\frac{I + \alpha Q}{\omega_d}\right)^2 + Q^2 = c^2,$$

or

$$I^2 + 2\alpha QI + \left(\alpha^2 + \omega_d^2\right) Q^2 = \omega_d^2 c^2.$$

According to the criterion below, this implicit equation for the trajectory is an ellipse.

---

**Ellipses**: An implicit equation involving two independent variables $x$ and $y$ of the form

$$Ax^2 + Bxy + Cy^2 = 1$$

defines an ellipse provided $B^2 - 4AC < 0$.

---

Which way around the closed orbit does the phase point move? There are several ways this can be determined:

- plot two or three locations of the phase point;
- evaluate the tangent vector at a phase point;
- evaluate the "velocity" of the phase point at some locations in the phase plane.

For example, the tangent vector is drawn in Fig. 5.7 for $c = 1$ at the place the trajectory crosses the positive $I$-axis.

To determine the orientation of the ellipse and the lengths of its major and minor axis, consider the distance $D$ from the origin – the center of the ellipse,

$$D^2(\theta) = c^2 \left[(1 + \alpha^2)\cos^2(\theta) + 2\alpha\omega_d\cos(\theta)\sin(\theta) + \omega_d^2\sin^2(\theta)\right] \qquad (5.2.25)$$

where we have introduced $\theta = \omega_d t - \phi$ as a new independent variable for convenience. The parameter $c$ merely selects the size of the ellipse and so we may select a particular ellipse with the choice $c = 1$.

The maximum and minimum values of $D(\theta)$ determine the length of the major and minor axes of the ellipse. The value of $\theta$ that determines the maximum value for $D(\theta)$ can then be used to locate the phase point that lies at the end of the major axis and that will determine the orientation of the ellipse. The maximum or minimum of $D(\theta)$ also occurs at the maximum or minimum of $D^2(\theta)$. Consequently, it is easier to locate the maximum and minimum of $D^2(\theta)$ by finding zero values for the derivative of Eq. (5.2.25).

$$\frac{dD^2}{d\theta}(\theta) = 2\left(\omega_d^2 - 1 - \alpha^2\right)\sin(\theta)\cos(\theta) + 2\alpha\omega_d\left(\cos^2(\theta) - \sin^2(\theta)\right)$$

$$= \left(\omega_d^2 - 1 - \alpha^2\right)\sin(2\theta) + 2\alpha\omega_d\cos(2\theta).$$

Zeros occur at

$$\tan(2\theta) = \frac{2\,\alpha\,\omega_{\mathrm{d}}}{1 + \alpha^2 - \omega_{\mathrm{d}}^2}.$$

For $\alpha = 0.1$ and $\omega_{\mathrm{d}} = \sqrt{6}\,/5$, the chosen values in Fig. 5.7, the maximum occurs when $\theta = 0.063$ and $\mathbf{z} = (0.998, -0.131)$; the length of the major axis is 1.01 and the ellipse is rotated by an angle $-0.131$ radians. The minimum occurs at $\theta = \pi/2 - 0.063$ and $\mathbf{z} = (-0.063, -0.483)$; the minor axis has length 0.487.

Let us return to the solution vector $\mathbf{y}(t) = \exp(-\alpha t)\,\mathbf{z}(t)$ and examine the effect the presence of the exponential $\exp(-\alpha t)$ has on the solution trajectories. Instead of the phase point completing the closed circuit defined by $\mathbf{z}(t)$, it steadily shortens in time; it does not return to its starting point, but falls short. It falls short every time it goes around; thus the spirals as shown in Fig. 5.8 for each choice of $c$. The

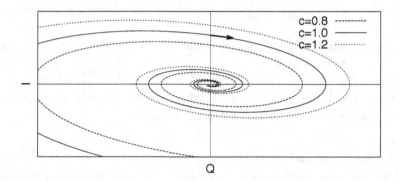

Fig. 5.8   Trajectories near the equilibrium point $(0, 0)$.

benefit of drawing the ellipses first becomes apparent; the orientation and aspect ratio of the spirals mimic those of the ellipse.

One final observation. As the phase point traverse the trajectory, its projection onto the $Q$-axis describes how the solution behaves. It swings back and forth with an ever decreasing amplitude. There are two time scales in this motion: the apparent period and the rate of decay in the exponential. If the rate of decay is low compared to the period, then the spirals has many turns – called a tightly wound spiral. Otherwise the turns are hardly noticeable – called a loosely wound spiral. The spirals shown in Fig. 5.8 are somewhere between tightly and loosely bound spirals.

### 5.2.4   *Abstract viewpoint*

A homogeneous system of $N$ linear differential equations with constant coefficients takes the form,

$$\frac{\mathrm{d}y_i}{\mathrm{d}t}(t) = \sum_{j=1}^{N} A_{ij}\, y_j(t) \quad \text{for } i = 1, \ldots, N. \tag{5.2.26}$$

There are $N$ equations to determine the $N$ unknown functions $y_i(t)$. Written in terms of vectors and matrices, the system has the simple form,

$$\frac{d\mathbf{y}}{dt}(t) = \mathcal{A}\mathbf{y}(t).$$  (5.2.27)

Additionally we need initial conditions, $\mathbf{y}(0) = \mathbf{y}_0$, to obtain a specific solution.

No surprise that the solution is constructed in the usual way. A guess is made of the form

$$\mathbf{y}_h(t) = \mathbf{a}\,e^{\lambda t}.$$  (5.2.28)

Obviously, we do not yet know the values for $\lambda$ and the coefficient vector $\mathbf{a}$, but expect to find them after we have substituted our guess into the system of differential equations (5.2.27).

$$\lambda\,\mathbf{c}\,e^{\lambda t} = \mathcal{A}\,\mathbf{c}\,e^{\lambda t}, \quad \text{or} \quad \mathcal{A}\mathbf{c} = \lambda\,\mathbf{c}.$$  (5.2.29)

We have an eigenvalue problem – see Sec. C.4. There are non-trivial solutions for $\mathbf{a}$ only for specific choices of $\lambda$.

Fortunately, the two-dimensional case ($N = 2$) provides most of the important examples that arise in the general case, so that is where we will start. Where we go from here depends on the nature of the eigenvalues.

## Real roots

Let $\lambda_1$ and $\lambda_2$ be real roots to the characteristic equation Eq. (C.4.4) with associated eigenvectors $\mathbf{a}_1$ and $\mathbf{a}_2$. That means there are two solutions of the form Eq. (5.2.28) and the general homogeneous solution is composed of a linear combination of them.

$$\mathbf{y}_h(t) = c_1\,\mathbf{a}_1\,e^{\lambda_1 t} + c_2\,\mathbf{a}_2\,e^{\lambda_2 t}.$$  (5.2.30)

The coefficients $c_1$ and $c_2$ are available to satisfy the initial condition $c_1\,\mathbf{a}_1 + c_2\,\mathbf{a}_2 = \mathbf{y}_0$. Since there are two components to the vectors, the initial condition gives two equations in two unknowns.

The nature of the phase diagram, and hence of the general solution, depends on whether the eigenvalues are positive or negative. There are three cases to consider:

(1) Both negative, $\lambda_1 < \lambda_2 < 0$. We have already seen an example, the chained chemical reactions in Sec. 5.2.2. Since $\lambda_1 < \lambda_2$, $\exp(\lambda_1 t) < \exp(\lambda_2 t)$ and the contribution to the sum Eq. (5.2.30) of the vector $\mathbf{a}_1$ diminishes more rapidly than the contribution form $\mathbf{a}_2$. The motion of the phase point is to move parallel to the line whose direction is given by $\mathbf{a}_1$ until it approaches the line whose slope is given by $\mathbf{a}_2$. Then the point swerves and approaches the origin in alignment with $\mathbf{a}_2$. Figure 5.6 gives a good example. The equilibrium point is said to be a *stable node*.

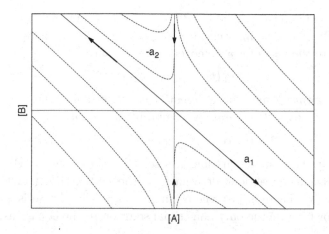

Fig. 5.9   Trajectories near the equilibrium point $(0, 0)$.

(2) Both positive, $0 < \lambda_2 < \lambda_1$. This case is the reverse of the previous
case in the following sense. Replace $\lambda_1 \to -\lambda_1$, $\lambda_2 \to -\lambda_2$ and $t \to -t$
and observe that the solution is the same as (5.2.30) except that the new
eigenvalues are both negative and $\lambda_1 < \lambda_2$. The phase diagram is the
same as shown in Fig. 5.6. But there is a crucial difference; time runs
backwards! The phase point moves away from the origin in the direction
of $a_2$ but soon swerves and then moves parallel to the line defined by $a_1$.
Simply imagine the arrows in Fig. 5.6 reversed. The equilibrium point is
said to be an *unstable node*.

(3) One positive and one negative, $\lambda_2 < 0 < \lambda_1$. The contribution to the
solution from $a_2$ decays in time, while the contribution from $a_1$ increases
exponentially. The phase point moves at first towards the line defined
by $a_1$ but then veers and moves parallel to it as shown in Fig. 5.9. The
equilibrium point is said to be a *saddle point*.

**Double roots**

There is only one eigenvalue and the nature of the solution depends on whether
there is one eigenvector or two. Suppose there are two independent eigenvectors. Then the general homogeneous solution is

$$\mathbf{y}_h(t) = c_1\,\mathbf{a}_1\,e^{\lambda t} + c_2\,\mathbf{a}_2\,e^{\lambda t} = \big(c_1\mathbf{a}_1 + c_2\mathbf{a}_2\big)\,e^{\lambda t}. \qquad (5.2.31)$$

Because the eigenvectors are independent, the initial condition will determine
$c_1$ and $c_2$. The example Eq. (C.4.8) indicates how this case might arise. Once
the initial condition has determined $c_1$ and $c_2$, the solution vector remains the
same combination of the eigenvectors $a_1$ and $a_2$ except that its length changes
by the exponential factor $\exp(\lambda t)$. The consequence is that the phase point
moves along a line, outwards if $\lambda > 0$ and inwards if $\lambda < 0$. The phase diagram
looks like spikes emanating from the origin. The equilibrium point is said to
be a *star node*, stable if $\lambda < 0$ and unstable if $\lambda > 0$.

If there is only one eigenvector, we need another contribution to the solution, and, as the checklist in Sec. 1.6.8 suggests, we should add a contribution $t \exp(\lambda t)$ to the guess for a homogeneous solution.

$$\mathbf{y}_{\mathrm h}(t) = (c_1 \mathbf{a} + c_2 \mathbf{b}\, t)\, \mathrm{e}^{\lambda t}. \tag{5.2.32}$$

Upon substituting into the differential equation and balancing terms with $\exp(\lambda t)$ and $t \exp(\lambda t)$ separately, we obtain two equations

$$\mathcal{A}\mathbf{b} = \lambda\mathbf{b}, \quad \text{and} \quad c_1 \mathcal{A}\mathbf{a} = c_1 \lambda\mathbf{a} + c_2 \mathbf{b}. \tag{5.2.33}$$

To illustrate how these equations are solved, let us consider the case of the chained chemical reactions described by the system Eq. (5.2.11) but with $k_1 = k_2 = k$. There is just one eigenvalue $\lambda = -k$. There is also only one eigenvector because the only non-trivial solution to the equations,

$$-k\, b_1 = -k\, b_1,$$
$$k\, b_1 - k\, b_2 = -k\, b_2,$$

is $b_1 = 0$ and $b_2$ undetermined. The simplest choice is $b_2 = 1$; it is the choice we made before in Eq. (5.2.15). The second equation in Eq. (5.2.33) becomes

$$c_1 \begin{bmatrix} -k & 0 \\ k & -k \end{bmatrix} \begin{pmatrix} a_1 \\ a_2 \end{pmatrix} = -k\, c_1 \begin{pmatrix} a_1 \\ a_2 \end{pmatrix} + c_2 \begin{pmatrix} 0 \\ 1 \end{pmatrix}.$$

The first equation is identically satisfied, and the second equation requires $c_1 k\, a_1 = c_2$; $a_2$ is undetermined.

The general homogeneous solution has been determined:

$$\mathbf{y}_{\mathrm h}(t) = \left[ \frac{1}{k} \begin{pmatrix} c_2 \\ k\, a_2 \end{pmatrix} + c_2 \begin{pmatrix} 0 \\ 1 \end{pmatrix} t \right] \mathrm{e}^{-kt}. \tag{5.2.34}$$

We still have two unknown coefficients $a_2$ and $c_2$ available to satisfy the initial condition. The choice of initial condition we used before is $[\mathrm{A}](0) = [\mathrm{A}_0]$ and $[\mathrm{B}](0) = 0$ and so $c_2 = k\,[\mathrm{A}]_0$ and $a_2 = 0$. Note that the solution agrees with Eq. (1.4.19).

The phase diagram, generated with different choices for $a_2$ and $c_2$ in Eq. (5.2.34), is shown in Fig. 5.10 for the case of a negative eigenvalue. There are several intriguing features to this phase diagram. The trajectories all swirl towards the origin, with the line defined by $\mathbf{b}$ – the eigenvector of $\mathcal{A}$ – separating two regions. We may obtain an equation for the trajectory by using the first component to express $t$ in terms of $[\mathrm{A}]$ provided $c_2 \neq 0$, and then substituting the result into the second component to obtain

$$[\mathrm{B}] = \frac{ka_2}{c_2} [\mathrm{A}] - \ln\left( \frac{k[\mathrm{A}]}{c_2} \right) [\mathrm{A}].$$

The slope of the trajectory become infinity as $[\mathrm{A}]$ approaches zero, meaning that all the trajectories have a vertical slope at the origin. Stated generally, all trajectories approach the origin with a tangent that lies in the direction of $\mathbf{b}$. The equilibrium point is said to be a *degenerate node*.

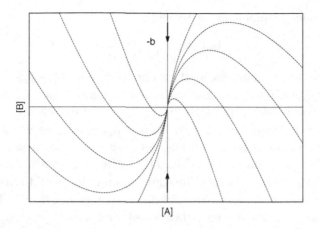

Fig. 5.10   Trajectories near the equilibrium point (0, 0).

## Complex roots

Write the eigenvalues as $\lambda_r \pm i\,\lambda_i$ and the eigenvectors as $\mathbf{a} \pm i\,\mathbf{b}$. The general homogeneous solution is formed as in Eq. (5.2.30) (except $c_2 = \bar{c}_1$) to produce a real-valued solution.

$$\mathbf{y_h}(t) = e^{\lambda_r t}\left[c_1\,(\mathbf{a} + i\,\mathbf{b})\,e^{i\,\lambda_i t} + \bar{c}_1\,(\mathbf{a} - i\,\mathbf{b})\,e^{-i\,\lambda_i t}\right].\qquad(5.2.35)$$

The complex coefficient $c_1$ has real and imaginary parts providing two real quantities available to satisfy the initial condition. Normally, it is more convenient to express $c_1$ in a polar form, $c_1 = c/2\,\exp(-i\phi)$ where $c$ and $\phi$ are real. Then Eq. (5.2.35) may be rewritten as

$$\mathbf{y_h}(t) = c\,e^{\lambda_r t}\left[\mathbf{a}\cos(\lambda_i t - \phi) - \mathbf{b}\sin(\lambda_i t - \phi)\right].$$

If $\lambda_r = 0$, then the phase plane contains ellipses; an example is given by Eq. (5.2.22) with the corresponding ellipses shown in Fig. 5.7. The orientation of the ellipses can be determined by examining the length of the solution vector $D$. Since

$$D^2(\theta) = \mathbf{y_h} \cdot \mathbf{y_h} = c^2\left[\mathbf{a} \cdot \mathbf{a}\cos^2(\theta) - 2\,\mathbf{a} \cdot \mathbf{b}\cos(\theta)\sin(\theta) + \mathbf{b} \cdot \mathbf{b}\sin^2(\theta)\right]$$

where $\mathbf{x} \cdot \mathbf{y}$ is the scalar (dot) product of $\mathbf{x}$ and $\mathbf{y}$ and $\theta = \lambda_i t - \phi$. The length reaches it largest value at the end of the major axis of the ellipse – and its lowest value at the end of the minor axis. We can determine the maximum and minimum values of $D$, or more conveniently of $D^2$, in the standard way by setting the derivative with respect to $\theta$ to zero.

$$\frac{\mathrm{d}D^2}{\mathrm{d}\theta}(\theta) = c^2\left[(\mathbf{b} \cdot \mathbf{b} - \mathbf{a} \cdot \mathbf{a})\sin(2\theta) - 2\,\mathbf{a} \cdot \mathbf{b}\cos(2\theta)\right] = 0.$$

The maximum or minimum occurs when

$$\tan(2\theta) = \frac{2\,\mathbf{a}\cdot\mathbf{b}}{\mathbf{b}\cdot\mathbf{b} - \mathbf{a}\cdot\mathbf{a}}. \tag{5.2.36}$$

Let $\theta_{\max}$ be the maximum value; the minimum value will be $\theta_{\min} = \theta_{\max} + \pi/2$. Then $\mathbf{y}_h(\theta_{\max})$ and $\mathbf{y}_h(\theta_{\min})$ give the major and minor axes, respectively. This is exactly how we determine the ellipses given by Eq. (5.2.22).

There is an important special case. The ellipses are just circles. This case arises when $\mathbf{a}$ is orthogonal to $\mathbf{b}$ and both have the same length. Mathematically, $\mathbf{a} \cdot \mathbf{b} = 0$ and $\mathbf{a} = \pm\mathbf{b}$. Then $D^2 = c^2\mathbf{a}\cdot\mathbf{a}$, a constant in $\theta$. In other words, a circle with radius $D$. The equilibrium point is said to be a *center*.

If $\lambda_{\mathrm{r}} \neq 0$, then the ellipses become spirals, as shown in Fig. 5.8. The trajectories move away from the origin when $\lambda_{\mathrm{r}} > 0$ and towards it when $\lambda_{\mathrm{r}} < 0$. The best way to sketch the spirals is to first determine the orientation and aspect ratio of the ellipses when $\lambda_{\mathrm{r}} = 0$, and then draw the trajectory as the length of the solution vector changes exponentially. If the length changes rapidly – $|\lambda_{\mathrm{r}}|$ is large compared to $|\lambda_{\mathrm{i}}|$ – then the spiral is loosely wound; if the change is slow, then the spiral is tightly wound. The equilibrium point is said to be a *spiral*, stable if $\lambda_{\mathrm{r}} < 0$ and unstable if $\lambda_{\mathrm{r}} > 0$.

One final observation. If an eigenvalue is zero, then the system is flawed. The equations are not linearly independent. This situation occurs for example when the equations are derived by using the same information twice, usually inadvertently.

---

### Principle 28

*The eigenvalues and eigenvectors of the matrix $\mathcal{A}$ in a homogeneous system of linear differential equations with constant coefficients determine the features of the trajectories in the phase space, especially near the origin which is an equilibrium point.*

---

### 5.2.5   *Exercises*

**These are some basic exercises to help develop and improve skill.**

(1) Suppose the following expressions are solutions to a two-dimensional system of equations. Draw some trajectories of the solution with different choices for the coefficients $c_1$ and $c_2$.

(a)

$$\mathbf{y}(t) = c_1 \begin{pmatrix} 1 \\ 0 \end{pmatrix} e^{-2t} + c_2 \begin{pmatrix} 0 \\ 1 \end{pmatrix} e^{-t}$$

(b)

$$\mathbf{y}(t) = c_1 \begin{pmatrix} 1 \\ 0 \end{pmatrix} e^{2t} + c_2 \begin{pmatrix} 0 \\ 1 \end{pmatrix} e^{t}$$

(c)

$$\mathbf{y}(t) = c_1 \begin{pmatrix} 1 \\ 0 \end{pmatrix} e^{-t} + c_2 \begin{pmatrix} 0 \\ 1 \end{pmatrix} e^{-t}.$$

(2) Construct the general solution to the two-dimensional system of equations

$$\frac{d\mathbf{y}}{dt}(t) = \mathcal{A}\mathbf{y}(t)$$

where

(a)

$$\mathcal{A} = \begin{bmatrix} 1 & 3 \\ 1 & -1 \end{bmatrix}$$

(b)

$$\mathcal{A} = \begin{bmatrix} 4 & -2 \\ 3 & -1 \end{bmatrix}$$

(c)

$$\mathcal{A} = \begin{bmatrix} -1 & -1 \\ 1 & -1 \end{bmatrix}.$$

Be sure to sketch the phase diagram by indicating some of the trajectories.

**Some problems that arise in applications.**

(3) Reconsider the chained chemical reactions as given in Eq. (5.2.11) but with the choice $k_1 = 0.2$ and $k_2 = 1.0$. Draw the phase diagram and relate its features to the solution displayed in Fig. 1.13.

(4) Reconsider the LCR-circuit as given in (5.2.17) and (5.2.18) but with the choice $\alpha = 1$ and $\omega_0^2 = 0.25$. Construct the general solution and draw the phase diagram. Compare the solution to the results shown in Fig. 1.17.

(5) The equation of motion for a pendulum is derived in Exercise 5 of Sec. 1.5.5. When the deflection is small, we obtain the linear approximation,

$$\frac{d^2\theta}{dt^2}(t) + \frac{g}{l}\theta(t).$$

It is possible to write this equation in the form of a system by a clever trick. Define

$$y_1(t) = \theta(t) \quad \text{and} \quad y_2(t) = \frac{d\theta}{dt}(t)$$

and note that the derivatives are

$$\frac{dy_1}{dt}(t) = y_2(t) \quad \text{and} \quad \frac{dy_2}{dt}(t) = \frac{d^2\theta}{dt^2}(t) = -\frac{g}{l}y_1(t).$$

Construct the solution to the system and compare the results to the solution
obtained previously. Draw the phase diagram and comment on its features.

(6) Only two of the system of equations (5.2.5) and (5.2.6) are used in the example
of chained chemical reactions. Suppose the other one (1.4.3) is also included;
the system is now three-dimensional ($N = 3$). Compute the eigenvalues and
eigenvectors. One of the eigenvalues is zero. Construct the general solutions
and discuss the consequence of this zero eigenvalue.

### 5.2.6  *Higher dimensional systems*

Both the examples of chained chemical reactions and of LCR-circuits are two-
dimensional and the calculation of the eigenvalues and eigenvectors is reasonably
easy. What happens if the system is higher dimensional as expressed in Eq. (5.2.26)?
Fortunately, the question is not as difficult as it might seem. The eigenvalues satisfy
the characteristic polynomial, the order of which is the dimension of the system, for
example, it is a quadratic when the system has dimension four. Since the roots of
a polynomial are either real or occur in complex conjugate pairs, we know precisely
how to express the general solution.

Let $\mathcal{A}$ have real eigenvalues $\lambda_j$ for $j = 1, \ldots, n$ and complex eigenvalues $\lambda_{r,k} \pm$
$i\,\lambda_{i,k}$ for $k = 1, \ldots, m$. Assume that the eigenvalues are all distinct, no repeated
roots. Then $n + 2m = N$, the dimension of the system. Let the corresponding
eigenvectors to the real eigenvalues be $\mathbf{a}_j$ and $\mathbf{b}_{r,k} \pm i\,\mathbf{b}_{i,k}$ for the complex eigenvalues.
Then the general homogeneous solution is

$$\mathbf{y}(t) = \sum_{k=1}^{m} d_k\, e^{\lambda_{r,k}t} \left[ \mathbf{b}_{r,k} \cos(\lambda_{i,k}t - \phi) - \mathbf{b}_{i,k} \sin(\lambda_{i,k}t - \phi) \right]$$

$$+ \sum_{j=1}^{n} c_j\, \mathbf{a}_j\, e^{\lambda_j t}.$$

The challenge now is to imagine what the phase diagram might look like. The
best we can do is to put together the various components of the general solution.
For example, suppose $N = 3$ and there is one negative eigenvalue and two complex
conjugate eigenvalues with $\lambda_r < 0$. Then the eigenvectors $\mathbf{b}_r$ and $\mathbf{b}_i$ define a plane
at some orientation in three-dimensional space in which a trajectory will spiral
into the origin. The eigenvector $\mathbf{a}$ gives the direction in which a trajectory moves
towards this plane. Putting the two pictures of the trajectories together gives us
the composite picture. The trajectory spirals inwards as it approaches the plane
defined by $\mathbf{b}_r$ and $\mathbf{b}_i$. Something like the spinal made when frozen ice-cream is
swirled onto a cone.

Another example worth noting occurs when there are only real eigenvalues,
but some are positive and others negative. The eigenvectors associated with the
negative eigenvalues define a hyperplane. Initially, a trajectory moves in a direction
parallel to this hyperplane towards the other hyperplane defined by the eigenvectors

associated with the positive eigenvalues. As soon as the trajectory gets close to the hyperplane defined by the positives eigenvalues it moves outwards in the direction parallel to this hyperplane. Just imagine Fig. 5.6 with the vectors $\mathbf{a}_1$ and $\mathbf{a}_2$ replaced by hyperplanes.

### 5.2.6.1 *Higher-order differential equations*

There is a trick that converts higher-order differential equations into systems of first-order equations. A simple example is the third-order differential equation,

$$\frac{\mathrm{d}^3 y}{\mathrm{d}x^3}(x) + a\frac{\mathrm{d}^2 y}{\mathrm{d}x^2}(x) + b\frac{\mathrm{d}y}{\mathrm{d}x}(x) + c\,y(x) = 0\,.$$

Define

$$y_1(x) = y(x)\,, \quad y_2(x) = \frac{\mathrm{d}y}{\mathrm{d}x}(x)\,, \quad y_3(x) = \frac{\mathrm{d}^2 y}{\mathrm{d}x^2}(x)\,,$$

and note that their derivatives are

$$\frac{\mathrm{d}y_1}{\mathrm{d}x}(x) = y_2(x)\,, \quad \frac{\mathrm{d}y_2}{\mathrm{d}x}(x) = y_3(x)\,, \frac{\mathrm{d}y_3}{\mathrm{d}x}(x) = -a\,y_3(x) - b\,y_2(x) - c\,y_1(x)\,.$$

Thus the equivalent system of first-order equations is

$$\frac{\mathrm{d}}{\mathrm{d}x}\begin{pmatrix} y_1(x) \\ y_2(x) \\ y_3(x) \end{pmatrix} = \begin{bmatrix} 0 & 1 & 0 \\ 0 & 0 & 1 \\ -c & -b & -a \end{bmatrix}\begin{pmatrix} y_1(x) \\ y_2(x) \\ y_3(x) \end{pmatrix}\,.$$

The system has dimension three corresponding to the order of the original differential equation. It is easy to adapt the trick to even higher-order differential equations. Note also that we expect three initial equations for a third-order differential equation, and these initial conditions translate into initial conditions for $\mathbf{y}(0)$:

$$y(0) = y_1(0) = \alpha\,, \quad \frac{\mathrm{d}y}{\mathrm{d}t}(0) = y_2(0) = \beta\,, \quad \frac{\mathrm{d}^2 y}{\mathrm{d}t^2}(0) = y_3(0) = \gamma\,.$$

A special case of some importance is the second-order differential equation (1.4.25). The equivalent system is

$$\frac{\mathrm{d}}{\mathrm{d}t}\begin{pmatrix} y_1(t) \\ y_2(t) \end{pmatrix} = \begin{bmatrix} 0 & 1 \\ -b & -a \end{bmatrix}\begin{pmatrix} y_1(t) \\ y_2(t) \end{pmatrix}$$

where $y_1(t) = y(t)$ and $y_2(t)$ is the derivative of $y(t)$.

The eigenvalues of this system must satisfy the characteristic equation Eq. (C.4.4),

$$\lambda^2 + a\,\lambda + b = 0\,,$$

the same equation as Eq. (1.4.27). The corresponding eigenvectors follow easily from Eq. (C.4.5). Then the general solution is

$$\mathbf{y}(t) = \begin{pmatrix} y_1(t) \\ y_2(t) \end{pmatrix} = c_1\begin{pmatrix} 1 \\ \lambda_1 \end{pmatrix}e^{\lambda_1 t} + c_2\begin{pmatrix} 1 \\ \lambda_2 \end{pmatrix}e^{\lambda_2 t}$$

which is identical to Eq. (1.4.30).

By applying the initial conditions Eq. (1.4.31) we are led to

$$\begin{bmatrix} 1 & 1 \\ \lambda_1 & \lambda_2 \end{bmatrix} \begin{pmatrix} c_1 \\ c_2 \end{pmatrix} = \begin{pmatrix} \alpha \\ \beta \end{pmatrix},$$

and there is a unique solution for the coefficients provided the determinant Eq. (C.3.4) of the matrix is not zero. In other words, $\lambda_2 - \lambda_1 \neq 0$; the eigenvalues are distinct.

---

**Reflection**

The roots of the characteristic polynomial of a higher-order differential equation must be the same as the eigenvalues of the equivalent system of first-order differential equations.

---

## 5.3   Inhomogeneous linear equations

As scientists and engineers, we relish the opportunity to study systems that are forced, either by our own hand or by some external agent. Forcing reveals specific information about systems that allows deepening understanding and the prospects of improved design if the system is of our own making. Forcing can arise mathematically in many ways, through inhomogeneous terms in differential equations, through boundary terms and in some cases through initial conditions, and we have seen several examples throughout this book. Here we will concentrate on the forcing that arises as inhomogeneous terms in a system of first-order linear differential equations and the importance of initial conditions as a forcing or as a control. The typical question we address is "how do we start up a system to achieve a desired outcome?"

We allow each differential equation to be forced in some way,

$$\frac{dy_i}{dt}(t) = \sum_{j=1}^{N} B_{ij}\, y_j(t) + r_i(t) \quad \text{for } i = 1, \dots, N$$

or in terms of vectors and matrices

$$\frac{d\mathbf{y}}{dt}(t) = \mathcal{B}\, \mathbf{y}(t) + \mathbf{r}(t). \tag{5.3.1}$$

Aside from special cases, there are only two strategies for constructing particular solutions that take care of the forcing term $\mathbf{r}(t)$. One is the method of variation of parameters, a version of which we explored in Sec. 1.6.7. Details will be provided at the end of this section. What emerges in this approach is the need to perform integrals and in most cases the anti-derivatives are unknown. The other approach is the method of undetermined coefficients, and this is where we will start, of course, by way of an example.

### 5.3.1 *LCR circuit with constant applied voltage*

This is a problem we considered in Sec. 1.6.1. This time we use the system based on Eq. (1.5.23) and Eq. (1.5.24). The related homogeneous system is stated in Eq. (5.2.17) and Eq. (5.2.18).

$$\frac{dQ}{dt}(t) = I(t),$$

$$\frac{dI}{dt}(t) = -\frac{R}{L}I(t) - \frac{1}{LC}Q(t) + \frac{V_0}{L}.$$

Written as Eq. (5.3.1),

$$\frac{d}{dt}\begin{pmatrix} Q(t) \\ I(t) \end{pmatrix} = \begin{bmatrix} 0 & 1 \\ -\omega_0^2 & -2\alpha \end{bmatrix}\begin{pmatrix} Q(t) \\ I(t) \end{pmatrix} + \begin{pmatrix} 0 \\ V_0/L \end{pmatrix} \tag{5.3.2}$$

where the circuit parameters $\alpha$ and $\omega_0$ are defined in Eq. (1.5.30).

The construction of a particular solution starts with the observation that the forcing vector is a constant in time. By following the spirit of Table 1.4, the guess for the particular solution is a constant vector $\mathbf{y}_p(t) = \mathbf{d}$. Substitute into Eq. (5.3.2),

$$\begin{bmatrix} 0 & 1 \\ -\omega_0^2 & -2\alpha \end{bmatrix}\begin{pmatrix} d_1 \\ d_2 \end{pmatrix} = -\begin{pmatrix} 0 \\ V_0/L \end{pmatrix}.$$

Clearly, $d_2 = 0$ and $d_1 = V_0/(\omega_0^2 L)$. Notice that the particular solution is also an equilibrium solution to Eq. (5.3.2).

The general solution has two parts, $\mathbf{y}(t) = \mathbf{y}_p(t) + \mathbf{y}_h(t)$ where the homogeneous contribution satisfies Eq. (5.2.19) and is given in Eq. (5.2.21) for the case where $\alpha^2 < \omega_0^2$. Thus the general solution is

$$\mathbf{y}(t) = \mathbf{d} + c\,e^{-\alpha t}\left[\mathbf{a}\cos(\omega_d t - \phi) - \mathbf{b}\sin(\omega_d t - \phi)\right]. \tag{5.3.3}$$

The two coefficients $c$ and $\phi$ are determined by the initial conditions but we can infer what must happen to the solution in the long term by noting that the homogeneous solution decays and the general solution approaches the equilibrium point given by $\mathbf{d}$.

The system Eq. (5.3.2) is autonomous – no explicitly dependency on $t$ in the right hand side – and so we can draw a phase diagram. The difficult work has already been done because the phase diagram is just that in Fig. 5.8 but with the origin displaced to the equilibrium point $\mathbf{d}$. What does this mean? It means that no matter what the initial conditions are, in the course of time the phase point approaches the equilibrium point as a spiral. No matter how we start up the circuit, the charge $Q(t)$ on the capacitor will approach the value $V_0/(\omega_0^2 L)$.

Incidentally, there are times that the LCR circuit is designed with $\alpha^2 > \omega_0^2$ so that the eigenvalues are real and negative,

$$\lambda_1 = -\alpha + \sqrt{\alpha^2 - \omega_0^2} \quad \text{and} \quad \lambda_2 = -\alpha - \sqrt{\alpha^2 - \omega_0^2}.$$

The general solution is now

$$\mathbf{y}(t) = \mathbf{d} + c_1\, \mathbf{a}_1\, e^{\lambda_1 t} + c_2\, \mathbf{a}_1\, e^{\lambda_2 t}, \tag{5.3.4}$$

much like Eq. (5.2.16): the eigenvectors $\mathbf{a}_1$ and $\mathbf{a}_2$ correspond to the eigenvalues $\lambda_1$ and $\lambda_2$, respectively. The phase diagram is similar to Fig. 5.6 except that the origin is shifted to $\mathbf{d}$. As a consequence, the charge on the capacitor eventually approaches the equilibrium point $\mathbf{d}$ parallel to $\mathbf{a}_1$ since that is the eigenvector that corresponds to the eigenvalue with the slowest decay rate.

## 5.3.2   *LCR circuit with alternating applied voltage*

This is a problem we considered in Sec. 1.6.2 where the external voltage is $A\sin(\omega t)$. The system is similar to Eq. (5.3.2).

$$\frac{\mathrm{d}}{\mathrm{d}t}\begin{pmatrix} Q(t) \\ I(t) \end{pmatrix} = \begin{bmatrix} 0 & 1 \\ -\omega_0^2 & -2\alpha \end{bmatrix}\begin{pmatrix} Q(t) \\ I(t) \end{pmatrix} + \begin{pmatrix} 0 \\ (A/L)\sin(\omega t) \end{pmatrix}. \tag{5.3.5}$$

The guess for the particular solution follows the standard approach.

$$\mathbf{y}_{\mathrm{p}}(t) = \mathbf{d}\cos(\omega t) + \mathbf{e}\sin(\omega t).$$

Upon substitution into (5.3.5),

$$-\omega\begin{pmatrix} d_1 \\ d_2 \end{pmatrix}\sin(\omega t) + \omega\begin{pmatrix} e_1 \\ e_2 \end{pmatrix}\cos(\omega t) = \begin{bmatrix} 0 & 1 \\ -\omega_0^2 & -2\alpha \end{bmatrix}\begin{pmatrix} d_1 \\ d_2 \end{pmatrix}\cos(\omega t)$$

$$+ \begin{bmatrix} 0 & 1 \\ -\omega_0^2 & -2\alpha \end{bmatrix}\begin{pmatrix} e_1 \\ e_2 \end{pmatrix}\sin(\omega t) + \begin{pmatrix} 0 \\ A/L \end{pmatrix}\sin(\omega t).$$

Balance the terms with $\sin(\omega t)$:

$$\begin{bmatrix} 0 & 1 \\ -\omega_0^2 & -2\alpha \end{bmatrix}\begin{pmatrix} e_1 \\ e_2 \end{pmatrix} + \omega\begin{pmatrix} d_1 \\ d_2 \end{pmatrix} = -\begin{pmatrix} 0 \\ A/L \end{pmatrix}. \tag{5.3.6}$$

Balance the terms with $\cos(\omega t)$:

$$\begin{bmatrix} 0 & 1 \\ -\omega_0^2 & -2\alpha \end{bmatrix}\begin{pmatrix} d_1 \\ d_2 \end{pmatrix} - \omega\begin{pmatrix} e_1 \\ e_2 \end{pmatrix} = \begin{pmatrix} 0 \\ 0 \end{pmatrix}. \tag{5.3.7}$$

We have ended up with two systems of algebraic equations for the two unknown vectors $\mathbf{d}$ and $\mathbf{e}$.

There are a few ways to solve these two algebraic systems but the easiest is to use Eq. (5.3.7) to replace the vector $\mathbf{e}$ in Eq. (5.3.6).

$$\begin{bmatrix} 0 & 1 \\ -\omega_0^2 & -2\alpha \end{bmatrix}\begin{bmatrix} 0 & 1 \\ -\omega_0^2 & -2\alpha \end{bmatrix}\begin{pmatrix} d_1 \\ d_2 \end{pmatrix} + \omega^2\begin{pmatrix} d_1 \\ d_2 \end{pmatrix} = -\omega\begin{pmatrix} 0 \\ A/L \end{pmatrix}.$$

After simplification,

$$\begin{bmatrix} \omega^2 - \omega_0^2 & -2\alpha \\ 2\alpha\,\omega_0^2 & \omega^2 - \omega_0^2 + 4\alpha^2 \end{bmatrix} \begin{pmatrix} d_1 \\ d_2 \end{pmatrix} = \omega \begin{pmatrix} 0 \\ A/L \end{pmatrix}.$$

The solution is given by Eq. (C.1.7) and (C.1.9): $d_1 = -2\alpha\omega A/(LD)$ and $d_2 = -(\omega^2 - \omega_0^2)\,\omega A/(LD)$ where $D$ is given by Eq. (1.6.12). Subsequently, Eq. (5.3.6) provides $e_1 = -(\omega^2 - \omega_0^2)\,A/(LD)$ and $e_2 = (\omega_0^2 + 2\alpha)\,(\omega^2 - \omega_0^2)\,A/(LD)$. Altogether,

$$\mathbf{y}_{\mathrm{p}}(t) = -\frac{\omega A}{LD}\begin{pmatrix} 2\alpha \\ \omega^2 - \omega_0^2 \end{pmatrix}\cos(\omega t) + \frac{(\omega^2 - \omega_0^2)\,A}{LD}\begin{pmatrix} -1 \\ \omega_0^2 + 2\alpha \end{pmatrix}\sin(\omega t). \qquad (5.3.8)$$

This result agrees with the previous result in Eq. (1.6.13).

By including the homogeneous solution Eq. (5.2.21), the general solution is

$$\mathbf{y}(t) = \mathbf{d}\cos(wt) + \mathbf{e}\sin(\omega t) + c\,e^{-\alpha t}\left[\mathbf{a}\cos(\omega_{\mathrm{d}}t - \phi) - \mathbf{b}\sin(\omega_{\mathrm{d}}t - \phi)\right].$$

Technically, the system Eq. (5.3.4) is not autonomous so the phase diagram loses its meaning, but we notice that the homogeneous solution decays and eventually, the solution is just that given by the particular solution and it has the form of an ellipse Eq. (5.2.22).

---

**Reflection**

Provided the forcing term $\mathbf{r}(t)$ has a certain form, we can use the method of undetermined coefficients for a linear system with constant coefficients by making an appropriate guess for the particular solution and using unknown coefficient vectors.

---

### 5.3.3  *Stability*

The solutions of the previous two subsections promote the introduction of the notion of stability. In one case, a constant applied voltage drives the circuit to the eventual state where the charge on the capacitor reaches a certain fixed value. Presumably, this is exactly the consequence that the engineer has designed the circuit to achieve. In the other case, an alternating voltage is applied and the eventual result is a steady alternating charge on the capacitor. Again, we imagine that this steady state pattern is exactly what the engineer hoped for.

Success in achieving the desired outcome in both cases rests on a fundamental property of the LCR circuit, its natural behavior in the absence of forcing as described by the homogeneous solution. And the most important property of the homogeneous solution is that it decays to zero because either both eigenvalues are real and negative as in Eq. (5.3.4) or the real part of the complex eigenvalues is negative as in Eq. (5.3.3) and Eq. (5.3.8). Since the homogeneous solution accommodates the initial conditions, the fact that it decays to zero means that the eventual behavior is reached no matter what the initial conditions are, a circumstance referred to as *global stability* or *asymptotic stability* as mathematicians refer to it.[5]

More generally, stability refers to the property that the solution does not deviate very far from a desired steady state. An example different from the asymptotically stable example Eq. (5.3.3) is the solution when the homogeneous solution is an ellipse,

$$\mathbf{y}(t) = \mathbf{d} + c\left[\mathbf{a}\cos(\omega_0 t - \phi) - \mathbf{b}\sin(\omega_0 t - \phi)\right]. \tag{5.3.9}$$

Obviously, for this to occur there must be no resistance ($\alpha = 0$) in the circuit. The meaning in the result Eq. (5.3.9) is as follows: imagine we have reached the equilibrium point $\mathbf{d}$ but that subsequently a small perturbation – $c$ is small – occurs with the consequence that the new solution exhibits a small oscillation around the equilibrium point. The solution never deviates very far from the equilibrium point.

To illustrate with a more concrete example, suppose a stationary structure, a building or a bridge for example – described by $\mathbf{d}$ – receives a hit so that it is disturbed by a small amount – $c$ is small – and subsequently it vibrates by a small amount – the solution is given by Eq. (5.3.9) – then we might be satisfied with the design of the structure even if it vibrates for a long time as long as the vibration is small. We consider the situation as stable enough, and refer to it as *neutrally stable*. Neutral stability is identified mathematically by complex roots to the characteristic equation for the homogeneous solution that have zero real parts. Of course, the structure may shake violently and even collapse if it receives a large hit, a tornado for example: stability refers to the consequences of small perturbations to the steady state.

The final case is that of instability. What do we mean by instability? Intuitively, we would regard the situation where a slight tap on a structure causes it to wobble and then collapse as an example of instability. Mathematically, the stationary structure is represented by $\mathbf{d}$ but now the homogeneous solution has positive real eigenvalues, $0 < \lambda_2 < \lambda_1$. The general solution is still of the form Eq. (5.3.4) but the consequences are very different. Even if $c_1$ and $c_2$ are small – the structure receives a small hit – the growth of the exponentials in the homogeneous solution soon exaggerates the disturbance to the structure. The largest eigenvalue $\lambda_1$ selects the exponential with the largest growth and the solution becomes dominated by $c_1\,\mathbf{a}_1\exp(\lambda_1 t)$. At some point the exponential becomes so large that the assumptions underlying the formulation of the mathematical model are no longer valid; the structure collapses.

Clearly, if both eigenvalues are real and positive there is no uncertainty about the instability that arises, but what happens if only one eigenvalue is positive. In other words, suppose $\lambda_2 < 0 < \lambda_1$. The solution Eq. (5.3.9) is unchanged, and if we can set the initial conditions so that $c_1 = 0$, then the solution decays to the

---

[5]Asymptotic here means the dominant behavior as time goes by, and it is the exponential decay of the homogeneous solution that leaves the particular solution as the dominant behavior.

equilibrium point **d**. In terms of the phase diagram, we start on the line defined by $\mathbf{a}_2$ – see Fig. 5.9 – and stay on the line. But wait! That may be impossible to do; it might require specifying an initial condition with infinite precision. So it is generally accepted that it is practically impossible to achieve $c_1 = 0$ and just the smallest amount of $c_1$ will cause the solution to grow exponentially; the equilibrium point is unstable.

If the eigenvalues are complex conjugates and the real part is positive, then the phase diagram is an outward growing spiral. The solution exhibits an oscillation that continues to grow. We have all had the experience of an unbalanced washing machine when it begins to make a growing thunking sound during a spin cycle. We rush to the machine to stop it lest the instability causes damage.

---

**Reflection**

If the particular solution to a system of linear differential equations is a constant corresponding to a fixed point or has a steady oscillatory pattern, then its stability is determined by the nature of the homogeneous solution.

---

### 5.3.4 *Abstract view*

The solution to a system of linear differential equations,

$$\frac{d\mathbf{y}}{dt}(t) = \mathcal{A}\mathbf{y}(t) + \mathbf{r}(t) \tag{5.3.10}$$

is constructed in the standard way by expressing the solution in two parts,

$$\mathbf{y}(t) = \mathbf{y}_\mathrm{p}(t) + \mathbf{y}(t),$$

where the particular solution $\mathbf{y}_\mathrm{p}(t)$ takes care of the forcing term and the homogeneous solution $\mathbf{y}_\mathrm{h}(t)$ takes care of the initial conditions.

The homogeneous solution follows from the standard approach of using an exponential as a guess, leading to an eigenvalue problem Eq. (5.2.29). Let $\lambda_1$ and $\lambda_2$ be the eigenvalues with $\mathbf{a}_1$ and $\mathbf{a}_2$ the associated eigenvectors. The assumption here is that the eigenvalues are distinct. Then the homogeneous solution is given by Eq. (5.2.30).

The particular solution can be constructed by variation of parameters and the details are presented in Sec. 5.3.6. Here we proceed on the basis that $\mathbf{r}(t)$ contains components that are polynomials, sines and cosines, and exponentials because we can then guess the form of the particular solution based on Table 1.4. Some examples are:

**Polynomials.** For example, suppose $\mathbf{r}(t) = \mathbf{r}\,t^2$ where $\mathbf{r}$ is some constant vector. Then the guess for the particular solution should be $\mathbf{y}_\mathrm{p}(t) = \mathbf{a} + \mathbf{b}\,t + \mathbf{c}\,t^2$ where $\mathbf{a}$, $\mathbf{b}$ and $\mathbf{c}$ are constant vectors. Substitute into Eq. (5.3.10) and balance the different power of $t$. As a result there are three equations for the three unknown vectors,

$$\mathcal{A}\mathbf{c} = -\mathbf{r}, \quad \mathcal{A}\mathbf{b} = \mathbf{c}, \quad \mathcal{A}\mathbf{a} = \mathbf{b}. \tag{5.3.11}$$

Obviously, we solve these equation in the sequence $\mathbf{c}$, $\mathbf{b}$, $\mathbf{a}$ by using Gaussian elimination as presented in Appendix C.3.

**Exponentials.** For example, suppose $\mathbf{r}(t) = \mathbf{r}\exp(\alpha t)$ where $\mathbf{r}$ is a constant vector. Then the guess for the particular solution should be $\mathbf{y}_\mathrm{p}(t) = \mathbf{a}\exp(\alpha t)$ where $\mathbf{a}$ is a constant vector. Substitute into Eq. (5.3.10) and balance.

$$\mathcal{A}\mathbf{a} - \alpha\mathbf{a} = -\mathbf{r}. \tag{5.3.12}$$

Thus we must solve a system of algebraic equation, written for convenience as $\mathcal{B}\mathbf{a} = \mathcal{A}\mathbf{a} - \alpha\mathbf{a} = -\mathbf{r}$. The solution to $\mathcal{B}\mathbf{a} = -\mathbf{r}$ follows by Gaussian elimination.

**Sines and cosines.** For example, suppose $\mathbf{r}(t) = \mathbf{r}\cos(\alpha t) + \mathbf{s}\sin(\alpha t)$ where $\mathbf{r}$ and $\mathbf{s}$ are constant vectors. Then the guess for the particular solution should be $\mathbf{y}_\mathrm{p}(t) = \mathbf{a}\cos(\alpha t) + \mathbf{b}\sin(\alpha t)$ where $\mathbf{a}$ and $\mathbf{b}$ are constant vectors. Substitute into Eq. (5.3.10) and balance cosines and sine separately.

$$\mathcal{A}\mathbf{a} - \alpha\mathbf{b} = -\mathbf{r}, \quad \mathcal{A}\mathbf{b} + \alpha\mathbf{a} = -\mathbf{s}.$$

Multiply the second equation by $\alpha$ and replace $\alpha\mathbf{b}$ from the first equation. Then the two systems may be rewritten as

$$\mathcal{A}^2\mathbf{a} + \alpha^2\mathbf{a} = -\mathcal{A}\mathbf{r} - \alpha\mathbf{s}, \quad \alpha\mathbf{b} = \mathcal{A}\mathbf{a} + \mathbf{r}. \tag{5.3.13}$$

For convenience express the first equation as $\mathcal{B}\mathbf{a} = \mathcal{A}^2\mathbf{a} + \alpha^2\mathbf{a} = \mathbf{c} = -\mathcal{A}\mathbf{r} - \alpha\mathbf{s}$ and solve $\mathcal{B}\mathbf{a} = \mathbf{c}$ by Gaussian elimination.

These results require several observations:

- The equations in Eq. (5.3.11) will have unique solutions provided $\det(\mathcal{A}) \neq 0$ – see Appendix C.3. If $\det(\mathcal{A}) = 0$, then the system of differential equations Eq. (5.3.10) is not linearly independent and the mathematical model is suspect.
- Equation (5.3.12) may have no solution if $\alpha$ is an eigenvalue of $\mathcal{A}$ because the consequence is that $\det(\mathcal{B}) = 0$. The reason is that $\exp(\alpha t)$ is a homogeneous solution – see Exercises 12 and 13 in Sec. 1.2.5. If there is no solution, the remedy is to change the guess for the particular solution to $\mathbf{y}_\mathrm{p}(t) = (\mathbf{a} + \mathbf{b}t)\exp(\alpha t)$. The possibilities are explored in the exercises.
- Similarly, the solution to Eq. (5.3.13) will be unique if $\det(\mathcal{B}) \neq 0$. If $\det(\mathcal{B}) = 0$, then the factors for $\mathcal{B} = (\mathcal{A} + i\alpha)(\mathcal{A} - i\alpha)$ imply that $\mathcal{A}$ has complex eigenvalues $\pm i\alpha$ and the conditions are right for resonance.
- Of course, we may have combinations of polynomials, exponentials and cosines and sines in $\mathbf{r}(t)$. All we need to do is identify all the possible functions of time necessary for the guess of the particular solution and combine them with unknown constant vectors as coefficients.

Now that we have constructed the general solution, we turn to the consequences. Often scientists and engineers force systems to achieve a desirable outcome. That outcome is expressed by the particular solution and typically has one of two possible forms, a constant state corresponding to an equilibrium point in the phase plane or a steady periodic state expressed as a closed orbit in the phase plane.[6] The important question then is the outcome stable? Will the outcome actually occur?

The question of stability is settled by the nature of the homogeneous solutions. Recall that the homogeneous solution is determined by the substitution of $\mathbf{c}\,\exp(\lambda t)$ into the homogeneous version of Eq. (5.3.10). As a consequence we are led to an eigenvalue problem,

$$\mathcal{A}\mathbf{c} - \lambda\mathbf{c} = 0.$$

Then we have the following possibilities, stated for a system with $N$ equations:

(1) All real eigenvalues are negative and all complex eigenvalues have negative real parts. Consequently, the homogeneous solution decays to zero and the particular solution, the steady state, persists no matter what the initial conditions. We call this *globally or asymptotically stable* and is usually the desired situation for engineering design.

(2) All real eigenvalues are negative and there are some complex eigenvalues that have zero real part. Consequently, parts of the homogeneous solution decay, leaving behind a periodic solution. Only if the amplitude of this periodic solution is small, in other words, the initial conditions are close to steady state solution, do we consider the steady state as *neutrally stable*. The steady state solution is contaminated by a persistent small oscillation. The challenge for scientists and engineers is to start up the process described by the system Eq. (5.3.10) so that the solution eventually comes as close to the steady state solution as possible.

(3) Some real eigenvalues are positive or some complex eigenvalues have a positive real part. Consequently, there are parts of the homogeneous solution that grow in time, destroying the possibility of a long term steady state. The state is *unstable*.

It is wonderful that mathematics provides such precise definitions of stability, but alas that precision may be deceiving. Suppose that one eigenvalue is real and positive but the eigenvalue is very, very small, $\lambda = 10^{-10}$ per year. It would take about $10^{10}$ years before any appreciable change to the steady state would be observed. Would we not be tempted to say that the steady state is stable enough?

---

[6]Scientists and engineers often ramp up a system from rest to reach the desired steady state. Mathematically, the forcing is modeled by a ramp function, such as an exponential. This means that the particular solution is time dependent, but it is its long term behavior that is important and that occurs when the ramp is finished and the forcing is just a constant.

### Principle 29

*The stability of a steady state solution, whether a constant or a periodic solution, to a system of linear differential equations with a constant coefficient matrix is determined entirely by the nature of the homogeneous solution.*

### 5.3.5   Exercises

**These are some basic exercises to help develop and improve skill.**

(1) Construct the specific solution solution to the two-dimensional system of equations

$$\frac{d\mathbf{y}}{dt}(t) = \mathcal{A}\mathbf{y}(t) + \mathbf{r}(t)$$

where

(a)

$$\mathcal{A} = \begin{bmatrix} 1 & 3 \\ 1 & -1 \end{bmatrix}, \quad \mathbf{r}(t) = \begin{pmatrix} -1 - 3t \\ t \end{pmatrix} \quad \text{and} \quad \mathbf{y}(0) = \begin{pmatrix} 0 \\ 1 \end{pmatrix}.$$

**Answer:** $y_1(1) = 0.865$, $y_2(1) = 1.135$.

(b)

$$\mathcal{A} = \begin{bmatrix} 4 & -2 \\ 3 & -1 \end{bmatrix}, \quad \mathbf{r}(t) = \begin{pmatrix} \sin(t) - 4\cos(t) \\ \sin(t) - 2\cos(t) \end{pmatrix} \quad \text{and} \quad \mathbf{y}(0) = \begin{pmatrix} 2 \\ 1 \end{pmatrix}.$$

**Answer:** $y_1(0.1) = 2.22$, $y_2(0.1) = 1.32$.

(c)

$$\mathcal{A} = \begin{bmatrix} -1 & -1 \\ 1 & -1 \end{bmatrix}, \quad \mathbf{r}(t) = -\begin{pmatrix} 0 \\ e^{-t} \end{pmatrix} \quad \text{and} \quad \mathbf{y}(0) = \begin{pmatrix} 1 \\ 1 \end{pmatrix}.$$

**Answer:** $y_1(1) = 0.677$, $y_2(1) = 0.199$.

Note that Exercise 2 in Sec. 5.2.5 asks for the construction of the homogeneous solutions to these exercises and the results may be used here.

**Some problems that arise in applications.**

(2) Assume that the circuit with a constant applied voltage is initially dead – see Eq. (1.6.3); the system of equations is given by Eq. (5.3.2) and the solution is given by Eq. (5.3.4) under the assumption that $\alpha^2 > \omega_0^2$. Determine the specific solution and draw the trajectory in the phase diagram for the choice $\alpha = 1$ and $\omega = 0.5$.

**Answer:** This choice is also used in Exercise 4 of Sec. 5.2.5 and the two answers should agree.

(3) In Sec. 5.3.1, the mathematical model for the application of a constant voltage to an LCR-circuit is autonomous and leads to a specific phase diagram. Consider now the difference if the applied voltage is ramped up exponentially. The system Eq. (5.3.2) is replaced with

$$\frac{d}{dt}\begin{pmatrix} Q(t) \\ I(t) \end{pmatrix} = \begin{bmatrix} 0 & 1 \\ -\omega_0^2 & -2\alpha \end{bmatrix}\begin{pmatrix} Q(t) \\ I(t) \end{pmatrix} + \begin{pmatrix} 0 \\ V_0/L \end{pmatrix}(1 - e^{-rt}).$$

**Answer:** This problem is also considered in Exercise 2(a) of Sec. 1.6.6 with the choice $\alpha = 0.1$ and $\omega_0 = 0.5$; the two answers should agree.

**The case when the forcing term is a copy of the homogeneous solution.**

(4) Imagine the situation where we force the chained chemical reaction that is presented in (5.2.5) and Eq. (5.2.6) – see also Eq. (1.4.1) and Eq. (1.4.2). The equations are written as a system in Eq. (5.2.11).

$$\frac{d\mathbf{y}}{dt}(t) = \mathcal{A}\mathbf{y}(t) + \mathbf{r}(t).$$

Construct the particular solution for the two cases:

(a)

$$\mathbf{r}(t) = \begin{pmatrix} 0 \\ 1 \end{pmatrix} e^{-k_1 t}.$$

(b)

$$\mathbf{r}(t) = \begin{pmatrix} 1 \\ 0 \end{pmatrix} e^{-k_1 t}.$$

Then apply the initial conditions $y_1(0) = 0$ and $y_2(0) = 0$.
**Answer:** With the choice $k_1 = 1$ and $k_2 = 2$,

$$\text{(a)} \quad \mathbf{y}(1) = \begin{pmatrix} 0 \\ 0.865 \end{pmatrix} \quad \text{(b)} \quad \mathbf{y}(1) = \begin{pmatrix} 0.368 \\ 0.135 \end{pmatrix}.$$

### 5.3.6 *General forcing term*

The particular solution to Eq. (5.3.10) can be constructed by the method of variation of parameters, a natural extension of the method described in Sec. 1.2.6 – see also Principle 7. This method replaces the coefficients in Eq. (5.2.30) by functions;

$$\mathbf{y}_p(t) = c_1(t)\,\mathbf{a}_1\,e^{\lambda_1 t} + c_2(t)\,\mathbf{a}_2\,e^{\lambda_2 t}. \tag{5.3.14}$$

The choice assumes that the eigenvalues are real, but the approach also works when the eigenvalues are complex. The difference is that the complex conjugate of $\mathbf{a}_2$ must be $\mathbf{a}_1$ and the complex conjugate $c_2(t)$ must be $c_1(t)$.

Upon substitution into Eq. (5.3.10),

$$\frac{dc_1}{dt}(t)\,\mathbf{a}_1\,e^{\lambda_1 t} + \frac{dc_2}{dt}(t)\,\mathbf{a}_2\,e^{\lambda_2 t} + c_1\,\lambda_1\,\mathbf{a}_1\,e^{\lambda_1 t} + c_2\,\lambda_2\,\mathbf{a}_2\,e^{\lambda_2 t}$$

$$= c_1\,\mathcal{A}\,\mathbf{a}_1\,e^{\lambda_1 t} + c_2\,\mathcal{A}\,\mathbf{a}_2\,e^{\lambda_2 t} + \mathbf{r}(t)\,.$$

By using the facts that $\mathbf{a}_1$ and $\mathbf{a}_2$ are eigenvectors, $\mathcal{A}\,\mathbf{a}_1 = \lambda_1\,\mathbf{a}_1$ and $\mathcal{A}\,\mathbf{a}_2 = \lambda_2\,\mathbf{a}_2$, this result simplifies further.

$$\frac{dc_1}{dt}(t)\,\mathbf{a}_1\,e^{\lambda_1 t} + \frac{dc_2}{dt}(t)\,\mathbf{a}_2\,e^{\lambda_2 t} = \mathbf{r}(t)\,.$$

At this stage we appear stuck, until we realize that we should express $\mathbf{r}(t)$ in term of $\mathbf{a}_1$ and $\mathbf{a}_2$,

$$\mathbf{r}(t) = r_1(t)\,\mathbf{a}_1 + r_2(t)\,\mathbf{a}_2\,.$$

Now we balance the terms containing $\mathbf{a}_1$ and $\mathbf{a}_2$ separately with the result,

$$\frac{dc_1}{dt}(t)\,e^{\lambda_1 t} = r_1(t) \quad\text{and}\quad \frac{dc_2}{dt}(t)\,e^{\lambda_2 t} = r_2(t)\,, \tag{5.3.15}$$

and we are left with two integrals to perform to obtain $c_1(t)$ and $c_2(t)$.

### 5.3.6.1 *Example*

This is a good opportunity to show the connections between the variation of parameters suggested here with the approach of Sec. 1.6.7 by using the same example,[7]

$$\frac{d^2 y}{dx^2}(x) + \frac{dy}{dx}(x) - 6y(x) = \sin(2x)\,.$$

First, convert this second-order differential equation into a system by defining

$$y_1(x) = y(x) \quad\text{and}\quad y_2(x) = \frac{dy}{dx}(x)\,.$$

As a result,

$$\frac{d}{dx}\begin{pmatrix} y_1(x) \\ y_2(x) \end{pmatrix} = \begin{bmatrix} 0 & 1 \\ 6 & -1 \end{bmatrix}\begin{pmatrix} y_1(x) \\ y_2(x) \end{pmatrix} + \begin{pmatrix} 0 \\ \sin(2x) \end{pmatrix}\,.$$

The eigenvalues and eigenvectors of the matrix – see Appendix C.4 – are

$$\mathbf{a}_1 = \begin{pmatrix} 1 \\ -3 \end{pmatrix} \text{ for } \lambda_1 = -3, \quad \mathbf{a}_2 = \begin{pmatrix} 1 \\ 2 \end{pmatrix} \text{ for } \lambda_2 = 2\,.$$

So the guess for the particular solution is

$$\mathbf{y}_\mathrm{p}(x) = c_1(x)\begin{pmatrix} 1 \\ -3 \end{pmatrix} e^{-3x} + c_2(x)\begin{pmatrix} 1 \\ 2 \end{pmatrix} e^{2x}\,.$$

The next step is to express $\mathbf{r}(x)$ in term of $\mathbf{a}_1$ and $\mathbf{a}_2$.

$$r_1(x)\begin{pmatrix} 1 \\ -3 \end{pmatrix} + r_2(x)\begin{pmatrix} 1 \\ 2 \end{pmatrix} = \begin{pmatrix} 0 \\ \sin(2x) \end{pmatrix}\,,$$

---

[7]Do not let the change of independent variable confuse you!

which has the result $r_1(x) = -\sin(2x)/5$ and $r_2(x) - \sin(2x)/5$. Consequently, the guess Eq. (5.3.14) for the particular solution leads to Eq. (5.3.15) where

$$\frac{dc_1}{dx}(x) = -\frac{\sin(2x)}{5}\,e^{3x} \quad \text{and} \quad \frac{dc_2}{dx}(x) = \frac{\sin(2x)}{5}\,e^{-2x}\,.$$

After integration,

$$c_1(x) = -\frac{e^{3x}}{65}\left[3\sin(2x) - 2\cos(2x)\right],$$

$$c_2(x) = -\frac{e^{-2x}}{20}\left[cos(2x) + \sin(2x)\right],$$

and the particular solution is

$$\mathbf{y}_{\mathrm{p}}(x) = -\frac{1}{260}\begin{pmatrix} 5\cos(2x) + 25\sin(2x) \\ 32\cos(2x) + 17\sin(2x) \end{pmatrix},$$

a result that agrees with that in Sec. 1.6.7.

## 5.4   Nonlinear autonomous equations

The general form for two first-order differential equations in two unknown functions, Eq. (5.2.1) and Eq. (5.2.2), is given at the start of Sec. 5.2. As acknowledged there, there is little hope of constructing an analytic solution: in most cases, numerical methods present the only possible means of constructing solutions.

If the system is autonomous, Eq. (5.2.3) and Eq. (5.2.4), then we do have the opportunity to gain an understanding of the general features of the solution by drawing a phase diagram. Unlike linear equations where the general solution can be constructed and used to draw the phase diagram, nonlinear equations require us to embrace Principle 27. First locate all equilibrium points in the phase plane. Perform a linear approximation near each equilibrium point and use the solution to establish its stability. Consequently, the phase diagram in the vicinity of each equilibrium point can be drawn and what remains is the effort to link up the trajectories from one equilibrium point to another. As usual a specify example helps clarify the process.

### 5.4.1   *Predator-prey model*

The logistic equation describes the evolution of the population of a species suffering resource limitations. What happens if they are also the prey of some predator. Take the case of antelopes and lions. Let the population size of the antelopes be given by $N_1(t)$ and that of the lions by $N_2(t)$. We expect the population size of the antelopes to follow the logistic equation, but in addition we must account for the loss of antelopes due to kills by the lions. A reasonable assumption is that the number of kills is proportional to the product of the populations of antelopes and lions since the product indicates the likelihood of lions catching antelopes. Thus the evolution equation for the population size of the antelopes takes the form,

$$\frac{dN_1}{dt}(t) = \sigma N_1(t)\left(1 - \frac{N_1(t)}{K}\right) - k_1 N_1(t)\, N_2(t)\,. \tag{5.4.1}$$

The situation for the lions is different. Without antelopes, the lions will just die. Their population will grow in size only if they make kills. So

$$\frac{dN_2}{dt}(t) = -r N_2(t) + k_2 N_1(t)\, N_2(t)\,. \tag{5.4.2}$$

We now have two equations, Eq. (5.4.1) and Eq. (5.4.2), that govern the evolution of the antelopes and lions, and they form a two-dimensional system of autonomous nonlinear differential equations.

### 5.4.1.1   *Locate equilibrium points*

An equilibrium point occurs when the rate of change of all quantities, in this case the population sizes of the antelopes and lions, is zero. That means they do not change in time and the solution is a constant state. Mathematically, we seek values, $(\overline{N}_1, \overline{N}_2)$, so that the right hand sides of Eq. (5.4.1) and Eq. (5.4.2) are zero.

$$\sigma \overline{N}_1 \left(1 - \frac{\overline{N}_1}{K}\right) - k_1 \overline{N}_1\, \overline{N}_2 = 0\,, \tag{5.4.3}$$

$$-r\overline{N}_2 + k_2 \overline{N}_1\, \overline{N}_2 = 0\,. \tag{5.4.4}$$

Start with the second equation Eq. (5.4.4). It has solutions $\overline{N}_2 = 0$ or $\overline{N}_1 = r/k_2$.

If $\overline{N}_2 = 0$, then the first equation Eq. (5.4.3) has solutions $\overline{N}_1 = 0$ or $\overline{N}_1 = K$. These are the solutions we had before in the absence $(\overline{N}_2 = 0)$ of any lions – see Eq. (5.1.3).

For the other case, $\overline{N}_1 = r/k_2$, Eq. (5.4.3) becomes

$$\frac{\sigma}{k_1}\left(1 - \frac{r}{Kk_2}\right) - \frac{rk_1}{k_2}\overline{N}_2 = 0\,, \quad \text{or} \quad \overline{N}_2 = \frac{\sigma}{k_1}\left(1 - \frac{r}{Kk_2}\right).$$

As a result, we have found three equilibrium points:

$$(0,0)\,, \quad (K,0)\,, \quad \left(\frac{r}{k_2}, \frac{\sigma}{k_1}\left(1 - \frac{r}{Kk_2}\right)\right). \tag{5.4.5}$$

### 5.4.1.2   *Stability*

So far so good, but we want to know whether these equilibrium points are stable or not, and what trajectories near them do. The only way we can find out is by making a linear approximation to the equations near each equilibrium point and studying what the solutions do. We anticipate that the procedure at each equilibrium point will be the same, but differ only in details.

We consider the process in general by setting

$$N_1(t) = \overline{N}_1(t) + \hat{N}_1(t) \quad N_2(t) = \overline{N}_2(t) + \hat{N}_2(t),$$

and substituting into Eq. (5.4.1) and Eq. (5.4.2). We consider $\hat{N}_1(t)$ and $\hat{N}_2(t)$ to be small and neglect any powers or products of them. They are often called the perturbed quantities. Thus Eq. (5.4.1) becomes

$$\frac{d\hat{N}_1}{dt}(t) = \sigma \overline{N}_1 \left(1 - \frac{\overline{N}_1}{K}\right) - k_1 \overline{N}_1 \overline{N}_2$$

$$+ \left(\sigma - \frac{2\sigma}{K}\overline{N}_1 - k_1\overline{N}_2\right)\hat{N}_1(t) - k_1\overline{N}_1\hat{N}_2(t)$$

$$= \left(\sigma - \frac{2\sigma}{K}\overline{N}_1 - k_1\overline{N}_2\right)\hat{N}_1(t) - k_1\overline{N}_1\hat{N}_2(t), \qquad (5.4.6)$$

where we use Eq. (5.4.3) to simplify the result.

We follow the same approach for Eq. (5.4.2). The result is

$$\frac{d\hat{N}_2}{dt}(t) = -r\overline{N}_2 + k_2\overline{N}_1\overline{N}_2 + k_2\overline{N}_2\hat{N}_1(t) + \left(k_2\overline{N}_1 - r\right)\hat{N}_2(t)$$

$$= k_2\overline{N}_2\hat{N}_1(t) + \left(k_2\overline{N}_1 - r\right)\hat{N}_2(t), \qquad (5.4.7)$$

where we use Eq. (5.4.4) to simplify the result.

Let us take stock of what we have achieved. The equations Eq. (5.4.6) and Eq. (5.4.7) are linear with constant coefficients since $\overline{N}_1$ and $\overline{N}_2$ are known constants. These equations are homogeneous since there are no forcing terms: they are referred to as *perturbation equations*. The process of obtaining this linear approximation is usually referred to as *linearization*.[8] The analysis of the stability of the equilibrium points is made complicated by the presence of so many parameters, so let us simplify the construction of the solutions to Eq. (5.4.6) and Eq. (5.4.7) by setting the parameters to some known values. The choice $\sigma = k_2 = 2$ and $K = k_1 = r = 1$ means that the equilibrium points Eq. (5.4.5) are at

$$(0,0), \quad (1,0), \quad (1/2,1).$$

Now use the perturbation equations Eq. (5.4.6) and Eq. (5.4.7) to study the stability of these equilibrium points.

(1) The first case $(0,0)$ leads to the perturbation equations,

$$\frac{d\hat{N}_1}{dt}(t) = 2\,\hat{N}_1(t), \qquad (5.4.8)$$

$$\frac{d\hat{N}_2}{dt}(t) = -\hat{N}_2(t). \qquad (5.4.9)$$

These equations are uncoupled and the solutions are

$$\hat{N}_1 = \varepsilon_1\,e^{2t}, \quad \hat{N}_2 = \varepsilon_2\,e^{-t}.$$

---

[8]The term linearization is used whenever nonlinear equations are approximated by linear ones.

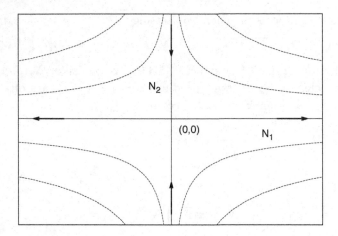

Fig. 5.11   Trajectories near the equilibrium point $(0, 0)$.

But it is also useful to view the equations Eq. (5.4.8) and Eq. (5.4.9) as a system,

$$\frac{d}{dt}\begin{pmatrix} \hat{N}_1(t) \\ \hat{N}_2(t) \end{pmatrix} = \begin{bmatrix} 2 & 0 \\ 0 & -1 \end{bmatrix}\begin{pmatrix} \hat{N}_1(t) \\ \hat{N}_2(t) \end{pmatrix}.$$

The eigenvalues are 2, $-1$ and the associate eigenvectors are easily determined by Eq. (C.4.5). The general solution is

$$\begin{pmatrix} \hat{N}_1(t) \\ \hat{N}_2(t) \end{pmatrix} = \varepsilon_1 \begin{pmatrix} 1 \\ 0 \end{pmatrix} e^{2t} + \varepsilon_2 \begin{pmatrix} 0 \\ 1 \end{pmatrix} e^{-t}. \tag{5.4.10}$$

Since there is one positive and one negative eigenvalue, the phase diagram near the equilibrium point $(0, 0)$ looks like Fig. 5.11. Trajectories are shown outside the area of physical relevance – $N_1 \geq 0$ and $N_2 \geq 0$ – to be sure we gain a complete understanding of the phase diagram near the equilibrium point.

The consequences of the solution Eq. (5.4.10) are easily understood. If the population size of the lions, although small under the assumptions of linearization, is larger than that of the antelopes, the initial trend is for the lions to die away because there are too few antelopes to sustain their population size. Once the size of the lion population has decreased sufficiently, the antelopes are less likely to be killed and in the presence of abundant food, their population grows in size rapidly.

(2) The second case $(1, 0)$ leads to the perturbation equations written as a system,

$$\frac{d}{dt}\begin{pmatrix} \hat{N}_1(t) \\ \hat{N}_2(t) \end{pmatrix} = \begin{bmatrix} -2 & -1 \\ 0 & 1 \end{bmatrix}\begin{pmatrix} \hat{N}_1(t) \\ \hat{N}_2(t) \end{pmatrix}.$$

The eigenvalues are $-2$, 1 and the general solution is

$$\begin{pmatrix} \hat{N}_1(t) \\ \hat{N}_2(t) \end{pmatrix} = \varepsilon_1 \begin{pmatrix} 1 \\ 0 \end{pmatrix} e^{-2t} + \varepsilon_2 \begin{pmatrix} 1 \\ -3 \end{pmatrix} e^{t}. \tag{5.4.11}$$

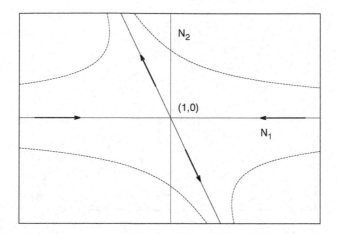

Fig. 5.12   Trajectories near the equilibrium point (1, 0).

In this case too, the eigenvalues have opposite signs and the trajectories near (1, 0) are shown in Fig. 5.12.

The consequences of the solution Eq. (5.4.11) are easily understood. In the absence of lions ($N_2 = 0 \to \varepsilon_2 = 0$), the number of antelopes reaches a balance with the availability of food and the population size reaches $K = 1$. The trajectories in the phase diagram move towards the equilibrium point $K = 1$. But when lions are present, their population size initially small will grow rapidly from the availability of many antelopes. Of course, the size of the antelope population will begin to decrease. Hence the trajectories track parallel to the line that represents the unstable part of the solution.

(3) The third case $(1/2, 1)$ has the perturbation equations,

$$\frac{d}{dt}\begin{pmatrix} \hat{N}_1(t) \\ \hat{N}_2(t) \end{pmatrix} = \begin{bmatrix} -1 & -1/2 \\ 2 & 0 \end{bmatrix}\begin{pmatrix} \hat{N}_1(t) \\ \hat{N}_2(t) \end{pmatrix}.$$

This time the eigenvalues are complex $-1/2 \pm i\,\sqrt{3}\,/2$ with complex eigenvectors,

$$\mathbf{a} \pm i\,\mathbf{b} = \begin{pmatrix} 1 \\ -1 \end{pmatrix} \pm i \begin{pmatrix} 0 \\ -\sqrt{3} \end{pmatrix}.$$

The general solution is – see Eq. (5.2.20),

$$\begin{pmatrix} \hat{N}_1(t) \\ \hat{N}_2(t) \end{pmatrix} = \varepsilon\, e^{-t/2}\left[\mathbf{a}\cos\left(\frac{\sqrt{3}\,t}{2} - \phi\right) - \mathbf{b}\sin\left(\frac{\sqrt{3}\,t}{2} - \phi\right)\right]. \qquad (5.4.12)$$

The result indicates the presence of a spiral whose aspect ratio and orientation can be determined through Eq. (5.2.36). The major and minor axes of the underlying elliptical shape correspond to $\theta = 2.22$ and $\theta = 0.65$, respectively. The aspect ratio, the ratio of the lengths of the major and minor axes, is 2.45 and the major axis of the ellipse is oriented with the angle 107 degrees. Some of the spirals are shown in Fig. 5.13.

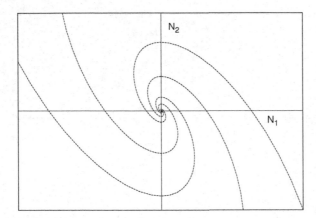

Fig. 5.13    Trajectories near the equilibrium point $(1/2,\ 1)$.

The consequences of the solution Eq. (5.4.12) are easily understood. The populations of the antelopes and lions are almost in balance but they are almost out of phase. When the population size of the antelopes is higher than their equilibrium value, that of the lions is lower. Since abundant food is available to the lions they prosper and increase in number rapidly while the antelopes are killed off more rapidly, until the size of the lion population becomes larger than their equilibrium value and the size of the antelope population drops lower. The situation is reversed, and more lions start to starve and die off leading to an increase in the number of antelopes, and so on. Eventually, the pattern settles to a steady state of just the right balance in the numbers of lions and antelopes.

### 5.4.1.3 *Compose the phase diagram*

The starting point in composing the phase diagram is to mark in all the equilibrium points and draw the local trajectories as revealed by their stability. These local trajectories are restricted to nearby regions of the equilibrium point as a consequence of invoking a linear approximation. The challenge now is to extend these trajectories to fill in the phase diagram and in particular to consider how these trajectories connect the different equilibrium points. There are several guidelines that govern the process of drawing trajectories. They are stated on the assumption that the right hand sides of Eq. (5.2.3) and Eq. (5.2.4) contain well-defined functions, as is the case for Eq. (5.4.1) and Eq. (5.4.2).

- Obviously, each starting point marks out a unique trajectory. Different trajectories cannot cross each other because there would be a common point, the intersection point, that would have two different trajectories starting from the same initial point and that is not possible.
- When will the trajectories stop? One possibility is never; they go out to infinity. Another is that they head towards and terminate at an equilibrium point, just as

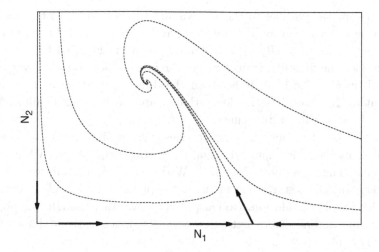

Fig. 5.14    Trajectories in the phase plane.

we saw in the one-dimensional case. Of course it takes an infinite amount of time to reach the equilibrium point because the trajectories approach the equilibrium point exponentially. Finally, it is possible for the trajectory to be a closed orbit. That means a periodic solution as a phase point moves around the orbit endlessly.

- Trajectories can start in the vicinity of an unstable equilibrium point and move away from it. Trajectories can also start very far away, effectively at infinity, before moving into the central regions of the phase space.

Let us see how these guidelines help us draw the phase diagram for Eq. (5.4.1) and Eq. (5.4.2). The result is shown in Fig. 5.14 for the choice we made before, $\sigma = k_2 = 2$ and $K = k_1 = r = 1$. We proceed to build up the phase diagram as follows:

(1) We place the equilibrium points on the phase space – Eq. (5.4.5).
(2) We draw the arrows that mark the stable and unstable directions at the equilibrium points $(0, 0)$ and $(1, 0)$ – see Figs. 5.11 and 5.12.
(3) Consider the trajectories near $(0, 0)$. Observe that if the initial condition has $N_1 = 0$ (no antelope), then Eq. (5.4.1) predicts that $N_1(t) = 0$ always. The lions simply die away as predicted by Eq. (5.4.2). Thus there is a trajectory that goes straight down the positive $N_2$-axis. Now consider the initial condition $N_2 = 0$ (no lions): Eq. (5.4.1) reduces to the logistic equation and the antelope population grows if the initial value is less than $K$. There is a trajectory starting at the origin that moves along the positive $N_1$-axis.
(4) The trajectory that starts at the origin and moves along the positive $N_1$-axis will eventually approach the equilibrium point $(K, 0)$. If the initial condition has $N_1 > K$ and $N_2 = 0$ (no lions), then the antelope population decreases in size until it stabilizes at $(K, 0)$. But if $N_2$ has a small positive value, then the

trajectory moves parallel to the $N_1$-axis until it gets close to the equilibrium point where it veers away and becomes parallel to the unstable trajectory, the trajectory defined initially by $\varepsilon_1 = 0$ and $\varepsilon_2$ small in Eq. (5.4.11).

(5) Where does the unstable trajectory go, the one defined initially by $\varepsilon_1 = 0$ and $\varepsilon_2$ small in Eq. (5.4.11)? It heads in the direction of the stable equilibrium point at $(1/2,1)$, and it seems likely that it approaches that equilibrium point as one of the spirals in its vicinity.

(6) The last step is to draw other trajectories, bearing in mind that they cannot cross one another and that there is only one stable equilibrium where they can terminate. Thus, we obtain Fig. 5.14. With Fig. 5.14 in hand, we can imagine the likely trajectory starting from any initial point, and that allows us to sketch the behavior of the solution. No matter where we start though, the population sizes of lions and antelopes settle to a perfect balance at the stable equilibrium point $(1/2, 1)$.

One final comment. The trajectories in Fig. 5.14 are obtained by solving Eq. (5.4.1) and Eq. (5.4.2) with numerical methods. The position of the equilibrium points and their stability guides the choice of initial conditions to make sure trajectories are determined that fill in the phase diagram in a useful way.

### 5.4.2 *Abstract view*

While it is easy to state the general form for an autonomous system of nonlinear differential equations,

$$\frac{d\mathbf{y}}{dt}(t) = \mathbf{f}\big(\mathbf{y}(t)\big), \tag{5.4.13}$$

the details of phase diagram can be very complicated, especially if the dimension $N$ of the system is large. One reason why the examples studied are all two-dimensional is that drawing the phase plane is tractable. Nevertheless, the strategy in drawing a general phase diagram follows the steps we used in drawing the phase plane shown in Fig. 5.14. So it is useful to restate the strategy, pointing out where difficulties might arise.

#### 5.4.2.1 *Locate equilibrium points*

Equilibrium points are detected by the requirement that the solution does not change in time. Thus we must find all values $\overline{\mathbf{y}}$ where

$$\mathbf{f}(\overline{\mathbf{y}}) = 0. \tag{5.4.14}$$

Normally, the values for $\overline{\mathbf{y}}$ are distinct; they mark a point in the phase diagram. But there is no guarantee this is so. As an example, consider

$$\frac{d}{dt}\begin{pmatrix} y_1(t) \\ y_2(t) \end{pmatrix} = \begin{pmatrix} y_1(t)\big(1 - y_2(t)\big) \\ y_1(t) \end{pmatrix}$$

and clearly $y_1 = 0$ defines an equilibrium line! Fortunately, cases such as this one rarely arise in science and engineering.

Assume, then, that the equilibrium points are distinct.

### 5.4.2.2 *Stability*

The trajectories in the vicinity of each equilibrium point is determined by linearization. Set $\mathbf{y}(t) = \overline{\mathbf{y}} + \hat{\mathbf{y}}(t)$ and substitute into Eq. (5.4.13),

$$\frac{d\hat{\mathbf{y}}}{dt}(t) = \mathbf{f}\left(\overline{\mathbf{y}} + \hat{\mathbf{y}}(t)\right).$$

If we perform a multi-dimensional Taylor series expansion about the equilibrium point, we obtain an approximation for each equation,

$$f_i\left(\overline{y}_1 + \hat{y}_1(t), \overline{y}_2 + \hat{y}_2(t), \dots\right) = f_i\left(\overline{y}_1, \overline{y}_2, \dots\right)$$
$$+ \frac{\partial f_1}{\partial y_1}\left(\overline{y}_1, \overline{y}_2, \dots\right)\hat{y}_1(t) + \frac{\partial f_1}{\partial y_1}\left(\overline{y}_1, \overline{y}_2, \dots\right)\hat{y}_2(t) + \cdots.$$

Note first that Eq. (5.4.14) requires $f_i\left(\overline{y}_1, \overline{y}_2, \dots\right) = 0$. For linearization, we ignore all products of hatted quantities, and the system takes the approximate form

$$\frac{d}{dt}\begin{pmatrix} \hat{y}_1(t) \\ \hat{y}_2(t) \\ \dots \\ \hat{y}_N(t) \end{pmatrix} = \begin{bmatrix} f_{11} & f_{12} & \cdots & f_{1N} \\ f_{21} & f_{22} & \cdots & f_{2N} \\ & & \dots\dots & \\ f_{N1} & f_{N2} & \cdots & f_{NN} \end{bmatrix}\begin{pmatrix} \hat{y}_1 \\ \hat{y}_2 \\ \dots \\ \hat{y}_N \end{pmatrix}, \qquad (5.4.15)$$

where $f_{ij} = \dfrac{\partial f_i}{\partial y_j}\left(\overline{\mathbf{y}}\right)$ are the elements of a matrix $\mathcal{J}(\overline{\mathbf{y}})$ called the *Jacobian matrix*. As a result of linearization, we obtain a system of linear, homogeneous equations (5.4.15) with a constant coefficient matrix that is evaluated at each equilibrium point in turn.

The eigenvalues of $\mathcal{J}(\overline{\mathbf{y}})$ determine the stability of the equilibrium point and the nature of the local trajectories. Let $\alpha_i$ be the positive real roots with corresponding eigenvector $\mathbf{a}_i$ $(i = 1, \dots, N_1)$; $-\beta_i$ the negative real eigenvalues with corresponding eigenvector $\mathbf{b}_i$ $(i = 1, \dots, N_2)$; and $\gamma_i \pm i\,\delta_i$ complex eigenvalues with corresponding eigenvectors $\mathbf{c}_i \pm i\,\mathbf{d}_i$ $(i = 1, \dots, N_3)$. Assume the eigenvalues are all distinct, then the number of eigenvalues will be $N_1 + N_2 + 2N_3 = N$ and the general solution is

$$\hat{\mathbf{y}}(t) = \sum_i^{N_1} \varepsilon_{1,i}\,\mathbf{a}_i\,e^{\alpha_i t} + \sum_i^{N_2} \varepsilon_{2,i}\,\mathbf{b}_i\,e^{-\beta t}$$
$$+ \sum_i^{N_3} \varepsilon_{3,i}\,e^{\gamma_i t}\left[\mathbf{c}_i\cos(\delta_i t - \phi_i) - \mathbf{d}_i\sin(\delta_i t - \phi_i)\right]. \qquad (5.4.16)$$

Each sum is present in the solution only if there are eigenvalues of that type. If there are eigenvalues of multiplicity more than one, then the general solution must be modified as done in Eq. (5.2.32), for example.

The consequences of the nature of the general solution Eq. (5.4.16) may be summarized as follows:

(1) Suppose that there are no positive real eigenvalues ($N_1 = 0$) and that all $\gamma_i < 0$, then every contribution of the general solution decays exponentially: The equilibrium point is stable.

(2) There is at least one positive real eigenvalue or at least one complex eigenvalue with $\gamma_i > 0$, then there are contributions to the general solution that grow in size: The equilibrium point is unstable. Do not forget that the solutions are only valid in the vicinity of the equilibrium point and as the unstable components grow in size we must continue the calculation of the solution by numerically methods.

(3) Unfortunately, the situation is borderline when the equilibrium point appears stable ($N_1 = 0$) except that at least one of the real eigenvalues or at least one of the real parts of the complex eigenvalues are zero. The most common case is where the real part of a complex eigenvalue is zero. Then the neglect of higher-order terms during linearization is suspect.

A simple example illustrates the difficulties in determining the correct behavior of the trajectories in the phase diagram under these circumstances. Consider

$$\frac{dy_1}{dt}(t) = y_2(t) + a\left[y_1^3(t) + y_1(t)\,y_2(t)^2\right],$$

$$\frac{dy_2}{dt}(t) = -y_1(t) + a\left[y_2^3(t) + y_1^2(t)\,y_2(t)\right],$$

where $a$ is a parameter. The linearized version is

$$\frac{dy_1}{dt}(t) = y_2(t),$$

$$\frac{dy_2}{dt}(t) = -y_1(t),$$

which has eigenvalues $\pm i$ and general solution,

$$y_1(t) = \varepsilon \cos(t - \phi),$$

$$y_2(t) = -\varepsilon \sin(t - \phi).$$

The trajectories are just circles $r^2 = y_1^2 + y_2^2 = \varepsilon^2$ in the phase diagram. If $a = 0$ then this is a legitimate solution and the phase diagram is correct, but when $a \neq 0$ it is not. To determine what the trajectory is when $a \neq 0$, note that

$$\frac{dr^2}{dt}(t) = 2\,y_1(t)\,\frac{dy_1}{dt}(t) + 2\,y_2(t)\,\frac{dy_2}{dt}(t)$$

$$= 2\,a\left[y_1^4(t) + 2y_1^2(t)\,y_2^2(t) + y_2^4(t)\right]$$

$$= 2\,a\,r^4.$$

Clearly, if $a > 0$, then $r(t)$ increases and the trajectories must be growing spirals. On the other hand, if $a < 0$ then $r(t)$ decreases and the trajectories must be inward collapsing spirals.

Of course, this example is contrived to make the point. In general, we can try to include the next terms in the expansion beyond just the linear terms and hope that we can determine the nature of the trajectories near the equilibrium point. All kinds of exciting situations can arise, one of them is that the circular orbits found under linearization remain closed: then we have nonlinear periodic orbits. The solutions to the equations for a pendulum exhibit this behavior.

### 5.4.2.3  Compose the phase diagram

Once the equilibrium points and the trajectories in their vicinity are known, the challenge is to complete the phase diagram with just the right amount of detail. Normally, this must be done through numerical methods, but there are some guidelines that will help.

(1) Follow the trajectories that start in the unstable directions at the equilibrium points. Determine where they go. Sometimes they approach stable equilibrium points or they may head out to infinity. They may also approach closely to a closed orbit.
(2) Follow the trajectories that start in the stable directions, except proceed backwards in time. In the words, find out from where the trajectories started.
(3) If there remain parts of the phase diagram that remain empty, pick some initial starting places and compute the subsequent trajectories. It can be great fun filling in the phase diagram like a jigsaw puzzle. At the end of the day, we want to understand the general nature of the solutions and how better than to assess how the trajectories must go in the phase diagram.

### 5.4.3  *Exercises*

(1) Newton's laws often provide systems of equations. For example, consider a body moving vertically under the influence of gravity. Let $y(t)$ mark its location above a reference point, the ground say. Let $v(t)$ be its velocity recorded as positive in the upward direction consistently with the positive upward location $y(t)$. Then Newton's law of motion tells us

$$\frac{dy}{dt} = v,$$
$$\frac{dv}{dt} = -g - \frac{k}{m} v|v|,$$

where $k$ is a drag resistance coefficient and $m$ is the mass of the body. Draw the phase plane for $y > 0$. What is the long term motion of the body? You may want to compare the phase plane with the phase diagram in Exercise 7 of Sec. 5.1.3.

(2) A simple model for military action against terrorist forces is

$$\frac{dg}{dt} = r - k_1 gm,$$

$$\frac{dm}{dt} = (T - m) - k_2 g,$$

where $g$ and $m$ record the number of terrorists and the number of military personnel, respectively. The unit of time is months. The parameter $r$ gives the number of terrorist recruits per month ($r = 100$ say); the parameter $T$ is the total force maintained on a monthly basis ($T = 10,000$ say); $k_1$ is a coefficient determining the number of deaths of terrorists ($k_1 = 10^{-5}$ say, corresponding to about 100 terrorists killed per month); and $k_2$ is a coefficient that determines the number of successful terrorist attacks ($k_2 = 10^{-1}$ say corresponding to about 100 deaths among the military per month). Determine the likely stable equilibrium state. Estimate the time it will take for the equilibrium point to be nearly reached.

(3) Consider the situation where two species are in competition. Imagine they are in a resource limited environment and their population sizes would satisfy the logistic equation Eq. (5.1.2) in the absence of competition. A simple model for the competition is to assume that both populations suffer a reduction in the rate of change in their population sizes proportionally to the product of the two population sizes. Let $N_1(t)$ and $N_2(t)$ be the two population sizes. The system of equations is

$$\frac{dN_1}{dt}(t) = N_1(t)\left(a_1 - b_1 N_1(t) - c_1 N_2(t)\right),$$

$$\frac{dN_2}{dt}(t) = N_2(t)\left(a_2 - b_2 N_2(t) - c_2 N_1(t)\right).$$

The parameters $b_1$ and $b_2$ represent resource limitation; the parameters $c_1$ and $c_2$ represent competition. Determine the phase diagram for the two cases and discuss the implications:

(a) The effects of competition are stronger than resource limitations, $c_1 c_2 > b_1 b_2$. In particular, make the choice, $a_1 = 7$, $b_1 = 3$, $c_1 = 2$, $a_2 = 4$, $b_2 = 1$ and $c_2 = 2$.

(b) The effects of competition are weaker than resource limitations, $c_1 c_2 < b_1 b_2$. In particular, make the choice $a_1 = 7$, $b_1 = 2$, $c_1 = 3$, $a_2 = 4$, $b_2 = 2$ and $c_2 = 1$.

# Appendix:
# The Exponential Function

## A.1 A review of its properties

The most important functions in science and engineering are the polynomials, the trigonometric functions and exponentials. Nowhere is the exponential function more important than in the solution of differential equations. There are two standard ways to write the exponential function, $\exp(x)$, usually when appearing in line text, or $e^x$, when its properties are to be used. The following list gives the properties of the exponential function:

- The inverse:

$$y = e^x, \quad x = \ln(y), \quad e^{\ln \alpha} = \alpha. \tag{A.1.1}$$

- The reciprocal:

$$\frac{1}{e^x} = e^{-x}. \tag{A.1.2}$$

- Multiplication:

$$e^\alpha e^\beta = e^{\alpha+\beta}. \tag{A.1.3}$$

- Powers:

$$\left(e^\alpha\right)^\beta = e^{\alpha\beta}. \tag{A.1.4}$$

- Derivative:

$$\frac{d}{dt} e^{\alpha t} = \alpha e^{\alpha t}. \tag{A.1.5}$$

- Special values:

$$e^0 = 1, \quad e^\infty = \infty, \quad e^{-\infty} = 0. \tag{A.1.6}$$

The exponential $\exp(x)$ is shown in Fig. A.1 and its reciprocal $\exp(-x)$ in Fig. A.2. If the independent variable is $t$ instead of $x$, then $\exp(t)$ shows growth (increasing), while $\exp(-t)$ shows decay (decreasing).

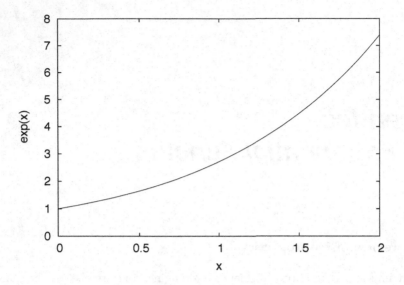

Fig. A.1    The exponential function, $\exp(x) = e^x$.

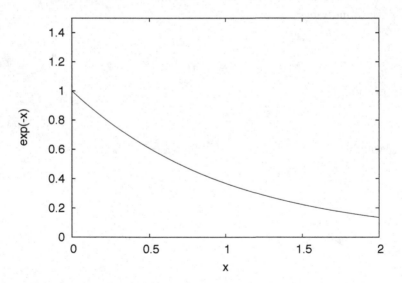

Fig. A.2    The reciprocal of the exponential function, $\exp(-x)e^{-x}$.

# Appendix:
# The Taylor Series

B

## B.1  A derivation of the Taylor series

There are a few basic functions that are our "very close friends," polynomials, sines and cosines and exponentials, because we know everything about them, including their properties, such as derivatives, integrals, etc. Not surprising then, we try to use them as much as possible. We even use them to approximate other more complex functions. In this book you will see two such attempts, the Taylor series and the Fourier series. In this appendix, we will revisit the Taylor series and review its most important properties.

Suppose we have a function $f(x)$ and we would like to approximate it with a polynomial, for example, a quadratic $a + bx + cx^2$ where the constants $a$, $b$, $c$ will be chosen to make a match with $f(x)$ as best as possible. Obviously the match will not be exact but only approximate. What principle should we invoke to attempt this match? There are several different properties we could use, but the Taylor series is created by matching information at a single location $x = 0$.

Starting with

$$f(x) \approx a + bx + cx^2, \tag{B.1.1}$$

the function and the quadratic must have the same value at $x = 0$. Thus

$$f(0) = a \tag{B.1.2}$$

and one of the constants (often called a coefficient) has been determined. Next, we differentiate both sides to obtain

$$\frac{\mathrm{d}f}{\mathrm{d}x}(x) = b + 2cx.$$

Now we demand that these derivatives match at $x = 0$. Thus,

$$\frac{\mathrm{d}f}{\mathrm{d}x}(0) = b, \tag{B.1.3}$$

and another coefficients has been determined. We continue the process once more because we still have one undetermined coefficient. We find

$$\frac{\mathrm{d}^2 f}{\mathrm{d}x^2}(0) = 2\,c\,. \tag{B.1.4}$$

By using Eq. (B.1.2), Eq. (B.1.3) and Eq. (B.1.4), we have constructed the approximation,

$$f(x) \approx f(0) + \frac{\mathrm{d}f}{\mathrm{d}x}(0)\,x + \frac{1}{2}\frac{\mathrm{d}^2 f}{\mathrm{d}x^2}(0)\,x^2\,.$$

An obvious thought strikes us. Just how good is this quadratic approximation? After all, we only match derivatives up to second order. Will the approximation improve if higher-order terms in the polynomial are included and matched with higher-order derivatives? We can address these concerns by considering a general polynomial with $N+1$ terms and then estimating the difference between the function and the polynomial.

The first step is to express the polynomial with $N + 1$ terms. Our previous expression with letters from the alphabet is not useful here because we might run out of letters. Rather, we introduce subscripts and write out a sum of terms.

$$f(x) \approx \sum_{n=0}^{N} a_n\, x^n = a_0 + a_1\, x \cdots + a_n\, x^n + \cdots\,. \tag{B.1.5}$$

Just as in Eq. (B.1.2), matching the function and the polynomial at $x = 0$ leads to

$$f(0) = a_0\,.$$

Now take the derivative,

$$\frac{\mathrm{d}f}{\mathrm{d}x}(x) \approx \sum_{n=1}^{N} n\, a_n\, x^{n-1} = a_1 + 2\, a_2\, x + \cdots + n\, a_n\, x^{n-1} + \cdots\,.$$

Clearly, we have used the fact that a derivative of a sum is the sum of the derivatives. Note that we express the result in two ways, one using the summation symbol and the other just writing out a typical term. Sometimes it helps to see what a typical term will be like. Just as before in Eq. (B.1.3), set $x = 0$ to obtain

$$\frac{\mathrm{d}f}{\mathrm{d}x}(0) = a_1\,.$$

Obviously we should proceed recursively. Let us try to figure out what the $m$th derivative will be. The $m$th derivative of any term $x^n$ where $n < m$ will obviously be zero. So the first term in the sum that is not zero sum will start with $n = m$. The $m$ derivatives of $x^m$ will produce the constant $m! = m\,(m-1)\,(m-2) \times \cdots \times 2 \times 1$. The next term $x^{m+1}$ will become $(m+1)\,(m-1)\ldots 2\,x$. Thus we expect

$$\frac{\mathrm{d}^m f}{\mathrm{d}x^m}(x) = m!\, a_m + (m-1)\, m\,(m-1)\ldots 2\, a_{m+1}\, x + \cdots\,.$$

Fortunately we do not have to worry about the precise nature of the higher-order polynomial terms when $n > m$ since they become 0 when evaluated at $x = 0$ and we just have the result

$$\frac{\mathrm{d}^m f}{\mathrm{d}x^m}(0) = m!\, a_m \,. \tag{B.1.6}$$

All of the coefficients $a_n$ can be evaluated this way and the approximation (B.1.5) becomes

$$f(x) \approx \sum_{n=0}^{N} \frac{1}{n!} \frac{\mathrm{d}^n f}{\mathrm{d}x^n}(0)\, x^n \,.$$

The procedure to determine the coefficients $a_n$ by matching the derivatives at $x = 0$ does not care how many terms $N$ there are in the sum. So why not go all the way and take $N \to \infty$. The hope is that the infinite sum which has required an infinite number of matching derivatives produces $f(x)$ exactly! This then is the Taylor series!

$$f(x) = \sum_{n=0}^{\infty} \frac{1}{n!} \frac{\mathrm{d}^n f}{\mathrm{d}x^n}(0)\, x^n \,. \tag{B.1.7}$$

## B.2   Accuracy of the truncated Taylor series

A surprising number of results in science and engineering stem from using just a few terms in the Taylor series, even just one or two. The reason will become clear in this section. Let us build up some understanding of some of the properties of a general sum,

$$S_N = \sum_{n=0}^{N} \alpha_n \,. \tag{B.2.1}$$

For the Taylor series $\alpha_n = a_n\, x^n$ depends on $x$ but that is not of concern at the moment; just imagine we have selected a specific $x$ and think of it as a constant. The important part is the pattern in $n$.

An important example of Eq. (B.2.1) is the geometric series,

$$S_N = \sum_{n=0}^{N} r^n = 1 + r + r^2 + \cdots + r^n + \cdots + r^N \tag{B.2.2}$$

where $r$ is some constant. Note that there is a very special pattern in the terms, in particular, $\alpha_{n+1} = r\, \alpha_n$. We exploit this property by multiplying $S_N$ with $r$,

$$r\, S_N = \sum_{n=0}^{N} r^{n+1} = r + r^2 + \cdots + r^{n+1} + \cdots + r^{N+1} \,,$$

and noticing most of the terms also appear in $S_N$. Indeed, if we subtract we find just two terms that do not cancel.

$$r\, S_N - S_N = r^{N+1} - 1 \,,$$

which leads to an explicit result for the sum,

$$S_N = \frac{r^{N+1} - 1}{r - 1}. \tag{B.2.3}$$

The result Eq. (B.2.3) contains several important features. First, we note that $r = 1$ causes a problem.[1] Second, we consider what happens when we take the limit $N \to \infty$. The result depends on whether $|r| > 1$ or $|r| < 1$. If $|r| > 1$, $r^{N+1} \to \infty$ and the sum does not converge. On the other hand, if $|r| < 1$ then $r^{N+1} \to 0$, and the sum converges to

$$S_\infty = \frac{1}{1 - r}.$$

In other words, if $|r| < 1$,

$$\frac{1}{1 - r} = \sum_{n=0}^{\infty} r^n. \tag{B.2.4}$$

Because we have an explicit result for the sum, we can also regard the accuracy in truncating the sum at $N$ terms. The difference

$$E_N = S_\infty - S_N = \frac{r^{N+1}}{1 - r}$$

measures the error in using $S_N$ in place of $S_\infty$. It has the bound,

$$|E_N| = \frac{|r|^{N+1}}{1 - r}.$$

If $|r|$ is small, then only a few terms are needed to have an accurate approximation. If $|r|$ is close to 1, then many terms will be needed.

With the insights gained from considering the geometric series Eq. (B.2.2) we can tackle obtaining an error estimate for the general sum Eq. (B.2.1). The error is given by

$$E_N = S_\infty - S_N = \sum_{n=N+1}^{\infty} \alpha_n.$$

To proceed, we imagine that we can find a relationship between the terms so that

$$|\alpha_{n+1}| \leq r\, |\alpha_n| \quad \text{when } n > N. \tag{B.2.5}$$

Of course you will notice the similarity with the pattern in the geometric series. The constant $r$ in Eq. (B.2.5) must not depend on $n$ and of course must be positive. As a consequence we can find a chain of bounds on the terms.

$$|\alpha_{N+2}| \leq r\, |\alpha_{N+1}|,$$

$$|\alpha_{N+3}| \leq r\, |\alpha_{N+2}| \leq r^2\, |\alpha_{N+1}|,$$

---

[1]In this case the sum becomes $\sum_{n=0}^{N} 1 = N + 1$. Note that the result is consistent with taking the limit of Eq. (B.2.3) as $r \to 1$.

and the pattern will be

$$|\alpha_{N+1+m}| \le r^m |\alpha_{N+1}| .$$  (B.2.6)

Then we can bound the error as follows:

$$|E_N| = \left| \sum_{n=N+1}^{\infty} \alpha_n \right|$$

$$\le \sum_{n=N+1}^{\infty} |\alpha_n|$$

$$\le \sum_{m=0}^{\infty} |\alpha_{N+1+m}| .$$  (B.2.7)

Now the bounds (B.2.6) can be used to give

$$|E_N| \le |\alpha_{N+1}| \sum_{m=0}^{\infty} r^m$$

$$\le \frac{|\alpha_{N+1}|}{1-r}$$  (B.2.8)

where we used the result Eq. (B.2.4) in the last step. As long as $r < 1$, the series converges and the error estimate Eq. (B.2.8) makes sense.

The application of the result Eq. (B.2.8) to the Taylor series is immediate since $\alpha_n = a_n x^n$. Thus, we hope to find $r$ so that $|a_{n+1}| |x|^{n+1} \le r |a_n| |x|^n$, or

$$\frac{|a_{n+1}|}{|a_n|} |x| \le r < 1 .$$  (B.2.9)

Usually, by making $|x|$ small enough, we can be sure that this bound is satisfied.

## B.3   Standard examples

The exponential function valid for all $x$:

$$e^x = \sum_{n=0}^{\infty} \frac{x^n}{n!} .$$  (B.3.1)

The cosine function valid for all $x$:

$$\cos(x) = \sum_{n=0}^{\infty} \frac{(-1)^n}{(2n)!} x^{2n} .$$  (B.3.2)

The sine function valid for all $x$:

$$\sin(x) = \sum_{n=0}^{\infty} \frac{(-1)^n}{(2n+1)!} x^{2n+1} .$$  (B.3.3)

The geometric series valid for $|x| < 1$:

$$\frac{1}{1-x} = \sum_{n=0}^{\infty} x^n .$$  (B.3.4)

To illustrate the use of the error estimate Eq. (B.2.8), consider the Taylor series for the exponential Eq. (B.3.1) where $a_n = 1/n!$. Thus Eq. (B.2.9) becomes

$$\frac{n!}{(n+1)!}\,|x| = \frac{|x|}{n+1}\,.$$

Provided we pick $N + 1 > |x|$, we estimate

$$r = \frac{|x|}{N+1} \qquad \text{and} \qquad |E_N| = \frac{1}{1-r}\frac{|x|^{(N+1)}}{(N+1)!}\,.$$

One way to interpret this result is to pick a range in $x$, for example, $|x| < 1$, and seek the value of $N$ that makes $E_N$ small enough. The choice $N = 4$ leads to $r = 0.2$ and $E_N = 0.01$.

# Appendix:
# Systems of Linear Equations

**C**

## C.1 Algebraic equations

The simplest example of a linear equation is

$$a\,x = r,\tag{C.1.1}$$

where $a$ is a parameter, $x$ an unknown variable and $r$ the forcing term. The solution is obvious,

$$x = \frac{r}{a}.\tag{C.1.2}$$

Of course this solution fails if $a = 0$ (the solution does not exist), but then Eq. (C.1.1) would read $0 = r$, which is nonsense unless the special case occurs where $r = 0$. If that is the case, then we would have no restrictions on $x$ and it would remain undetermined.

Surprising, this view of the simplest equation extends to systems of equations. For example, consider the system of two equations for two unknowns $x_1$ and $x_2$,

$$a_{11}x_1 + a_{12}x_2 = r_1,\tag{C.1.3}$$
$$a_{21}x_1 + a_{22}x_2 = r_2.\tag{C.1.4}$$

Here $a_{11}$, $a_{12}$, $a_{21}$ and $a_{22}$ are parameters and $r_1$, $r_2$ are forcing terms. A special notation has been introduced that helps keep track of the placement of the parameters and forcing terms. The first subscript in $a_{jk}$ refers to the $j$th equation, while the second subscript refers to which unknown $x_k$ is being multiplied by $a_{jk}$. The subscripts on $r$ refer to which equation the forcing term appears.

The standard approach to solving the two equations Eq. (C.1.3) and Eq. (C.1.4) is called the method of substitution. Take the first equation and write

$$x_1 = \frac{1}{a_{11}}\left(r_1 - a_{12}x_2\right),\tag{C.1.5}$$

then substitute this expression for $x_1$ into the second equation.

$$\frac{a_{21}}{a_{11}}\left(r_1 - a_{12}x_2\right) + a_{22}x_2 = r_2.$$

By design, this equation contains only the unknown $x_2$ and can be re-arranged to appear like Eq. (C.1.1).

$$\left(a_{22} - \frac{a_{12}a_{21}}{a_{11}}\right) x_2 = r_2 - \frac{a_{21}}{a_{11}} r_1 .$$

The solution for $x_2$ follows from Eq. (C.1.2) but it is more convenient to introduce an intermediate step to remove the division by $a_{11}$. Simply multiply the equation by $a_{11}$, leading to the equivalent equation.

$$(a_{11}a_{22} - a_{12}a_{21})\, x_2 = a_{11}r_2 - a_{21}r_1 . \tag{C.1.6}$$

Now the solution is obtained directly;

$$x_2 = \frac{a_{11}r_2 - a_{21}r_1}{a_{11}a_{22} - a_{12}a_{21}} . \tag{C.1.7}$$

The same concerns that arose in the solution Eq. (C.1.2) apply here. The denominator, $D$ say,

$$D = a_{11}a_{22} - a_{12}a_{21} \tag{C.1.8}$$

must not be zero! If it is, there is no solution for $x_2$ unless by good fortune the numerator $a_{11}r_2 - a_{21}r_1 = 0$ too, in which case $x_2$ is undetermined. Assuming the solution for $x_2$ exists, we proceed by substituting the result into Eq. (C.1.5).

$$x_1 = \frac{1}{a_{11}} \left( r_1 - a_{12} \frac{a_{11}r_2 - a_{21}r_1}{a_{11}a_{22} - a_{12}a_{21}} \right) .$$

After some tedious algebra, a simpler result is obtained;

$$x_1 = \frac{a_{22}r_1 - a_{12}r_2}{a_{11}a_{22} - a_{12}a_{21}} . \tag{C.1.9}$$

Once again, there are difficulties if $D = 0$ unless $a_{22}r_1 - a_{12}r_2 = 0$ too.

## C.2   Gaussian elimination

There is alternate way to solve Eq. (C.1.3) and Eq. (C.1.4) called Gaussian elimination; it is the preferred method to code on computers. It starts by multiplying Eq. (C.1.3) by $a_{21}$ and Eq. (C.1.4) by $a_{11}$.

$$a_{21}a_{11}x_1 + a_{21}a_{12}x_2 = a_{21}r_1 , \tag{C.2.1}$$
$$a_{11}a_{21}x_1 + a_{11}a_{22}x_2 = a_{11}r_2 . \tag{C.2.2}$$

Now by subtracting Eq. (C.2.1) from Eq. (C.2.2), we eliminate the unknown $x_1$.

$$(a_{11}a_{22} - a_{12}a_{21})x_2 = a_{11}r_2 - a_{21}r_1 ,$$

and we have arrived at the same equation as Eq. (C.1.6). To eliminate the other unknown $x_2$, simply multiply Eq. (C.1.3) by $a_{22}$ and Eq. (C.1.4) by $a_{12}$ and subtract.

$$(a_{11}a_{22} - a_{12}a_{21})x_1 = a_{22}r_1 - a_{12}r_2 ,$$

which leads to Eq. (C.1.9).

In summary, if $D \neq 0$, then Eq. (C.1.7) and Eq. (C.1.9) give the unique solution to Eq. (C.1.3) and Eq. (C.1.4). If $D = 0$, then there is no solution unless both $a_{11}r_2 - a_{21}r_1 = 0$ and $a_{22}r_1 - a_{12}r_2 = 0$ in which case $x_1$ and $x_2$ remain undetermined. To understand better why the latter case may occur, note that $D = 0$ implies

$$\frac{a_{21}}{a_{11}} = \frac{a_{22}}{a_{12}}.$$

In addition,

$$a_{11}r_2 - a_{21}r_1 = 0, \quad \text{implies} \quad \frac{a_{21}}{a_{11}} = \frac{r_2}{r_1},$$

and

$$a_{22}r_1 - a_{12}r_2 = 0, \quad \text{implies} \quad \frac{a_{22}}{a_{12}} = \frac{r_2}{r_1}.$$

If we now choose $a_{21}/a_{11} = R$, then all these ratios give the results,

$$a_{21} = R\,a_{11}, \quad a_{22} = R\,a_{12}, \quad r_2 = R\,r_1.$$

So what! Well, if we now use these expressions in the second equation Eq. (C.1.4), we find it becomes

$$R\,a_{11}x_1 + R\,a_{12}x_2 = R\,r_1.$$

In other words, the second equation is just the first equation Eq. (C.1.3) multiplied by $R$.

## C.3    Matrix form

The equations Eq. (C.1.3) and Eq. (C.1.4) are examples of a system of equations where the number of equations and unknowns is two. There are often situations where the number of unknowns and equations is much larger, $N$ say. Obviously, it is tedious and, in some cases, practically impossible to write down all the equations. Instead, the introduction of subscripts as done in Eq. (C.1.3) and Eq. (C.1.4) allow us to write the equations in a compact form:

$$\sum_{k=1}^{N} a_{jk}\, x_k = r_j, \quad \text{for } j = 1, \ldots, N. \tag{C.3.1}$$

An even simpler form is to introduce the notion of vectors and matrices. To illustrate, Eq. (C.1.3) and Eq. (C.1.4) (or Eq. (C.3.1) with $N = 2$) are written as

$$\begin{bmatrix} a_{11} & a_{12} \\ a_{21} & a_{22} \end{bmatrix} \begin{pmatrix} x_1 \\ x_2 \end{pmatrix} = \begin{pmatrix} r_1 \\ r_2 \end{pmatrix}. \tag{C.3.2}$$

All the parameters $a_{jk}$ have been organized in rows and columns, the rows listing the parameters for a single equation, while the columns list the parameters that multiply a given unknown. The unknowns and the forcing terms are listed as columns. There seems little advantage to expressing the equations in the form Eq. (C.3.2) except that by introducing the symbols,

$$\mathcal{A} = \begin{bmatrix} a_{11} & a_{12} \\ a_{21} & a_{22} \end{bmatrix} \quad \mathbf{x} = \begin{pmatrix} x_1 \\ x_2 \end{pmatrix}, \quad \mathbf{r} = \begin{pmatrix} r_1 \\ r_2 \end{pmatrix}$$

the similarity with (C.1.1) becomes obvious.

$$\mathcal{A}\mathbf{x} = \mathbf{r}. \tag{C.3.3}$$

The consequence of the form Eq. (C.3.3) is that we expect that certain choices for $\mathcal{A}$ are likely to reflect the difficulties associated with $a = 0$ when we solve Eq. (C.1.1). Those difficulties are identified by $\det(\mathcal{A}) = 0$ where $\det(\mathcal{A})$ is called the determinant of $\mathcal{A}$. Unless $N$ is small, $\det(\mathcal{A})$ is complicated to calculate. For $N = 2$, the determinant is given by

$$\det(\mathcal{A}) = D = a_{11}a_{22} - a_{12}a_{21}, \tag{C.3.4}$$

and now the connection to the results Eq. (C.1.7) and Eq. (C.1.9) is clear. As long as $\det(\mathcal{A}) \neq 0$, we are assured that there is a solution $\mathbf{x}$.

## C.4   Eigenvalues and eigenvectors

Can we find a special vector $\mathbf{e}$ that converts the multiplication by a matrix $\mathcal{A}$ into just a multiple of the vector? Stated mathematically, is there a $\lambda$ and a vector $\mathbf{e}$ so that

$$\mathcal{A}\mathbf{e} = \lambda\mathbf{e}? \tag{C.4.1}$$

This problem is called an *eigenvalue problem*. This problem is called homogeneous because the zero vector obviously satisfies the equation, so the first clarification that must be made is that we seek $\mathbf{e} \neq 0$, although $\lambda = 0$ is possible.

To proceed, write out the equations for $N = 2$.

$$(a_{11} - \lambda)\,e_1 + a_{12}e_2 = 0, \tag{C.4.2}$$
$$a_{21}e_1 + (a_{22} - \lambda)\,e_2 = 0. \tag{C.4.3}$$

This system is the same Eq. (C.1.3) and Eq. (C.1.4) except that $a_{11}$ is replaced by $a_{11} - \lambda$ and $a_{22}$ is replaced by $a_{22} - \lambda$, and there is no right hand side $r_1 = r_2 = 0$.

Use Gaussian elimination to obtain solutions for $e_1$ and $e_2$. Multiply Eq. (C.4.2) by $a_{22} - \lambda$ and Eq. (C.4.3) by $a_{12}$ and subtract the results.

$$\left[ (a_{11} - \lambda)\,(a_{22} - \lambda) - a_{12}a_{21} \right] e_1 = 0.$$

Alternatively, we can multiple Eq. (C.4.2) by $a_{21}$ and Eq. (C.4.3) by $a_{a_11} - \lambda$ and subtract the results.

$$\left[ (a_{11} - \lambda)\,(a_{22} - \lambda) - a_{12}a_{21} \right] e_2 = 0.$$

Clearly, if we want non-zero values for either $e_1$ and/or $e_2$, we must choose

$$(a_{11} - \lambda)\,(a_{22} - \lambda) - a_{12}a_{21} = 0. \tag{C.4.4}$$

This is a quadratic equation for $\lambda$! This result is nothing more than the statement that the determinant for the matrix obtained from Eq. (C.4.2) and Eq. (C.4.3) is zero. As a consequence we cannot expect unique solutions for $e_1$ and $e_2$.

In general, we expect two roots to Eq. (C.4.4). For each root, we use either Eq. (C.4.2) or Eq. (C.4.3) to obtain $e_1$ in terms of $e_1$ or vice versa. Because only the ratio $e_2/e_1$ is specified, an eigenvector can have any length. Note that if $\mathbf{e}$ satisfies Eq. (C.4.1), then any multiple of $\mathbf{e}$ also satisfies Eq. (C.4.1).

There are three cases to consider:

## Real roots

Let $\lambda_1$ and $\lambda_2$ be the real roots to Eq. (C.4.4). Each eigenvector is determined by using Eq. (C.4.2) and Eq. (C.4.3) with $\lambda$ replaced by $\lambda_1$ and $\lambda_2$ separately. As a result, the ratio of the components satisfy

$$\frac{e_2}{e_1} = -\frac{a_{11} - \lambda}{a_{12}} = -\frac{a_{21}}{a_{22} - \lambda} \tag{C.4.5}$$

for each choice of $\lambda$. The ratios are the same by virtue of Eq. (C.4.4), but sometimes one of the ratios is indeterminate and the other must be chosen. If $a_{12} \neq 0$, a convenient choice is $e_1 = a_{12}$ and then $e_2 = -(a_{11} - \lambda)$. Otherwise, choose $e_1 = -(a_{22} - \lambda)$ and $c_2 = a_{21}$.

## Complex roots

The roots appear as a complex conjugate pair $\lambda_r \pm i\lambda_i$. As a consequence, the eigenvectors also are complex. Just calculate the eigenvector for $\lambda_r + i\lambda_i$ and the other eigenvector is just the complex conjugate. By using Eq. (C.4.5),

$$\frac{e_2}{e_1} = -\frac{a_{11} - \lambda_r}{a_{12}} + i\frac{\lambda_i}{a_{12}}. \tag{C.4.6}$$

A convenient choice, then, is to take $e_1 = a_{12}$ and $e_2 = -(a_{11} - \lambda_r - i\lambda_i)$.

## Double roots

This case is the most tricky because there may be one or two eigenvectors. One eigenvector occurs when there is only one possibility for the ratio of components Eq. (C.4.5). Two eigenvectors occur when both components are arbitrary. Two simple examples show the different cases.

- **One eigenvector**
  The matrix,

$$\mathcal{A} = \begin{bmatrix} 1 & 0 \\ -1 & 1 \end{bmatrix} \tag{C.4.7}$$

has just one eigenvalue $\lambda = 1$ and one eigenvector $e_1 = 0, e_2 = 1$.

- **Two eigenvectors**

The matrix

$$\mathcal{A} = \begin{bmatrix} 1 & 0 \\ 0 & 1 \end{bmatrix} \tag{C.4.8}$$

has just one eigenvalue but two eigenvectors: $e_1 = 1$, $e_2 = 0$, and $e_1 = 0$, $e_2 = 1$.

# Appendix:
# Complex Variables

## D.1 Basic properties

Complex numbers are usually first seen when the square root of a negative number is needed. A special symbol is introduced

$$\mathrm{i} = \sqrt{-1}$$

because the square root of any negative number can be expressed in terms of a multiple of i. Suppose $a > 0$, then

$$\sqrt{-a} = \sqrt{(-1)\,a} = \mathrm{i}\,\sqrt{a}\,. \tag{D.1.1}$$

Just like all square roots, the result can be multiplied by $\pm 1$. Note that if $a < 0$, then

$$\sqrt{a} = \sqrt{(-1)\,(-a)} = \mathrm{i}\,\sqrt{-a}\,. \tag{D.1.2}$$

The new argument of the square root is now a positive number.

In general, a complex number $z$ has two parts, a real part $x$ and an imaginary part $y$, written as

$$z = x + \mathrm{i}\,y = \Re\{z\} + \mathrm{i}\,\Im\{z\}\,.$$

The addition of two complex numbers is straightforward; simply add the real parts and the imaginary parts separately. Also, multiplication of a complex number by a real number $\alpha$ is achieved by multiplying the real part and the imaginary part separately by the number $\alpha$. These two operations, the addition of complex numbers and multiplication by a real number, are the same as though $z$ is a two-dimensional vector with components $(x, y)$. This analogy introduces a different representation for a complex number as a position vector with length $r$ and angle $\theta$; see Fig. D.1. Since $x = r\,\cos(\theta)$, $y = r\,\sin(\theta)$ is the standard polar representation for a vector, we are led to writing the complex number as

$$z = r\,\cos(\theta) + \mathrm{i}\,r\,\sin(\theta) \tag{D.1.3}$$

where

$$r = \sqrt{x^2 + y^2}\,, \quad \text{and} \quad \tan(\theta) = \frac{y}{x}\,. \tag{D.1.4}$$

But a word of caution: $\theta$ is not unique! Any multiple of $2\pi$ can be added to $\theta$ without changing the values of $x$ or $y$, in other words, the complex number $z$. Normally, the value of $\theta$ is chosen in the range $-\pi < \theta \le \pi$.

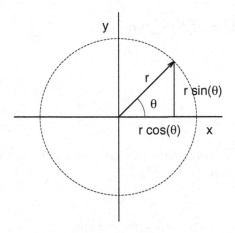

Fig. D.1   Polar representation for a complex variable.

There is a special and extremely useful representation for complex numbers that comes about from Eq. (D.1.3) and Euler's formula,

$$e^{i\theta} = \cos(\theta) + i\,\sin(\theta)\,. \tag{D.1.5}$$

An interesting way to derive Euler's formula from the solutions to a special differential equation is presented in Sec. 1.5.1. The consequences, though, are that

$$z = r\,e^{i\theta}\,, \tag{D.1.6}$$

and this form allows us to multiply complex numbers in a simple way.

One way to multiply two complex numbers, $z_1 = x_1 + i\,y_1$ and $z_2 = x_2 + i\,y_2$, is to apply the standard rules of algebra.

$$\begin{aligned} z_1 z_2 &= (x_1 + i\,y_1)\,(x_2 + i\,y_2) \\ &= (x_1 x_2 - y_1 y_2) + i\,(x_1 y_2 + x_2 y_1)\,. \end{aligned}$$

Parentheses are added to show clearly the resulting real and imaginary parts. The other way to multiply the two complex numbers is to use their polar forms.

$$z_1 z_2 = r_1 e^{i\theta_1} r_2 e^{i\theta_2} = r_1 r_2 e^{i(\theta_1 + \theta_2)}\,.$$

Obviously, the sum of the angles $\theta = \theta_1 + \theta_2$ may fall outside the range $-\pi < \theta \le \pi$, but by adding or subtracting a multiple of $2\pi$, $\theta$ can be adjusted to fall inside the normal range.

Division of complex numbers follows in a similar fashion.

$$\frac{z_1}{z_2} = \frac{r_1\,e^{i\theta_1}}{r_2\,e^{i\theta_2}} = \frac{r_1}{r_2}\,e^{i(\theta_1-\theta_2)}\,.$$

How can we express this result in terms of $x_1 + i\,y_1$ and $x_2 + i\,y_2$? We can rewrite the result as

$$\frac{z_1}{z_2} = \frac{\left(r_1\,e^{i\theta_1}\right)\left(r_2\,e^{-i\theta_2}\right)}{r_2^2} = \frac{(x_1 + i\,y_1)\,r_2\,e^{-i\theta_2}}{x_2^2 + y_2^2}\,.$$

We are almost there, but we need to relate $r_2\exp(-i\,\theta_2)$ to $z_2$. Since

$$\begin{aligned}
r_2\,e^{-i\theta_2} &= r_2\,\cos(-\theta_2) + i\,r_2\,\sin(-\theta_2) \\
&= r_2\,\cos(\theta_2) - i\,r_2\,\sin(\theta_2) \\
&= x_2 - i\,y_2\,,
\end{aligned}$$

we may write

$$\frac{z_1}{z_2} = \frac{(x_1 + i\,y_1)\,(x_2 - i\,y_2)}{x_2^2 + y_2^2}\,.$$

Since $x_2^2 + y_2^2 = (x_2 + i\,y_2)\,(x_2 - i\,y_2)$, we now understand the result in the following way,

$$\begin{aligned}
\frac{z_1}{z_2} &= \frac{x_1 + i\,y_1}{x_2 + i\,y_2} \\
&= \frac{(x_1 + i\,y_1)\,(x_2 - i\,y_2)}{(x_2 + i\,y_2)\,(x_2 - i\,y_2)} \\
&= \frac{(x_1 x_2 + y_1 y_2) + i\,(y_1 x_2 - x_1 y_2)}{x_2^2 + y_2^2}\,.
\end{aligned}$$

The trick then is to multiply both numerator and denominator by $x_2 - i\,y_2$, and a special symbol is introduced to record this quantity, called the complex conjugate,

$$\overline{z} = x - i\,y\,. \tag{D.1.7}$$

It is useful for several purposes:

$$r^2 = x^2 + y^2 = z\,\overline{z}\,,$$

$$\Re\{z\} = x = \frac{z + \overline{z}}{2}\,,$$

$$\Im\{z\} = y = \frac{z - \overline{z}}{2i}\,,$$

$$\overline{z} = r\,e^{-i\theta}\,.$$

It also has some properties that make working with the conjugate easy: If $z_1$ and $z_2$ are two complex variables, then the conjugate of a sum is a sum of the conjugates,

$$\overline{z_1 + z_2} = \overline{z_1} + \overline{z_2}\,, \tag{D.1.8}$$

and the conjugate of a product is the product of the conjugates,

$$\overline{z_1\, z_2} = \overline{z_1}\, \overline{z_2}\,. \tag{D.1.9}$$

Finally, it should be obvious the conjugate of the conjugate undoes the conjugation, $\overline{\overline{z}} = z$.

The result Eq. (D.1.9) suggests an appropriate way to raise a complex number to some power $p$;

$$z^p = (x + \mathrm{i}\,y)^p = \left(r\,\mathrm{e}^{\mathrm{i}\theta}\right)^p = r^p\,\mathrm{e}^{\mathrm{i}p\theta}\,, \tag{D.1.10}$$

but the possible multiple values for $\theta$ now make a difference. Since we can always add some integer multiple of $2\pi$ to $\theta$ without changing the value of $x + \mathrm{i}y$, Eq. (D.1.10) should be replaced by

$$z^p = (x + \mathrm{i}\,y)^p = r^p \mathrm{e}^{\mathrm{i}p(\theta + \mathrm{i}2n\pi)} = r^p \mathrm{e}^{\mathrm{i}p\theta} \mathrm{e}^{\mathrm{i}2pn\pi}\,. \tag{D.1.11}$$

This result is quite general and includes all possible cases by allowing $n$ to be any integer. Despite the infinite choices for the integer $n$, there are times when no new results occur because $pn$ can be an integer and then $\exp(\mathrm{i}2pn\pi) = 1$ and the result is the same is if $n = 0$. Some examples will illustrate the possibilities.

- Suppose $x + \mathrm{i}y = 1$ and $p = 1/2$. Then $r = 1$ and $\theta = 0 + 2n\pi$.

$$(1)^{1/2} = 1^{1/2}\mathrm{e}^{0 + \mathrm{i}n\pi}\,.$$

For $n = 0, 2, 4, \ldots$, $(1)^{1/2} = 1$, while for $n = 1, 3, 5, \ldots$, $(1)^{1/2} = -1$. Clearly, the only results are 1 and $-1$ as expected.

- Suppose $x + \mathrm{i}y = -1$ and $p = 1/2$. Then $r = 1$ and $\theta = \pi + 2n\pi$.

$$(-1)^{1/2} = 1^{1/2}\mathrm{e}^{\pi/2 + \mathrm{i}n\pi}\,.$$

For $n = 0, 2, 4, \ldots$, $(-1)^{1/2} = \mathrm{i}$, while for $n = 1, 3, 5, \ldots$, $(-1)^{1/2} = -\mathrm{i}$. Clearly, there only two results, i and $-\mathrm{i}$ as expected.

- Suppose $x + \mathrm{i}y = 1$ and $p = 1/4$. Then $r = 1$ and $\theta = 0 + 2n\pi$.

$$(1)^{1/4} = 1^{1/4}\mathrm{e}^{0 + \mathrm{i}n\pi/2}\,.$$

For $n = 0, 4, 8, \ldots$, $(1)^{1/4} = 1$; for $n = 1, 5, 9, \ldots$, $(1)^{1/4} = \mathrm{i}$; for $n = 2, 6, 10, \ldots$, $(1)^{1/4} = -1$; and for $n = 3, 7, 11, \ldots$, $(1)^{1/4} = -\mathrm{i}$. There are only four distinct choices, 1, $-1$, i, $-\mathrm{i}$.

- Suppose $x + \mathrm{i}y = -1$ and $p = 1/4$. Then $r = 1$ and $\theta = \pi + 2n\pi$,

$$(-1)^{1/4} = 1^{1/4}\mathrm{e}^{\mathrm{i}\pi/4 + \mathrm{i}n\pi/2}\,.$$

For $n = 0, 4, 8, \ldots$, $(-1)^{1/4} = (1 + \mathrm{i})/\sqrt{2}$; for $n = 1, 5, 9, \ldots$, $(-1)^{1/4} = (-1 + \mathrm{i})/\sqrt{2}$; for $n = 2, 6, 10, \ldots$, $(-1)^{1/4} = (-1 - \mathrm{i})/\sqrt{2}$; and for $n = 3, 7, 11, \ldots$, $(-1)^{1/4} = (1 - \mathrm{i})/\sqrt{2}$.

## D.2   Connections with trigonometric functions

Euler's formula is very useful in drawing the connection between exponentials with complex arguments and the trigonometric functions sine and cosine. For example, consider the statement of the multiplication of two exponentials;

$$e^{i\,(A+B)} = e^{iA}e^{iB}\,.$$

Replace the exponentials with Euler's formula. For the expression on the left,

$$e^{i\,(A+B)} = \cos(A+B) + i\,\sin(A+B)\,,$$

while the expression on the right is

$$
\begin{aligned}
e^{iA}e^{iB} &= \left[\cos(A) + i\,\sin(A)\right]\left[\cos(B) + i\,\sin(B)\right]\\
&= \left[\cos(A)\,\cos(B) - \sin(A)\,\sin(B)\right] + i\left[\sin(A)\,\cos(B) + \cos(A)\,\sin(B)\right].
\end{aligned}
$$

By equating the two expressions and balancing the real and imaginary parts, we obtain the addition angle formulas!

$$\cos(A+B) = \cos(A)\,\cos(B) - \sin(A)\,\sin(B)\,, \tag{D.2.1}$$

$$\sin(A+B) = \sin(A)\,\cos(B) + \cos(A)\,\sin(B)\,. \tag{D.2.2}$$

One of the consequences of the addition angle formulas is the ability to replace products of sines and cosines with sums of sines and cosines. To obtain the results, we write down the associated formulas for the difference in angles:

$$\cos(A-B) = \cos(A)\,\cos(B) + \sin(A)\,\sin(B)\,, \tag{D.2.3}$$

$$\sin(A-B) = \sin(A)\,\cos(B) - \cos(A)\,\sin(B)\,. \tag{D.2.4}$$

There is nothing really new in these results; we have simply replaced $B$ with $-B$ and used the properties of sines and cosines of negative arguments. By adding Eq. (D.2.1) and Eq. (D.2.3), we obtain

$$2\,\cos(A)\,\cos(B) = \cos(A+B) + \cos(A-B)\,. \tag{D.2.5}$$

By subtracting (D.2.1) from (D.2.3), we obtain

$$2\,\sin(A)\,\sin(B) = \cos(A-B) - \cos(A+B)\,. \tag{D.2.6}$$

By subtracting (D.2.4) from (D.2.2), we obtain

$$2\,\sin(A)\,\cos(B) = \sin(A+B) + \sin(A-B)\,. \tag{D.2.7}$$

# Index

transport phenomenon, 208
wave equation, 257
periodicity
definition, 119, 125
period, 119, 120, 123, 125
Principle 1, 14
Principle 2, 15
Principle 3, 17
Principle 4, 35
Principle 5, 37
Principle 6, 37
Principle 7, 44
Principle 8, 66
Principle 9, 68
Principle 10, 125
Principle 11, 126
Principle 12, 128
Principle 13, 144
Principle 14, 144
Principle 15, 145
Principle 16, 147
Principle 17, 150
Principle 18, 151
Principle 19, 161
Principle 20, 178
Principle 21, 185
Principle 22, 186
Principle 23, 187
Principle 24, 226
Principle 25, 241
Principle 26, 281
Principle 27, 299
Principle 28, 318
Principle 29, 330
propagating signals, 258

rate of change
absolute, 4
relative, 4
recursive relationships as precursors to
differential equations, 6, 10, 23, 169,
192, 208, 246, 267
resonance in LCR circuits, 103

second-order differential equations
boundary value problems, 184
characteristic equation, 66, 106
complex roots, 82, 89
imaginary roots, 78
real roots, 66

single root, 66
constant coefficients
homogeneous, 65, 89
inhomogeneous, 105
examples
chained chemical reactions, 59
chained chemical reactions with
forcing, 63
LCR circuit, 83
LCR circuit with alternating
voltage applied, 97
LCR circuit with constant applied
voltage, 96
LCR circuit with periodic forcing,
122
forcing by complex exponentials, 102
fundamental solutions, 75
general solution, 90, 106
homogeneous solutions, 105
initial conditions, 67, 74, 87, 90, 106
particular solutions
series solutions, 114
undetermined coefficients, 105
variation of parameters, 111
specific solutions, 107
uniqueness, 69
variable coefficients, 73
series solutions, 75
Wronskian, 68, 107
series solutions
first-order equations, 16, 51
second-order differential equations, 67
second-order equations, 114
systems of differential equations
autonomous, 304
examples
chemical reactions, 305
LCR circuit, 310
LCR circuit with forcing, 323, 324
higher dimensions
homogeneous solutions, 320
initial conditions, 314
linear
homogeneous, 305, 313
homogeneous solutions, 327
inhomogeneous, 322, 327
particular solutions, 327
undetermined coefficients, 327
variation of parameters, 331

Printed in the United States
By Bookmasters